Science and the Confucian Religion of Kang Youwei (1858–1927)

Ideas, History, and Modern China

Edited by

Ban Wang (*Stanford University*)
Wang Hui (*Tsinghua University*)

VOLUME 26

The titles published in this series are listed at *brill.com/ihmc*

Science and the Confucian Religion of Kang Youwei (1858–1927)

China before the Conflict Thesis

By

WAN Zhaoyuan

BRILL

LEIDEN | BOSTON

Portrait of Kang Youwei, circa 1905, by the American photographer Elmer Chickering (1857–1915), now in the public domain, image available from the United States Library of Congress under the digital ID cph.3a45114 (http://hdl.loc.gov/loc.pnp/cph.3a45114). The author would like to thank Mr. Zhang Qireng for the information on the source of this image.

Library of Congress Cataloging-in-Publication Data

Names: Wan, Zhaoyuan, 1978- author.
Title: Science and the Confucian religion of Kang Youwei (1858-1927) : China before the conflict thesis / by Zhaoyuan Wan.
Description: Leiden ; Boston : Brill, [2022] | Series: Ideas, history, and modern China, 1875-9394 ; volume 26 | Includes bibliographical references and index.
Identifiers: LCCN 2021026059 (print) | LCCN 2021026060 (ebook) | ISBN 9789004468214 (hardback) | ISBN 9789004468221 (ebook)
Subjects: LCSH: Kang, Youwei, 1858-1927. | Confucianists–China–Biography. | Religion and science–China–History. | Confucianism–China–History. | China–Religion. | China–Intellectual life–1644-1912.
Classification: LCC BL1875.K36 W36 2022 (print) | LCC BL1875.K36 (ebook) | DDC 299.5/12165–dc23
LC record available at https://lccn.loc.gov/2021026059
LC ebook record available at https://lccn.loc.gov/2021026060

Typeface for the Latin, Greek, and Cyrillic scripts: "Brill". See and download: brill.com/brill-typeface.

ISSN 1875-9394
ISBN 978-90-04-46821-4 (hardback)
ISBN 978-90-04-46822-1 (e-book)

This book is printed on acid-free paper and produced in a sustainable manner.

To Shoghi Effendi (1897–1957), whose writings have profoundly inspired me

In loving memory of my late father, Wan Baoshan 万宝山 *(1944–2016)*

Contents

Foreword

Scientism has been a core component of China's official ideology and mainstream intellectual culture for much of the past century. The iconic modernist author Lu Xun (1881–1936) is well known for his scathing critiques of religion and superstition in his novels and essays. The New Culture Movement, which marked the rise of modern Chinese thought, is associated with the worship of 'Mr. Science and Mr. Democracy', and its association with the 'anti-religious movement'. Since then, these ideas have been baked into the orthodoxy of modern Chinese thought, which often proudly claims that China is the only country without a religion.

But this is not the only way science and religion have been thought about in modern Chinese intellectual history. In this book, Wan Zhaoyuan challenges the mainstream narrative of modern Chinese intellectual history, which portrays an unchanging story of radical scientism and secularism over the past century, from the May Fourth Movement (1919) to the People's Republic (1949) until the present.

Indeed, the first wave of reformist intellectuals of the turn of the twentieth century – figures such as Kang Youwei (1858–1927), Tan Sitong (1865–1898) and Liang Qichao (1873–1929), grappled with the interplay between Western and Chinese thought, and between science and religion, with often striking subtlety and sophistication. This book focuses on Kang Youwei, who is known as a leading political reformer and Confucian thinker in the twilight of the Qing dynasty at the end of the 19th century and the turn of the 20th century, and as the mentor of several of China's leading modernizing intellectuals. Wan challenges the received view of Kang in both Chinese and Western historiography, which has focused on his political life and career, largely ignoring that he was a religious luminary, teacher and reformer, almost a self-proclaimed prophet. Kang has, it is true, featured prominently in a few recent accounts of modern constructions of religion in China, highlighting a previously ignored aspect of his influence – ranging from his role in the 'convert temples to schools' movement to his advocacy of the "Religion of Confucius" (*Kongjiao* 孔教) as China's national religion and his role in redemptive societies such as the Universal Morality Society 萬國道德會.[1] But the religious dimension of Kang's life and thought has, until now, not been the subject of an in-depth and systematic study.

1 Vincent Goossaert and David A. Palmer, *The Religious Question in Modern China* (Chicago: University of Chicago Press, 2011); Vincent Goossaert, '1898: The Beginning of the End for Chinese Religion?', *The Journal of Asian Studies* 65, no. 2 (May 2006): 307–35; Wei Leong Tay, 'Kang Youwei, The Martin Luther of Confucianism

Wan shows that Kang Youwei's religious thought evolved while he closely followed the latest developments in Western science and was a strong advocate of modern scientific methods and research, but that this did not stop him from engaging in sophisticated and creative reflections and proposals on the relationship between science and religion, offering more nuanced perspectives on the relationships between modern science and traditional Chinese cosmology and religion than the radical intellectuals of the next generation.

Drawing on a detailed study of the writings of Kang and his associates and disciples, presented in the context of historical and political changes as they affected their personal lives, Wan shows that the 'religious', the 'scientific' and the 'political' dimensions of Kang's life were inextricably intertwined, with events, experiences or new knowledge in one realm impacting on his thinking and action in the other realms – at a time when these categories, imported from the West, had not yet become reified and institutionalised within intellectual and political life. In addition to the life and thought of Kang Youwei himself, Wan also discusses the evolving thought of other figures in his intellectual network, such as Liang Qichao, Tan Sitong, and Chen Huanzhang (1880–1933).

This book is thus a significant contribution to an emerging field of research on the place of religion in the intellectual and political life of China in the late Qing and early Republican eras, which has challenged the conventional, secularist historiography.[2] It will also be a noted contribution to the growing literature on the transformations of Confucianism in the modern era.[3]

But, perhaps most importantly, this is a path-breaking work which opens the field of the history of science-religion relations in a country whose

---

and His Vision of Confucian Modernity and Nation', in *Secularization, Religion and the State*, ed. Haneda Masashi (Tokyo: The University of Tokyo Centre for Philosophy, 2010), 97–109.

2 Rebecca Nedostup, *Superstitious Regimes: Religion and the Politics of Chinese Modernity* (Cambridge, MA.: Harvard University Asia Center, 2009); Paul R. Katz, *Religion in China and Its Modern Fate* (Waltham, MA.: Brandeis University Press, 2014); Peter van der Veer, *The Modern Spirit of Asia: The Spiritual and the Secular in China and India* (Princeton: Princeton University Press, 2013); Prasenjit Duara, *The Crisis of Global Modernity: Asian Traditions and a Sustainable Future* (Cambridge: Cambridge University Press, 2016); Goossaert and Palmer, *Religious Question.*

3 Lionel M. Jensen, *Manufacturing Confucianism: Chinese Traditions & Universal Civilization* (Durham: Duke University Press, 1997); Hsi-yuan Chen, 'Confucianism Encounters Religion: The Formation of Religious Discourse and the Confucian Movement in Modern China' (Ph.D., Harvard University, 1999); Anna Sun, *Confucianism as a World Religion: Contested Histories and Contemporary Realities* (Princeton, N.J.: Princeton University Press, 2013); Sébastien Billioud, *The Varieties of Confucian Experience* (Leiden; Boston: Brill, 2018).

religious, scientific and ideological trajectories have been vastly different from the West's. Indeed, following the breakdown of the secularization thesis and its promise, there is a growing interest in thinking through alternative ways of conceiving of reality, as seen in, for example, the 'ontological turn' in anthropology, or the growing mainstreaming of 'alternative' practices of health, healing and meditation. The dualistic conflict between science and religion is now questioned in numerous quarters.

But this does not imply a simple surrender of 'science' to 'religion', both terms having become mutually constitutive and contrastive, through a history of tensions between European Christianity and the rise of Western Science, primarily from the 16th to 19th centuries. Other societies, including China and other Asian cultures, have long conceived of different ways of relating the spiritual and material, humanity and cosmos, inner and outer, body and mind, perception and understanding. How can the insights from these radically different ontologies relate to the modern categories of knowledge, including science and religion? This remains a vexing question that will undoubtedly exercise many minds into the future, as Asian and non-Western ways of thought reclaim their voices in the intellectual field. In this context, thinkers of the late 19th and early 20th centuries such as Kang Youwei, Liang Qichao, 'Abdu'l-Bahá (1844–1921), Swami Vivekananda (1863–1902), Mohandas Gandhi (1869–1948), Sri Aurobindo (1872–1950) or Chen Yingning (1880–1969), who all had one foot deeply rooted in traditional civilisations and cosmologies, and another foot well positioned in the rising currents of modern thought, have important insights for us to reflect on. This book, *Science and the Confucian Religion of Kang Youwei (1858–1927)*, will undoubtedly contribute to a widening reconsideration of the ever-evolving relationship between science and religion, in China, Asia and the world.

Professor David A Palmer (University of Hong Kong)

Notes and Abbreviations

In this book, Chinese characters have sometimes been provided to assist readers. In the main text, I have employed the traditional complex characters in accordance with the academic norm followed in studies on pre-1949 China, but in notes and the bibliography, I have retained the simplified characters in those titles that have been published in simplified characters. Translations in this book, unless otherwise stated, are all mine, and I take sole responsibility for any inaccuracy or infelicity of language.

KYWQJ *Kang Youwei quanji 康有为全集* [*Complete Works of Kang Youwei*]. Edited by Jiang Yihuan 姜义华 and Zhang Ronghua 张荣华. 12 vols. Beijing: Zhongguo remin daxue chubanshe, 2007.

LQCQJ *Liang Qichao quanji 梁启超全集* [*Complete Works of Liang Qichao*]. Edited by Zhang Pingxing 张品兴. 21 vols. Beijing: Beijing chubanshe, 1999.

TSTQJ *Tan Sitong quanji 譚嗣同全集* [*Complete Works of Tan Sitong*]. Edited by Cai Shangsi 蔡尚思 and Fang Xing 方行. Revised. 2 vols. Beijing: Zhonghua shuju, 1981.

Acknowledgements

This book is based on my doctoral thesis. From being a shepherd boy growing up in deprivation at a remote village located in Northwest China to becoming a Doctor of Philosophy at Oxford, UK, it has been a long journey for me.

Above all, I wish to express deepest gratitude to my supervisor, Professor Rob Iliffe, for guiding me to the completion of this journey through his insightful advice, inspiration and encouragement, and his gracious moral and financial support.

Heartfelt thanks are also due to Professor David Palmer (HKU), who encouraged me to take up this challenging topic in the first place, and from many discussions with whom I have greatly benefited; to Uncle William, who has gone to great pains to proofread my various drafts and offered me extremely helpful editorial suggestions; to Professor Zheng Dahua, my host at the Institute of Modern Chinese History, CASS, where I was a visiting scholar from March to July, 2015; to Professor Ruth Harris (Oxford) for examining my thesis, appreciating it, and advising me on turning it into a book; to Professor Bernard Lightman (York) for helping me finalise the book title as well as for commenting on the introduction; and to the reviewer who read my manuscript at Brill's invitation and offered me insightful suggestions for its final improvement.

The same appreciation goes to the following scholars and professors, named by reference to time of consultation: Zhuo Xinping (CASS), Ke Ping (NJU), Phyllis Chew (NIE), Siew-Huat Kong (UM), Yang Zijiang (LZJTU), Maurizio Marinelli (Sussex), Jim Liversey (Sussex), Jia Xiaoye (CASS), Du Jidong (CASS), Yu Weiya (PKU), Zhu Xiaoyuan (PKU), Chang Chihyun (SJTU), Han Qi (CAS), Yuan Jiangyang (UCAS), Liu Xiaoting (BNU), Zhang Xiang (CNU), Chen Na (Fudan), Huang K'o-wu (Sinica) and Tang Wenming (Tsinghua), as well as Oxford-based Lyndal Roper, Rana Mitter, Barend Ter Haar, Jane Garnett and Gordon Barret, for their insightful input and inspiring, sometimes highly critical, comments on my research proposals or thesis drafts.

My sincere thanks extend to the research coordinators in Sussex SHAHP and Oxford History Faculty as well as the librarians of Oxford China Centre, Linacre College (Oxford), the Needham Institute (Cambridge), the Institute of Modern Chinese History (CASS), and SOAS, in particular, Fiona Allan, Belinda Michaelides, Susanne Heinrich, Jackie Favarin, Fiona Richardson, John Moffett, Minh Chung and Trevor Langrish, for their support and efforts to help obtain research materials.

I also wish to thank Lanzhou Jiaotong University (N.W. China), NMPI (Macau), Oxford Museum of the History of Science, Sussex History Department, the Sisterhood and After Project (Sussex), and the Newton Project

(Oxford) for giving me various opportunities to earn bread; to the W. Wing Yip and Brothers (London), the Sussex Alumni Travel Grant and the UCCL (London), and the Linacre Domus Research Prizes (Oxford) for granting me a bursary, two travel fund grants and an award, respectively; and to the School of Philosophy, Beijing Normal University, for the financial support for the final revision and publication of this manuscript.

Of my friends and relatives who have helped me either financially, emotionally or intellectually throughout the course of my research and writing, special thanks go to May Farid, Molly Chuang, Lou Dajie, Afi and Sasan, Yanping and Shiping, Jian and Anmin, Jieyan and Hongtao, Xiayi and Allen, Zhang Xinyu, Qiao Liang, Wu Mengyu, Wu Zhengxuan, Liang Jianhua, Wei Leong Tay, Li Na, Li Hongpeng, Li Hongbin, Anca and Cornelis, Brandi and Tim, and my sisters Fengjun and Shujun.

This book would not have been possible without the professional assistance of Elizabeth You, Kayla Griffin, Qin Higley and Patricia Radder at Brill, and I am very grateful to each of them.

Last but not least, profound gratitude to my mother and late father, who persevered in giving me a school education despite dire poverty and extreme adversity; to my mother-in-law, who saved me a lot of time by help with the housework; and, last but foremost, to my dear wife, Qionghui, and our daughters, Holly and Lucy, the three angels of my life (I LOVE YOU THREE!)*, for their always being there during the seven-year odyssey of my doctoral research.

* This expression was burst out of the lips of Lucy, who was only three or four years old then, on hearing "I love you too (two)!" uttered by her sister. Since then, the expression has become commonly used in my family of four.

Introduction

> 'How could it be that Isaac Newton professed a religious belief?' a group of Grade 10 students asked their physics teacher.
> 'Well, he did not turn to religion until his declining years when he was not doing much scientific research.'
> Not fully satisfied with the answer, they turned to their politics teacher and received another explanation:
> 'Newton had a dual personality: he was a materialist during his scientific work, though an idealist in his private life.'

∵

This classroom episode occurred in 1953 at Nanjing No. 1 High School in the newly-founded People's Republic of China.[1] At that time students across the country were taught that Marxist materialism was the rational scientific worldview, while 'idealism', of which religion was regarded as an archetype, was a form of misguided thinking that had to be stoutly resisted. It followed that true knowledge could only be obtained under the guidance of dialectic materialism, not idealism. A believing scientist, therefore, was a contradiction in terms. That is why those students had been so puzzled when they first learned that Isaac Newton, one of the world's greatest scientists, had held a Christian faith. In response, the teachers had to explain away his religious faith in a way that conformed to orthodox ideology.

That dichotomous image of Newton persisted into the 1990s when I was a high school student. Newton was still being portrayed as an icon of science and reason during the prime of his life, eventually declining and becoming a deplorable victim of religion in his dotage. For this reason, Newton was an often-quoted example of how religion stands in the way of scientific endeavours. This Chinese view of Newton can largely be traced to Friedrich Engels'

---

1 An account of the anecdote was posted on 27 August 2012 by one of the students who had asked the question on a popular Chinese Internet forum, in relation to a discussion on why some scientifically-educated people in the West still hold religious beliefs. See Xiyang Wanchang 夕阳晚唱, 'Jiu Nanlaibeiwang de wenti xiangqide 就"南来北往"的问题想起的 [Nanlaibeiwang's Question Has Reminded Me of Something]', *Xici hutong 西祠胡同*, 27 August 2012, http://www.xici.net/d175186744.htm; Xiyang Wanchang 夕阳晚唱 and Wan Zhaoyuan 万兆元, 'Qingjiao 请教 [An Inquiry]', Email, 23 October 2012.

 | DOI: 10.1163/9789004468221_002

Dialectics of Nature, a work that has been extremely influential among twentieth-century Chinese Marxist intellectuals, and in which Newton was criticized for having 'in his old age busied himself with expounding on the *Revelation of St. John*'.[2] Imagine my surprise when, in 2008, I was introduced to a totally different Newton, as presented in a recent English biography of him, which I was invited to translate into Chinese:[3] a great scientist deeply engaged in religious studies throughout his life, who saw no contradiction between his interpreting the Bible and simultaneously decoding Nature, 'the other Book of God'![4] This surprise 'discovery' inspired me to embark on an exploration of the fascinating subject of science and religion, particularly in the Chinese context.

The Chinese dichotomous conception of Newton is a revealing example of the so-called 'Conflict Thesis', according to which science and religion are in a fundamental conflict. The thesis became part of China's intellectual orthodoxy during the iconoclastic New Culture Movement in the late 1910s and has held sway ever since.[5] My research, however, is not so much an examination of the Conflict Thesis as it is a search for its historical antecedents in the context of modern Chinese thought. Were there alternative configurations of the association between what are now generally categorized as science and religion before the Conflict Thesis took root? How did the immediate predecessors of the New Culture intellectuals understand the concepts of 'science' and 'religion' and their relationship? These questions focused my attention on the late Qing[6] reformers, especially Kang Youwei 康有為 (1858–1927), widely considered as a most representative thinker of the period of transition from Imperial

2 Friedrich Engels, 'Dialectics of Nature', in *Marx & Engels Collected Works*, trans. Clemens Dutt, vol. 25 (London: Lawrence & Wishart, 2010), 345.

3 Rob Iliffe, *Newton: A Very Short Introduction* (Oxford: Oxford University Press, 2007). Its Chinese version in simplified characters was published by the Yilin Press in Nanjing in 2009, which was reprinted in traditional Chinese characters by Oxford University Press (Hong Kong) in 2015.

4 For more on Newton's studies of the 'Two Books', see James E. Force and Richard H. Popkin, eds., *The Books of Nature and Scripture: Recent Essays on Natural Philosophy, Theology, and Biblical Criticism in the Netherlands of Spinoza's Time and the British Isles of Newton's Time* (Luxemburg: Springer Science+Business Media Dordrecht, 1994).

5 For an in-depth discussion on the rise of the Conflict Thesis in early twentieth-century Chinese thought, see Danny W. Y. Kwok, *Scientism in Chinese Thought: 1900–1950* (New Haven: Yale University Press, 1965).

6 In this book, the 'late Qing' is usually used to refer to the last years of the Qing dynasty from the 1894–5 Sino-Japanese war to the 1911 Revolution that overthrew the dynasty. Occasionally, the term is also used in its more general sense to cover the dynasty's last seven decades, starting from the first Opium War in the early 1840s.

to Republican China,[7] a period that has been characterised as the 'transitional era' (*zhuanxing shidai* 轉型時代) of modern Chinese intellectual history.[8] While Kang was chiefly known as the intellectual architect of the Reform of 1898 (*Wuxu bianfa* 戊戌變法), he was also a pioneering advocate of Western science in China and the pre-eminent promoter of a religious reform. Regarded by his disciples as 'the Martin Luther of Confucianism',[9] he endeavoured to restore Confucianism to its 'pristine' state, and he re-named his restatement of it as *Kongjiao* 孔教, the religion of Confucius.[10] In this book I will examine Kang's evolving concepts of 'science' and 'religion', and argue that his writings constitute one of the major early attempts to conceptualize their relationship in the Chinese context.

1 Definition of Terms

In many ways, the analysis of the so-called 'science and religion question' in twentieth-century China has reflected the stimulation and intellectual influence of the West. Indeed, modern Chinese concepts of 'science' and 'religion' have been shaped by their Western counterparts.

7 See, for example, the view of his major biographer in the West, Kung-ch'uan Hsiao, *A Modern China and a New World: K'ang Yu-Wei, Reformer and Utopian* (Seattle: University of Washington Press, 1975), 594.

8 The influential concept of the 'transitional era' was proposed by Hao Chang and echoed by a number of scholars, see Hao Chang, *Liang Ch'i-Ch'ao and Intellectual Transition in China, 1890–1907* (Cambridge, MA: Harvard University Press, 1971); Hao Chang, *Chinese Intellectuals in Crisis: Search for Order and Meaning (1890–1911)* (Berkeley: University of California Press, 1987); see also Wang Fansen 王汎森, *Zhongguo jindai sixiangshi de zhuanxing shidai* 中國近代思想史的轉型時代 [*The Transitional Era of Modern Chinese Intellectual History*] (Taipei: Linking Publishing, 2007).

9 Liang Qichao 梁启超, 'Nanhai Kangxiansheng zhuan 南海康先生传 [A Biography of Mr Kang of Nanhai (1901)]', in *LQCQJ*, vol. 2, 1999, 486.

10 *Kongjiao* (Kong being the surname of Confucius) was a new term employed by Kang to refer to his reformed Confucianism in contrast to orthodox Neo-Confucianism. In some English texts, the neologism 'Confucianity' has been proposed as a translation for *Kongjiao*, so as to highlight Kang's appropriation of Christian practices. See, for example, Lionel M. Jensen, *Manufacturing Confucianism: Chinese Traditions & Universal Civilization* (Durham: Duke University Press, 1997), 292; Anna Sun, *Confucianism as a World Religion: Contested Histories and Contemporary Realities* (Princeton, N.J.: Princeton University Press, 2013), 42–43. While cleverly coined, 'Confucianity' might lull a reader into overlooking the doctrinal differences between *Kongjiao* and Christianity.

Despite the existence of various definitions, 'science' remains difficult to define, and the term can be approached from different perspectives.[11] Science is sometimes defined as a 'system of behaviour' by which man attempts to master his environment, but more commonly it is defined as a 'system of knowledge' of the natural world.[12] In the Western context, the study of nature was traditionally denoted by the term 'natural philosophy', which sought to discover, without referring to the supernatural, the physical causes of natural phenomena. Though originally not concerned with mathematics, whose computations did not involve physical causes, natural philosophy achieved union with mathematics in the seventeenth century and thereby gave rise to the 'Scientific Revolution'.[13] Exact sciences such as chemistry, physics and biology had matured as independent disciplines by the early nineteenth century. Only then did natural philosophy come to be called 'science' in general, and the word 'science' was used as an umbrella term for a range of different sciences.[14] This 'modern science' is held to be essentially characterized by its methodology, which, according to one dictionary definition, consists in 'systematic observation, measurement, and experiment, and the formulation, testing, and modification of hypotheses'.[15]

To some extent, the term 'natural philosophy' carried a meaning similar to the classical Chinese idea of *gezhi* 格致, which means 'acquiring knowledge through the investigation of things'. So when a version of Western science was first introduced into China by the Jesuits in the late sixteenth and seventeenth centuries and then again by Protestant missionaries in the mid-nineteenth

11 For a well-informed discussion on the definition of science, see A. F. Chalmers, *What Is This Thing Called Science?*, 4th ed. (Queensland: University of Queensland Press, 2013); David C. Lindberg, *The Beginnings of Western Science: The European Scientific Tradition in Philosophical, Religious, and Institutional Context, Prehistory to A.D. 1450*, 2nd ed. (Chicago and London: The University of Chicago Press, 2007), 1–3.

12 J. G. Crowther, *The Social Relations of Science*, revised ed. (London: Cresset Press, 1967), 1; G. E. R. Lloyd, *Early Greek Science: Thales to Aristotle* (London: Chatto & Windus, 1970), 1.

13 It should be noted, however, that there have appeared different views on whether there had indeed been a 'Scientific Revolution', see Steven Shapin, *The Scientific Revolution* (Chicago: University of Chicago Press, 1998); cf. John Henry, *The Scientific Revolution and the Origins of Modern Science*, 3rd ed. (New York: Palgrave Macmillan, 2008).

14 Edward Grant, *A History of Natural Philosophy: From the Ancient World to the Nineteenth Century* (Cambridge; New York: Cambridge University Press, 2007), 307.

15 'Scientific Method', in *Oxford Dictionaries Online* (Oxford University Press), accessed 28 January 2014, http://www.oxforddictionaries.com/; see also Barry Gower, *Scientific Method: A Historical and Philosophical Introduction* (London and New York: Routledge, 1997). For an anarchistic view against the scientific method, see Paul K. Feyerabend, *Against Method: Outline of an Anarchistic Theory of Knowledge* (London: New Left Books, 1975).

century, *gezhi* was adopted as the term for natural science. Towards the end of the nineteenth century, however, a neologism *kexue* 科學 (which literally means systematized branches of learning) was invented by Japanese scholars to refer to the modern science introduced into Japan from the West, so as to distinguish it from the old term *gezhi*, which was reminiscent of Chinese traditional learning. The new term highlighted some key characteristics of modern science, notably its specialization and professionalization. This neologism was soon introduced into China from Japan (which had then become the main portal through which Chinese scholars acquired Western learning), and was used interchangeably with *gezhi* before it was eventually accepted as the standard Chinese equivalent for 'science' in the mid-1910s.[16] *Kexue* or science in the Chinese context is now generally understood as 'exact and systematized knowledge' of the world, which primarily refers to the natural sciences, but in its broad sense it also covers disciplines such as psychology, sociology and economics. Two features are often noticed in its Chinese usage. One is that science is conventionally approached from a pragmatic perspective, and therefore it is often used together or even elided with technology. The other is that, as is the case in the West but perhaps more forcibly, in the context of China, science is frequently employed as a standard for value judgment in various fields.[17]

The term 'religion' is even more problematic.[18] Its origin, the Latin *religiō*, has an obscure etymology, and Max Müller (1823–1900), the 'father of religious studies', concluded that it 'can only have meant respect, care, reverence' in its first conception.[19] In pre-modern West, the word 'religion' was rarely used to refer to sets of beliefs or practices (the meaning that is familiar to us), but it was often employed to denote a moral virtue such as piety. In this context, the expression 'true religion' was used by some of the Christian Church Fathers in

---

16 For a comprehensive discussion on the replacement of *gezhi* by *kexue*, see Fan Hongye 樊洪业, 'Cong gezhi dao kexue 从"格致"到"科学" [From "gezhi" to "Science"]', *Ziran bianzhengfa tongxun* 自然辩证法通讯 [*Journal of Dialectics of Nature*], no. 3 (1998): 39–50. It is noteworthy that at the turn of the century a similar process of redefining and retranslating the English term 'science' was also going on in the Arabic world, see Marwa S. Elshakry, 'Knowledge in Motion: The Cultural Politics of Modern Science Translations in Arabic', *Isis* 99, no. 4 (2008): 701–30.

17 For a summary of the use of *kexue* or science in the modern Chinese context, see Wu Guosheng 吴国盛, *Shenme shi kexue* 什么是科学 [*What Is Science?*] (Guangzhou: Guangdong renmin chubanshe, 2016), 1–17.

18 It has been widely acknowledged that there simply cannot be a universal definition of religion. See, for example, the discussion by Talal Asad, *Genealogies of Religion: Discipline and Reasons of Power in Christianity and Islam* (Baltimore & London: Johns Hopkins University Press, 1993), 27–55.

19 Max Müller, *Natural Religion* (London: Longmans, Green, & Co., 1889), 36.

the Middle Ages to refer to the pious worship of God rather than to doctrinal orthodoxy. This understanding of religion as an 'interior' virtue persisted until the sixteenth century, when the term began to be used in an 'exterior', generic sense, both as a body of doctrine to be affirmed and as a set of public practices to be distinguished from Islam, Judaism or even other branches of Christianity. This significant shift in the conception of religion is considered by the distinguished historian of science and religion Peter Harrison as a 'precondition' for a relationship between science and religion.[20] In the intellectual discourse in the West, religion in relation to science refers primarily and sometimes solely to Christianity, or in specific cases to either Protestantism or Roman Catholicism.[21] The modern concept of religion, in the sense of 'a systematic religious entity, conceptually identifiable and characterizing a distinct community', has come to be regarded as inadequate by such scholars as Wilfred Cantwell Smith, for the reason that it is a peculiarly European construct of a relatively recent origin, largely shaped by the Judeo-Christian monotheistic tradition, rather than a universally valid category.[22]

It is a more complicated story in China. When the Jesuits introduced into China the Roman Catholic religion as a structured system of beliefs and practices, organised for congregational worship in churches, there was no equivalent Chinese belief system with the same features.[23] The traditional concept that came closest in meaning was *jiao* 教, which literally means 'teaching', 'instruction' and 'education'. By the late nineteenth-century, China's religious landscape, which had traditionally been constituted by the institutionalized Three Teachings (*sanjiao* 三教), namely, Confucianism, Buddhism and Daoism, on the national level and by communal religions on the local level, had been altered by the emerging presence of 'foreign' Christianity, which had obtained legal status in China and was allowed to spread after the Opium Wars in the mid-nineteenth century.[24] In this context, Christianity was also considered a kind of *jiao*, and was referred to as *Tianzhujiao* 天主教 (Catholicism) or *Yesujiao* 耶穌教 (Protestantism), literally, the religion of the Lord of Heaven or of

20 Peter Harrison, *The Territories of Science and Religion* (Chicago: University of Chicago Press, 2015), 7–11.

21 See Gary B. Ferngren, *The History of Science and Religion in the Western Tradition: An Encyclopedia* (New York & London: Garland Publishing, 2000).

22 W. Cantwell Smith, *The Meaning and End of Religion* (Minneapolis, Minnesota: Fortress Press., 1991), 119.

23 Vincent Goossaert, 'The Concept of Religion in China and the West', *Diogenes* 52 (2005): 13–20.

24 Vincent Goossaert and David A. Palmer, *The Religious Question in Modern China* (Chicago: University of Chicago Press, 2011), 22–23.

Jesus, respectively. Like their Jesuit predecessors, who upheld Catholicism as the universal religion and concluded that China was mired in superstitions, heterodox beliefs, idolatry and atheism, nineteenth century Protestant missionaries also criticized Chinese traditional culture and claimed that China had no *true* religions.[25] In response, Chinese intellectuals started to contrast Chinese *jiao* with Christianity, thus initiating the 'religious question' into the intellectual discourse of modern China. Central to the discussion was the question whether the concept of *jiao* matched the notion of 'religion' introduced by the Christian missionaries.[26]

Many saw the difference between the two terms and opted for, once again, a Japanese neologism, *zongjiao* 宗教, which eventually, in the early twentieth century, became widely accepted as the Chinese equivalent for 'religion' in the Western sense outlined above.[27] Under this rubric were grouped Buddhism and Daoism, together with Islam, Protestantism and Catholicism, while, as I will show in this book, there were heated debates concerning whether Confucianism was a religion or not.[28] In the meantime, local cults and practices were now labelled as *mixin* 迷信 (superstitions), another new notion adopted from the West via Japan. Thus with the coming of such neologisms as *zongjiao*, *kexue* and *mixin* there appeared a triangle in Chinese modernist rhetoric, with science becoming the touchstone in differentiating 'religion' (compatible

25 See George Minamiki, *The Chinese Rites Controversy: From Its Beginning to Modern Times* (Chicago: Loyola University Press, 1985); Chen Huaiyu 陈怀宇, *Jindai chuanjiaoshi lun Zhongguo zongjiao: yi Muweilian Wujiao tongkao wei zhongxin* 近代传教士论中国宗教：以慕维廉五教通考为中心 [*Missionary Writings on Chinese Religions: William Muirhead and His Introduction to Five Religions in China*] (Shanghai: Shanghai renmin chubanshe, 2012).

26 For recent re-examinations of missionary notions of Chinese religion, see Zhang Zhigang 张志刚, *Zongjiao Zhongguohua yili chensi* 宗教中国化义理沉思 [*Sinocization of Religions: A Theoretical Study*] (Beijing: Zongjiao wenhua chubanshe, 2017), 128–40; Li Tiangang 李天纲, *Jinze: Jiangnan minjian jisi tanyuan* 金泽：江南民间祭祀探源 [*Jinze Town: Exploring the Origins of Folk Sacrificial Rituals in Jiangnan*] (Beijing: Sanlian shudian, 2017), 1–13, 508–25. For a discourse on *jiao* and 'religion' in the early twentieth century, see Hsi-yuan Chen, 'Confucianism Encounters Religion: The Formation of Religious Discourse and the Confucian Movement in Modern China' (Ph.D., Harvard University, 1999), 1–70.

27 See Goossaert, 'The Concept of Religion'; Vincent Goossaert, 'The Social Organization of Religious Communities in the Twentieth Century', in *Chinese Religious Life*, ed. David A. Palmer, Glenn Landes Shive, and Philip L. Wickeri (Oxford: Oxford University Press, 2011), 172–90.

28 Goossaert, 'The Social Organization of Religious Communities', 176; also see Chen, 'Confucianism Encounters Religion'.

with science) from 'superstition' (unscientific).[29] Kang, however, disapproved of those neologisms, in particular the usage of *zongjiao* for religion, arguing that the Chinese *jiao* represented the true sense of 'religion', a notion that in his view had been narrowed, if not distorted, in its scope by the misconceptions under which the protagonists, apologists and promoters of the Christian religion laboured.[30] His point was largely neglected by early twentieth century Chinese intellectuals.

Several decades later, however, an increasing number of scholars in Chinese religious studies have begun to argue that the concept of religion accepted in China, which has been largely shaped by its western counterpart, is inadequate to capture the quintessential nature of Chinese religious tradition. To understand Chinese religion, therefore, the category of 'religion' must be reconsidered.[31] Several attempts have been made in this direction, in particular to define the religious nature of Confucianism.[32] Rodney Taylor, for example, appropriated Frederick Streng's definition of religion as a means toward ultimate transformation in his analysis of Confucianism, positing that the Confucian concept of Heaven, which functioned as a religious absolute, has provided humanity with such a means – the pursuit of sagehood.[33] In contemporary China, the controversy over whether Confucianism was a religion has become revitalised again in the Chinese academia. While the mainstream view remains negative, there have appeared strong voices affirming the religious nature and status of Confucianism in imperial China.[34] This book adopts the

29 Goossaert and Palmer, *Religious Question*, 50–51.

30 Kang Youwei 康有为, 'Kongjiaohui xu (er) 孔教会序（二）[(A Second) Introduction to the Confucian Religion Association (1912)]', in *KYWQJ*, vol. 9, 2007, 345–46.

31 For a strong argument for this position, see Zhang, *Zongjiao zhongguohua*, 128–60; Li, *Jinze*, 1–13.

32 See, for example, C. K. Yang, *Religion in Chinese Society: A Study of Contemporary Social Functions of Religion and Some of Their Historical Factors* (Berkeley: University of California Press, 1961); Hans Küng and Julia Ching, *Christianity and Chinese Religions* (New York: Doubleday Religious Publishing Group, 1989); Wei-ming Tu, *Confucian Thought: Selfhood as Creative Transformation* (Albany, N.Y.: SUNY Press, 1985); Mou Zhongjian 牟钟鉴 and Zhang Jian 张践, *Zhongguo zongjiao tongshi 中国宗教通史* [*A General History of Chinese Religions*], revised ed. (Beijing: Zhongguo shehui kexue chubanshe, 2007).

33 Rodney L. Taylor, *The Religious Dimensions of Confucianism* (Albany: State University of New York Press, 1990), 2–4; see also Frederick J. Streng, *Understanding Religious Life* (Belmont: Wadsworth Publishing Company, 1985), 1–8.

34 The most vocal proponent of imperial Confucianism as a state religion were Li Shen 李申 and his mentor Ren Jiyu 任继愈. See Ren Jiyu 任继愈, ed., *Rujiao wenti zhenglunji 儒教问题争论集* [*Collected Essays on the Debate over Ruism/Confucianism as a Religion*] (Beijing: Zongjiao wenhua chubanshe, 2000); Li Shen 李申, *Zhongguo Rujiao shi 中国*

perspective that the religious dimension of Confucianism has now been widely acknowledged; without an understanding of this dimension, our appreciation of key concepts in Confucianism could, as the prominent contemporary exponent of Confucian thought Wei-ming Tu points out, be inadequate or even misleading.[35]

My research traces Kang's concepts of 'science' and 'religion' during the transitional era when he witnessed the replacement of Chinese traditional terminology (e.g. *jiao*) with modern terms (e.g. *zongjiao*) derived from Western/Japanese sources. In the following chapters, I examine the way Kang first framed those concepts, and eventually re-framed them in response to the unfolding intellectual and political discourse of his later years. It should be noted, however, that outside of the context of Kang's own body of work, when I discuss the matter in this book, I use both terms in a more general sense. I treat science and religion as two evolving systems of knowledge and practice, the former enabling people to pursue and apply knowledge of the natural and social world following a systematic, evidence-based methodology, and the latter investing people with a belief in a supernatural power or powers, which gives rise to private, communal, or institutional acts (including personal conduct) aimed at establishing a favourable relationship with the object of their belief.[36] This view of science can apply to discussions on scientific thoughts and practices in both traditional and modern society, and this concept of religion makes theist belief central to its definition, by which all the world's theistic religions are included, but secular systems, no matter how closely they resemble a religion, are not. By using the adjective 'evolving' in my treatment of science and religion, I stress that the two systems change and develop over time, and as a result, their relationship has not been static nor will it remain unchanged.

儒教史 A History of the Confucian Religion of China, 2 vols (Shanghai: Shanghai renmin chubanshe, 1999).

35 See Tu Wei-ming, *Zhongyong dongjian 中庸洞见 / An Insight of Chung-yung*, bilingual ed. (Beijing: People's Publishing House, 2008), 4.

36 In this I have drawn on the considered definition of 'science' by the Science Council – 'the pursuit and application of knowledge and understanding of the natural and social world following a systematic methodology based on evidence', and Morris Jastrow's broad yet useful definition of religion – 'the natural belief in a Power or Powers beyond our control, and upon whom we feel ourselves dependent; which belief and feeling of dependence prompt (1) to organisation, (2) to specific acts, and (3) to the regulation of conduct, with a view to establishing favourable relations between ourselves and the Power or Powers in question'. The Science Council, 'Our Definition of Science', 2009, https://sciencecouncil.org/about-science/our-definition-of-science/; Morris Jastrow, *The Study of Religion* (London: Walter Scott, 1901), 171–72.

2 Science and Religion

In an often-quoted statement in relation to the Conflict Thesis written about a century ago, the British philosopher, Alfred North Whitehead, considered science and religion to be 'the two most potent forces' of human life, and made an observation that withstands the passage of time: 'When we consider what religion is for mankind, and what science is, it is no exaggeration to say that the future course of history depends upon the decision of this generation as to the relations between them."[37]

The relationship between science and religion is both an old and a fresh issue in the West. In pre-modern Europe, a synthesis of knowledge and belief was created when Christian doctrine and ancient Greek cosmology were harmonized in the medieval period, and the study of nature was made subordinate to theology.[38] The pioneers of modern science, such as Galileo Galilei and Johannes Kepler, were either Roman Catholics or belonged to a Protestant church. In their efforts to expound the knowledge of astronomy, they entered the field of theology in order to explain how the Copernican system could be made consistent with Biblical exegesis, although in doing so, they had also disagreed with some of orthodox views of the divines.[39] For many seventeenth-century natural philosophers, like Robert Boyle and Isaac Newton, religion served as an incentive as well as a sanction to investigate the workings of a universe created by God.[40] While a differentiation between religion and natural philosophy did occur during the Scientific Revolution, their reintegration was later largely achieved by way of natural theology, which attempted to use scientific knowledge to prove that the natural order was the consequence both of the existence of God and of his specific attributes.[41]

Since the Scientific Revolution, generations of Western intellectuals have appraised the relationship between science and religion and have

---

37 Alfred N. Whitehead, *Science and the Modern World* (New York: The New American Library, 1925), 180; see also Alister E. McGrath, *Science and Religion: A New Introduction*, 2nd ed. (Malden, MA: Wiley-Blackwell, 2010), 1.

38 John H. Brooke, *Science and Religion: Some Historical Perspectives* (Cambridge: Cambridge University Press, 1991), 58–63; Robert H. Brown, *The Wisdom of Science: Its Relevance to Culture and Religion* (New York: Cambridge University Press, 1986), 148.

39 Herbert Butterfield, *The Origins of Modern Science: 1300–1800*, new ed. (London: G. Bell and Sons Ltd, 1957), 69–70.

40 Grant, *Natural Philosophy*, 298.

41 Brooke, *Science and Religion*, 24.

suggested a number of ways of conceptualizing this association.[42] Of these, the 'conflict' view or the 'irreconcilability' model was voiced by a number of eighteenth-century Enlightenment philosophers, but it did not begin to hold sway among European intellectuals until the late nineteenth century, when John Draper and Andrew White represented the history of science as a series of wars and conflicts between science and the Catholic church, or between men of science and theologians, whose celebrated examples include the cases of Galileo and Darwin.[43] Such a view became known as the Draper-White Thesis and was widely interpreted as a conflict between science and religion in general. This thesis exerted an enormous influence in the twentieth century, and the ensuing success of science and technology came to be held as a standing refutation of the authority and truth of religion.

Starting from the mid-twentieth century, however, the Draper-White Thesis has received substantial scholarly criticism. Alternative ways of relating science and religion have been proposed and critiqued. In his work on the origins of modern science, for example, the British historian Herbert Butterfield has suggested that in the long run it was Protestantism that became an ally with the scientific movement in the early modern period.[44] Ian Barbour has summarized four ways of relating science and religion, presenting three alternative models – independence, dialogue and integration – to the Conflict Thesis.[45] However, a number of historians of science such as John Brooke have demonstrated that science and religion have been so richly interrelated in different periods of history that no simple general model or master narrative on this subject is possible.[46] Peter Harrison has carried this 'complexity thesis' further and argued that the discrete categories of 'religion' and 'science' were intellectual constructs of the last three centuries, for in the remote past there were no clear-cut boundaries between their underlying semantic fields, both of which emphasized inner qualities.[47] In sum, Western scholarship on the subject over

42 Mikael Stenmark, 'Ways of Relating Science and Religion', in *The Cambridge Companion to Science and Religion*, ed. Peter Harrison (Cambridge: Cambridge University Press, 2010), 278–95; Ian G. Barbour, *Religion in an Age of Science* (San Francisco: Harper & Row, 1990), 3–10.

43 J. William Draper, *History of the Conflict between Religion and Science* (London: Henry S. King & Co., 1875); A. Dickson White, *A History of the Warfare of Science with Theology in Christendom*, 2 vols (New York: D. Appleton and Company, 1896).

44 Butterfield, *Origins of Modern Science*, 58.

45 Barbour, *Religion in an Age of Science*, 3–30.

46 Brooke, *Science and Religion*, 2–4; McGrath, *Science and Religion*.

47 Harrison, *The Territories*. For an anthropological perspective on the emergence of religion as a modern historical object, see also Asad, *Genealogies of Religion*, 27–82.

the past few decades has been extraordinarily rich in terms of its depth, diversity and popular appeal, and has demonstrated the immense historical complexity of science-religion relations.[48]

The vast majority of these studies, however, have been carried out within a Western context and are focused primarily on the relations between science and the Abrahamic faith traditions, in particular, Christianity, and their conclusions hold only for Western science and religion. In light of an emerging globalizing vision and at a time when increased efforts should be made to avoid provincialism in scholarly pursuits, it is highly desirable that other religious traditions and their non-Western contexts should also be considered and analysed.[49] For example, the 2011 collected volume on science and religion around the world represents a pioneering effort towards this direction by exploring the way nature has been understood and manipulated in such diverse regions as ancient China, India, and sub-Saharan Africa. In doing so, it demonstrates the value of plural perspectives for the construction and perception of the relationship between science and religion.[50]

Although in traditional China the line between religious reasoning and scientific thought was probably no less blurred than in the pre-industrial West, in current Chinese thought not only does the conflict model prevail, but also for political and ideological reasons, the tension between science and religion is much sharper than in the West. Having taken root during the New Culture Movement that upheld radical rationalism in the form of scientism, the Conflict Thesis held sway among intellectuals across the nation in the second half of the twentieth century, with regularly repeated endorsement from the official atheist ideology, and has remained the orthodox interpretation to this day.[51]

---

48 For a concise yet updated summary of the historiography on the historical inter-relations of science and religion over the past few decades, see Geoffrey Cantor, 'Science and Religion: From the Historian's Perspective', in *The Routledge Companion to Religion and Science*, ed. James W. Haag, Gregory R Peterson, and Michael L. Spezio (London: Routledge, 2012), 24–33.

49 Phee Seng Kang, 'Bridging Science and Religion in China: Emerging Opportunities for Global Dialogue', *Theology and Science* 4, no. 2 (2006).

50 See John H. Brooke and Ronald L. Numbers, eds., *Science and Religion around the World* (Oxford: Oxford University Press, 2011).

51 Zhuo Xinping 卓新平, 'Zhongguo zhishijie dui zongjiao yu kexue guanxi zhi lun 中国知识界对宗教与科学关系之论 [Discourse on the Relationship between Science and Religion among Chinese Intellectuals]', in *Qiao: kexue yu zongjiao 桥: 科学与宗教* [*Bridging Science and Religion*], ed. Kang, Phee Seng, Ted Peters, and Gaymon Bennett, 2nd ed. (Beijing: Zhongguo shehui kexue chubanshe, 2007), 230–44; Paul Shan, 'Science and Faith in China Today', in *Western Learning and Christianity in China: The Contribution*

According to this position, science, whose spirit is atheistic, cannot but contradict the theistic beliefs central to religion; the conflict between the two is often interpreted as the surrogate for that between atheism or theism or, similarly, for the tension between the materialist and idealist worldviews.[52] Accordingly, the views on the science-religion relations in Chinese academic discourse has been dominated by the Conflict Thesis.

Unlike the situation of scholarship in the West, the history of the complex relationship between science and religion in Chinese history is yet to be studied in depth. It must, indeed, be acknowledged that the subject has already received scholarly treatment in relevant studies on the history of science in traditional China, and in various works on the history of religion in China by modern scholars both at home and abroad.[53] In discussing his famous question as to why modern science was not developed in China, the renowned British sinologist Joseph Needham (1900–1995) has argued that Confucianism was a hindrance to science, while Daoism was more 'scientific' and had a closer connection with science.[54] The American sinologist Nathan Sivin has, however, challenged this claim, pointing out that there was no documented evidence that Daoist scientists or technologists were predominant in either quantity or quality.[55] Derk Bodde, the noted English translator of Fung Yu-lan's (馮友蘭, 1895–1990) *A History of Chinese Philosophy*, has joined in the discussion by concluding that the development of modern science in traditional China was inhibited by several aspects of the Chinese humanistic tradition, including what he has described as the non-supportive role of religion, the anti-scientific sentiment which resulted from a preoccupation with morals, and the organicist thinking in the Chinese approach to nature.[56] In general, the view that a Chinese intellectual tradition dominated by Confucianism was not favourable

---

and Impact of Johann Adam Schall von Bell, S.J. (1592–1666), ed. Roman Malek, vol. II (Nettetal: Steyler Verlag, 1998), 1204.

52 Li Shen 李申, ed., *Gaokeji yu zongjiao* 高科技与宗教 [*Advanced Science-Technology and Religion*] (Tianjin: Kexue jishu chubanshe, 2000), viii–ix.

53 See Yang, *Religion in Chinese Society*; Goossaert and Palmer, *Religious Question*; Mou and Zhang, *Zhongguo zongjiao*.

54 Joseph Needham, *Science and Civilisation in China: History of Scientific Thought*, vol. 2 (Cambridge: Cambridge University Press, 1956), 35, 103, 129.

55 Nathan Sivin, *Medicine, Philosophy and Religion in Ancient China: Researches and Reflections* (Aldershot: Variorum, 1995), 310.

56 Derk Bodde, *Chinese Thought, Society, and Science: The Intellectual and Social Background of Science and Technology in Pre-Modern China* (Honolulu: University of Hawaii Press, 1991).

to the growth of modern science has been widely shared by Chinese scholars in the twentieth century.[57]

On the other hand, the inclusive attitude regarding the relationship between natural knowledge and religion or faith in traditional Chinese thought has also been widely recognized. For in the Chinese way of holistic thinking, the two were basically regarded as one. This is evident in the afore-mentioned Confucian concept of *gezhi*, and in the relevant notion of *xiaxue shangda* 下學上達 (learning about what is below in order to penetrate to what is above).[58] As has been observed, the tendency of Chinese thought towards cosmic unity entailed the integration of science and faith.[59] Similarly, a recent case study has also stressed the role of religion in the thought of Zhang Heng 張衡 (78–139), a prominent Chinese mathematician, astronomer, and geographer from a Confucian background.[60]

There has been an impressive body of literature on what can be considered as a prelude to the science and religion question in modern China, the arrival of the Jesuit and Protestant missions in China. Led by Matteo Ricci (1552–1610),[61] the Jesuits introduced to late Ming China (1368–1644) a pre-modern version of Western science as well as the Christian doctrine as one package, under the comprehensive term, *tianxue* 天學, literally meaning 'learning from heaven'.[62] This integrated presentation was embraced by the Chinese converts,

---

57 See, for example, Yu-Lan Fung, 'Why China Has No Science - An Interpretation of the History and Consequences of Chinese Philosophy', *International Journal of Ethics* 32, no. 3 (1922): 237–63; Ren Hongjun 任鴻雋, 'Shuo Zhongguo wu kexue zhi yuanyin 說中國無科學之原因 [On the Reasons Why China Did Not Have Science]', *Kexue* 科學 [*Science*] 1, no. 1 (1915): 6–12; Ren Jiyu 任继愈, 'Rujiao de xingcheng 儒教的形成 [On the Formation of Confucian Religion]', *Zhongguo shehui kexue* 中国社会科学 [*Chinese Social Sciences*], no. 1 (1980): 10.

58 *Lunyu* (*Analects*) 14:37, in James Legge, trans., *The Chinese Classics I: Confucian Analects, The Great Learning, and The Doctrine of the Mean*, 2nd ed. (Oxford: The Clarendon Press, 1893), 288–89.

59 Shan, 'Science and Faith in China Today', 1202.

60 Lei Libai 雷立柏 [Leopold Leeb], *Zhang Heng, kexue yu zongjiao* 张衡,科学与宗教 [*Zhang Heng, Science and Religion*] (Beijing: Shehui kexue wenxian chubanshe, 2000).

61 A recent Chinese study has likened Ricci to a missionary-scientific Prometheus, bringing the fire of Western science and technology to China via the marine trade route. Liu Dachun 刘大椿, *Xixue dongjian* 西学东渐 [*Western Learning Spreading to the East*] (Beijing: Zhongguo renmin daxue chubanshe, 2018), 8.

62 Willard Peterson, 'Learning from Heaven: The Introduction of Christianity and Other Western Ideas into Late Ming China', in *The Cambridge History of China*, ed. Denis Twitchett and Frederick Mote, vol. 8 (Cambridge: Cambridge University Press, 1998), 789–839; see also Liam M. Brockey, *Journey to the East: The Jesuit Mission to China, 1579–1724* (Cambridge, MA.: Harvard University Press, 2007); Han Qi 韩琦, *Tongtian zhixue: Yesuhui shi*

most notably Xu Guangqi 徐光啟 (1562–1633, Christian name Paul Hsü), who eventually rose to a high position at the Ming court. Xu argued that the religious doctrine of *tianxue* could 'supplement Confucianism and correct Buddhism' (*buru yifo* 補儒易佛) while its scientific content was akin to the classical Chinese idea of 'fathoming the principles of things through investigating them' (*gewu qiongli* 格物窮理).[63] There were, however, also critics among the literati, who accused the Jesuits, and their Chinese collaborators as well, of touting Christianity through their introduction of Western scientific knowledge, as was evidently shown during the 1664 Calendar Case.[64]

Like their Jesuit predecessors, late nineteenth century Protestant missionaries introduced to China Western science and medicine with the covert, subtle intention to convey and bolster their Christian teachings. They presented a harmonious view of science and Christianity in their scientific translations, which were often informed by natural theology. This was evident, for instance, in Benjamin Hobson's (1816–1873 合信) medical texts, the *Shanghae Serial* printed by the London Missionary Society, and Alexander Williamson's (1829–1890 韋廉臣) works.[65] In hindsight, however, the tension between science and Christianity, which was gaining momentum in post Darwinian Europe and America, can still be discerned in the missionary writings in Chinese, but not so much in what was written about as in what was not. While the Jesuits failed to introduce to the Chinese the Copernican theory in a timely manner, the Protestants and their Chinese aides tended either to avoid mentioning in writing the Darwinian theory of natural selection or to present the theory as potential targets of criticism.[66] Furthermore, as we will see in Chapter Two, the missionary narrative of a Christian provenance of modern science was also

---

he tianwenxue zai Zhongguo de chuanbo 通天之学: 耶稣会士和天文学在中国的传播 [*Learning of Heaven: The Jesuits and the Spread of Astronomy in China*] (Beijing: Sanlian shudian, 2018).

63 Xu Guangqi 徐光啟, *Xu Guangqi ji* 徐光啟集 [*Collected Writings of Xu Guangqi*] (Beijing: Zhonghua shuju, 1963), 66; see also Catherine Jami, Peter M Engelfriet, and Gregory Blue, eds., *Statecraft and Intellectual Renewal in Late Ming China: The Cross-Cultural Synthesis of Xu Guangqi (1562–1633)* (Leiden: Brill, 2001), 155–70.

64 Pingyi Chu, 'Scientific Dispute in the Imperial Court: The 1664 Calendar Case', *Chinese Science*, no. 14 (1997): 12; Catherine Jami, 'Revisiting the Calendar Case (1664–1669): Science, Religion, and Politics in Early Qing Beijing', *Korean Journal of History of Science* 27, no. 2 (2015): 459–77; see also Han, *Tongtian zhixue*, 38–41, 116–26.

65 See Benjamin Elman, *A Cultural History of Modern Science in China* (Cambridge, MA.: Harvard University Press, 2006), 111–17, 130–31, 135–40.

66 See Benjamin Elman, *On Their Own Terms: Science in China, 1550–1900* (Cambridge, MA: Harvard University Press, 2005), 347; Xiaoxing Jin, 'The Evolution of Evolutionism in China, 1870–1930', *Isis* 111, no. 1 (2020): 50.

challenged by the Chinese literati, in particular by the reformers, in reference to the notion of Chinese origins of Western learning (*xixue zhongyuan* 西學中源).

The subject of science and religion has received renewed attention from the Chinese academia during the past four decades following China's reform and opening-up in the late 1970s. The new development has occurred alongside the emergence of a more neutral discourse on religion in Chinese society, and has also been stimulated by the rigorous studies in science and religion in the West. This revival of scholarly interest can be illustrated by the increase of published monographs and translated works on the subject,[67] and by the convening of a series of lectures or seminars on science and religion, such as the 2002 international forum jointly organized in Beijing by the Berkeley Centre for Theology and the Natural Sciences (CTNS) and the Chinese Academy of Social Sciences (CASS),[68] and the Templeton-sponsored Science and Religion Series (2005–2009) in collaboration with several top Chinese universities.[69] More importantly, the academic interest in science and religion is manifested in the large number of journal papers published on the topic over the recent decades. The Chinese Social Sciences Citation Index Database shows that an increasing number of papers on science and religion (i.e. those whose titles contain 'science and religion') have been published in its source journals from 1979 to 2018, listing 44 from the period of 1979–1989, 93 from the 1990s, 366 from the 2000s, and 296 from the 2010s.[70]

A general survey of this body of literature reveals that, as is to be expected, the prevailing position has been to justify the orthodox conflict thesis. Although attempts have been made to consider whether religion and science can be regarded as belonging to distinctive yet equally valid domains, this view has

67 Dozens of titles on science and religion by Western scholars, including Bertrand Russell, Ian Barbour, Edward Grant, Richard Olson, John Brooke and Peter Harrison, have been translated into Chinese during this period of time.

68 For more on this seminar forum that has been acclaimed as 'the first-ever dialogue between Chinese scientists and religious studies scholars in a public forum', see Phee Seng Kang, 'Bridging Science and Religion in China'.

69 See Melville Y. Stewart, *Science and Religion in Dialogue* (John Wiley & Sons, 2009); Zhang Xiuhua 张秀华, 'Kexue yu zongjiao guanxi tanjiu de xinjinlu: dui jinnian Zhongmei kexue yu zongjiao xueshu huiyi de fenxi 科学与宗教关系探究的新进路—对近年中美科学与宗教学术会议的分析 [New Paths for Inquiry of the Relation between Science and Religion: A Review of the China-America Academic Conferences on Science and Religion in Recent Years]', *Tsinghua daxue xuebao* 清华大学学报 (哲学社会科学) [*Journal of Tsinghua University* (*Philosophy and Social Sciences*)], no. 3 (2013): 88–95.

70 Data from CNKI (China National Knowledge Infrastructure) http://www.cnki.net/, accessed on 9 December 2018.

failed to receive wide recognition except among those intellectuals who professed a religious belief. Some of the papers published in the new century have also begun to consider the possibility of a harmonious relationship between science and religion, but that has proven to be a difficult and sensitive effort in the contemporary Chinese context.[71] As Zhuo Xinping, a leading Chinese scholar of religious studies, has observed, attempts by Chinese academics at building a new paradigm of the relationship between science and religion as an alternative to the deeply ingrained conflict model are encouraging, but they are not likely to make any breakthrough unless and until significant advance is made in the understanding of 'religion' in Chinese thought.[72] The 'religious question' in China, as Goossaert and Palmer have observed, remains 'as intractable as ever'.[73]

In general, there seems to be a tacit assumption in the literature that the question of science and religion, being chiefly a Western issue, did not become a subject of serious consideration in mainstream Chinese thought until the late 1910s with the rise of the New Culture intellectuals, who, from the outset, pronounced the relationship between the two fields to be antagonistic. Consequently, discussions of the history of the science and religion issue in the modern Chinese context normally start with the New Culture Movement.[74]

My study, however, demonstrates that earnest reflections on the subject had actually started at least one generation earlier with the late Qing reformers, who thrust themselves into the political and intellectual arenas in the mid-1890s and exerted significant and enduring influence for over a decade. Furthermore, the major issues concerning Chinese religion as well as the relationship between science and religion that are being discussed today had already been touched upon during that transitional era of modern Chinese history, and those considerations deserve our examination and appreciation.

3 'Studies of Half Kang'

The reformist thinkers were at a critical juncture in history, characterized by profound perplexity over the encounter between Chinese and Western systems

---

71 For a detailed analysis of the literature on science and religion over the recent decades, see Gou Wenzeng 勾文增, 'Dangdai kexue yu zongjiao yanjiu de yanjin 当代科学与宗教研究的演进 [A Review of Contemporary Research on Science and Religion in China]', *Nanjing ligong daxue xuebao* 南京理工大学学报 (社科版) [*Journal of Nanjing University of Science and Technology* (*Social Sciences*)], no. 2 (2012): 44–51.

72 Zhuo, 'Zhongguo zhishijie'.

73 Goossaert and Palmer, *Religious Question*, 401.

74 It should be noted, though, that Zhuo's insightful study starts with the early 1900s.

of knowledge and over the relationship between material and spiritual cultures.[75] They lived through the religious and scientific trends that gained momentum in later nineteenth century China. Partly to counter the influence of the 'intruding' foreign religion – Christianity, there appeared a revival of Buddhism among Chinese scholars who generally considered Buddhism more suitable for the Chinese mentality.[76] Meanwhile, there was a resurgence of the Gongyang School which emphasized the religious, mystical dimension of Confucius.[77] Accompanying these religious trends was the introduction of Western science and secular ideas, which progressed through different phases.[78] As a result, late Qing reformist thought contained both scientific and religious elements, a fact that is recognized even by Chinese Marxist scholars, although they assert that the former contradicted and overwhelmed the latter.[79] It is much to be desired, therefore, that more scholarly appreciation and treatment should be given to the ever-present religious dimension, and its relation to the scientific dimension, of the thought of the late Qing reformers.[80]

Among the reformist thinkers, Kang Youwei stands out as the leader of the reform campaign in the late 1890s. The campaign was triggered by China's defeat in the 1894–5 Sino-Japanese War and culminated in the short-lived Reform of 1898. Kang's ultimate vision was to transform the imperial government from a despotic to a constitutional monarchy through gradual reforms by drawing on the European precedents, so as to better prepare China for survival in its relations with the competing and often militarily aggressive nations.

75 Yunzhi Geng, *An Introductory Study on China's Cultural Transformation in Recent Times*, trans. Wang Huimin (Beijing & Heidelberg: Foreign Language Teaching and Research Press & Springer-Verlag, 2015), 1–2.

76 See Ch'i-ch'ao Liang, *Intellectual Trends in the Ch'ing Period*, trans. Immanuel C. Y. Hsü (Cambridge, MA: Harvard University Press, 1959), 116–17; Sin-wai Chan, *Buddhism in Late Ching Political Thought* (Hong Kong: Chinese University Press, 1985), 30–49.

77 Sun Chunzai 孫春在, *Qingmo de Gongyang sixiang 清末的公羊思想 [On the Thought of the Gongyang School in Late Qing]* (Taibei: Taiwan shangwu yinshuguan, 1985); Benjamin Elman, *Classicism, Politics, and Kinship: The Ch'ang-Chou School of New Text Confucianism in Late Imperial China* (Berkeley: University of California Press, 1990).

78 See Hao Chang, 'Intellectual Change and the Reform Movement, 1890–8', in *The Cambridge History of China*, ed. John K. Fairbank and Kwang-Ching Liu, vol. 11 (Cambridge: Cambridge University Press, 1980), 274–83; Elman, *A Cultural History*, 100–226.

79 See Mori Noriko, 'Liang Qichao, Late-Qing Buddhism, and Modern Japan', in *The Role of Japan in Liang Qichao's Introduction of Modern Western Civilization to China*, ed. Joshua A. Fogel (Berkeley: Institute of East Asian Studies, University of California Berkeley, 2004), 222–46; Li Zehou 李泽厚, *Kang Youwei Tan Sitong sixiang yanjiu 康有为谭嗣同研究 [A Treatise on the Thought of Kang Youwei and Tan Sitong]* (Shanghai: Renmin chubanshe, 1958), 6–7.

80 Goossaert and Palmer, *Religious Question*, 45; see also Li, *Kang Youwei*, 6–7.

Although the Reform was soon cut short by a coup, Kang remained committed to the vision during his ensuing fifteen years of exile in foreign lands. He refused to endorse the path of revolution and continued to advocate constitutional monarchy among overseas Chinese. Even after the Revolution of 1911, which overthrew the Manchu rule of the Qing dynasty and ushered in the Republican era, Kang's position did not change. He vigorously argued that a constitutional monarchy would be better suited to the needs of contemporary China than a republican form of government, because he was particularly alive to the crucial need for maintaining national unity. With that belief, he participated in the ill-fated Manchu Restoration of 1917 at a high cost to his political credibility.[81]

At the same time, Kang was the chief promoter of a religious reform, which aimed to 'restore', at both doctrinal and organizational levels, Confucianism to its 'pristine' state as a religion. Starting from the early 1890s, he embarked on a reinterpretation of the Confucian Classics, portraying Confucius as a founder of a *jiao* or religion as well as a reformer of political institutions. To materialize his vision, Kang presented to the Throne a proposal for the establishment of a National Confucian Church in the late 1890s, and in the early years of the Republican period, he played a central role in the organization the Confucian Religion Association (*Kongjiaohui* 孔教會) and in the campaign for the adoption of Confucianism as a state religion.[82] It should be added that, besides being a religious as well as political reformer, Kang was also an earnest student and advocate of Western science in China. An 'unhonoured prophet of a new faith' in science, he was one of the early voices and leading thinkers who called for saving China through science and technology.[83]

Most Chinese studies of Kang over the past century, however, have focused on his political thought and life, and these writings have been characterized by a dichotomous, 'Whiggish' view that is often underpinned by the tendency to

81 For a concise yet well-informed English introduction to Kang's life, see H. L. Boorman, ed., *Biographical Dictionary of Republican China*, vol. 2 (New York: Columbia University Press, 1968), 228–33.

82 Y. -P. Kuo, '"Christian Civilization" and the Confucian Church: The Origin of Secularist Politics in Modern China', *Past & Present* 218, no. 1 (2013): 235–64; Vincent Goossaert, 'Republican Church Engineering: The National Religious Associations in 1912 China', in *Chinese Religiosities: Afflictions of Modernity and State Formation*, ed. Mayfair Mei-hui Yang (Berkeley: University of California Press, 2008), 209–32.

83 Kung-ch'uan Hsiao, 'K'ang Yu-Wei's Excursion into Science', in *K'ang Yu-Wei: A Biography and a Symposium*, ed. Jung-pang Lo (Tucson: University of Arizona Press, 1967), 375–407. Kang has been unfairly labelled as a 'champion of the opposition' to science in O. Brière, *Fifty Years of Chinese Philosophy, 1898–1950* (Westport, Conn.: Greenwood Press, 1979), 15.

look back with hindsight about what appears progressive or politically correct in a subsequent phase of history.[84] Thus, Kang has been frequently depicted as 'progressive' before and during the Reform of 1898, but as 'conservative' and even 'reactionary' after it.[85] Accordingly, his work and thought in his later life have been considered to be lacking in conceptual value; therefore they have received less scholarly attention, a situation that has been aptly described as 'studies of half Kang'.[86]

In treating Kang primarily as a political thinker, and a utopian writer as well,[87] many scholars have tended to either downplay or overlook his religious convictions, considering his advocacy of a Confucian religion as being merely politically motivated, a cynical device for achieving a political and personal agenda. According to this reductivist view, his religious ideas were simply an expression of political interests.[88] Kang's Chinese-language biographers, when having to deal with Kang's religious discourse, especially his treatise on 'God', would normally attempt to offer a materialist analysis.[89] Hsiao Kung-ch'uan, Kang's main English-language biographer, has given due attention to Kang's religious thought, but has concluded that his subject's 'Confucian religion' ended up being a moral philosophy.[90] This view has itself been challenged by Lauren Pfister, who has argued that Kang's version of religion was based on principles that

84 For a critical account of Whig historiography, see Herbert Butterfield, *The Whig Interpretation of History*, new ed. (New York: W. W. Norton & Company, 1965).

85 For the most representative version of this view, see Tang Zhijun 汤志钧, *Kang Youwei yu Wuxu bianfa 康有为与戊戌变法* [*Kang Youwei and the Reform of 1898*] (Beijing: Zhonghua shuju, 1984).

86 Ma Honglin 马洪林, 'Kang Youwei yanjiu bainian huigu yu zhanwang 康有为研究百年回顾与展望 [A Review of Studies of Kang Youwei over the Past Century and Prospects for Future Research]', *Dongfang luntan 东方论坛*, no. 5 (2008): 83–92.

87 Recent studies in English of Kang's utopian thought include Federico Brusadelli, 'A Tale of Two Utopias: Kang Youwei's Communism, Mao Zedong's Classicism and the "Accommodating Look" of the Marxist Li Zehou', *Asian Studies* 5, no. 1 (2017): 103–22; Ban Wang, ed., *Chinese Visions of World Order: Tianxia, Culture, and World Politics* (Durham: Duke University Press, 2017), 49–64, 87–105.

88 For more on this view, see Geng Yunzhi 耿云志, 'Kang Youwei de shengren qingjie jiqi yi Kongjiao wei guojiao shuo 康有为的"圣人"情结及其以孔教为国教说 [Kang Youwei's Sage Complex and His Views on Confucianism as a State Religion]', in *Kang Youwei yu jindai wenhua 康有为与近代文化* [*Kang Youwei and Modern Culture*], ed. Fang Zhixin 方志钦 and Wang Jie 王杰 (Kaifeng: Henan daxue chubanshe, 2006).

89 See for example Ma Honglin 马洪林, *Kang Youwei pingzhuan 康有为评传* [*A Critical Biography of Kang Youwei*] (Nanjing: Nanjing University Press, 2009), 206–11.

90 Hsiao, *K'ang Yu-Wei*, 114.

can be recognized as legitimate religious claims.[91] The Taiwan-based scholar Zheng Zhiming has re-evaluated Kang's efforts to restore the 'ancient Chinese belief in *Shangdi* 上帝 (the Supreme Ruler) in his 'religious reinterpretation' of Confucianism.[92] Tang Wenming from the Chinese Mainland has argued that Kang's religious thought is still of value for modern-day China, and Zhang Xiang has examined Kang's unique approach to establishing, in light of 'universal principles' rather than a Western form of religion, a cosmopolitan religion with Confucius as its founder.[93] Similarly, in their much acclaimed work on Chinese religion, Vincent Goossaert and David A. Palmer have considered Kang to be 'one of the first and most vocal proponents of a new way of thinking about the religious question in China'.[94] It is now increasingly recognized among relevant scholars that Kang's actions and ideas cannot be fully understood without taking their spiritual-religious dimensions into consideration.[95]

A transcultural dimension has also emerged in studies of Kang. It has been stressed that Kang's 'reformation' of Confucianism in the late nineteenth century can be better appreciated when it is duly placed in the context of China-Europe cultural exchanges initiated by Ricci and his fellow Jesuits almost three centuries earlier.[96] Scholarly efforts have already been made to

---

91 Lauren Pfister, 'A Modern Ruist Religious Vision of a Global Unity: Kang Youwei's Utopian Vision and Its Humanistic Religious Refraction in European Sinology', in *Globalization and the Making of Religious Modernity in China* (Leiden and Boston: Brill, 2014).

92 Zheng Zhiming 郑志明, 'Kang Youwei de zongjiaoguan (shang) 康有为的宗教观 (上) [On Kang Youwei's Religious Views I]', *Ehu* 鹅湖 23, no. 3 (1997): 17–28; Zheng Zhiming 郑志明, 'Kang Youwei de zongjiaoguan (xia) 康有为的宗教观 (下) [On Kang Youwei's Religious Views II]', *Ehu* 鹅湖 23, no. 4 (1997): 22–31.

93 Tang Wenming 唐文明, *Fujiao zaikuan: Kang Youwei Kongjiao sixiang shenlun* 敷教在宽: 康有为孔教思想申论 [*Teaching with Tolerance: A Study of Kang's Thought on Confucian Religion*] (Beijing: Zhongguo remin daxue chubanshe, 2012); Zhang Xiang 张翔, 'Cong li gongli zhixue dao yi datong lijiao 从立公理之学到以大同立教 [From the Learning of Universal Principles to a Religion Based on the Great Unity]', *Zhexue dongtai* 哲学动态 [*Philosophical Trends*], no. 3 (2015): 13–22.

94 Goossaert and Palmer, *Religious Question*, 46.

95 Wei Leong Tay, 'Kang Youwei, The Martin Luther of Confucianism and His Vision of Confucian Modernity and Nation', in *Secularization, Religion and the State*, ed. Haneda Masashi (Tokyo: The University of Tokyo Centre for Philosophy, 2010), 97–109. see also Takeuchi Hiroyuki 竹内弘行, *Chūgoku no jukyō kindaika ron* 中国の儒教近代化論 [*The Modernization of Chinese Confucianism*] (Tokyo: Ken bunshuppan, 1995), 112–65.

96 Zhang Xiang 张翔, 'Ruxueshi xushu de fenduan yu Kongzi zhiyi de bifushi quanshi 儒学史叙述的分断与孔子之义的比附式诠释 [The Demarcation of the History of Confucianism and the Analogous Interpretation of the Meaning of Confucius]', *Zhongguo zhexueshi* 中国哲学史 [*History of Chinese Philosophy*], no. 6 (2019): 120–28.

compare Kang and other religious figures of the late nineteenth century. The Harvard-educated Indian scholar Krishna Prakash Gupta, for example, has examined the different forms of 'religious evolution' proposed by Kang and Swami Vivekananda (1863–1902), the Hindu spiritual leader and reformer.[97] More recently, in his monumental history of the nineteenth century, the German historian Jürgen Osterhammel places Kang, as well as Bahá'u'lláh (Mirza Husayn-'Alí, 1817–1892), the Persian religious leader and founder of the Bahá'í Faith, among 'the major thinkers' who 'cut across cultural boundaries'.[98] Similarly, in tracing the spread of religious cosmopolitanism in early twentieth century China, David Palmer and I have discussed the interactions between members of the Confucian Religion Association in Shanghai and the prominent American Bahá'í teacher Martha Root in the early 1920s, when Kang was still alive.[99]

Since the turn of the new millennium, as cultural conservatism gains momentum in China, the dichotomous view Kang has gradually been discredited, and the unsatisfactory situation of 'studies of half Kang' has been significantly remedied. Not only have different dimensions of Kang's intellectual output, including even his aesthetic pursuits and ecological thinking,[100] received increasing scholarly attention, but many of his political and religious proposals, including those put forward in his later years, which had been ridiculed or ignored over the course of the last century, have also begun to be treated with increasing sympathy and have received scholarly reappraisal.[101]

97 Krishna Prakash Gupta, 'Religious Evolution in Premodern Asia: A Comparative Study of K'ang Yu-Wei and Vivekananda' (Ph.D., Cambridge, MA., Harvard University, 1976).

98 Jürgen Osterhammel, *The Transformation of the World: A Global History of the Nineteenth Century*, trans. Patrick Camiller (Princeton and Oxford: Princeton University Press, 2014), 896–97.

99 Zhaoyuan Wan and David A. Palmer, 'The Cosmopolitan Moment in Colonial Modernity: The Bahá'í Faith, Spiritual Networks and Universalist Movements in Early Twentieth Century China', *Modern Asian Studies* 54, no. 6 (2020): 1787–1827, https://doi.org/10.1017/S0026749X19000210.

100 Aida Yuen Wong, *The Other Kang Youwei: Calligrapher, Art Activist, and Aesthetic Reformer in Modern China* (Leiden: Brill, 2016); Ban Wang, 'Confucianism and Nature: Ecological Motifs in Kang Youwei's Great Community', *Telos* 2018, no. 183 (2018): 47–67.

101 Major book-length studies include Tang, *Fujiao zaikuan*; Zeng Yi 曾亦, *Gonghe yu junzhu: Kang Youwei wanqi zhengzhi sixiang yanjiu 共和与君主: 康有为晚期政治思想研究* [*Republic versus Monarchy: A Study of Kang's Political Thought in His Later Years*] (Shanghai: Shanghai renmin chubanshe, 2010); Zhu Yitian 朱忆天, *Maixiang jindai wenming guojia de tansuo: Kang Youwei houqi sixiang yanjiu 迈向近代文明国家的探索: 康有为后期思想研究* [*Towards a Modern Civilized Country: A Study of Kang's Later Thought*] (Shanghai: Shanghai renmin chubanshe, 2016); Gan Chunsong 干春松, *Baojiao liguo: Kang Youwei de xiandai fanglue 保教立国：康有为的现代方略* [*To Establish the*

Kang's important position in modern Chinese thought as well as the Confucian tradition has become more widely acknowledged.[102] Hsiao's claim, made in the 1970s, that Kang had launched a new stage in the development of Confucianism[103] has been echoed by a number of prominent contemporary scholars on Chinese intellectual history. Tu, for example, considers Kang a precursor to the third epoch of Confucianism (the first two being classical Confucianism and Neo-Confucianism), that is, New Confucianism (*Xinrujia* 新儒家), which arose in the 1920s and has continued until this century.[104] However, the early New Confucianists, and their successors based in Taiwan and Hong Kong, who have prioritized 'inner self-cultivation' over the institutional and political dimensions of Confucianism, would normally not associate themselves with the name of Kang Youwei.[105] In contrast, the New Confucianists on the Chinese Mainland have advocated 'a return to Kang Youwei' and revisited many of his ideas, such as his unflinching advocacy of a titular monarchical republic and his endeavours to formally install Confucianism as a state religion, in their discussion on the future reintegration of Confucianism with the state.[106] This 'political turn' as well as its modernist approaches has been controversial, however.[107]

State through Preserving the Faith: Kang's Strategies for Modernization] (Beijing: Sanlian shudian, 2015); Zhang Yongle 章永乐, *Wanguo jingzheng: Kang Youwei yu Weiyena tixi de shuaibian 万国竞争: 康有为与维也纳体系的衰变* [*Nations as Rivals: Kang and the Decline of the Vienna System*] (Beijing: Shangwu yinshuguan, 2017); Wei Leong Tay, 'Secularization as Sacralization: Religion and the Formation of Modern Chinese Nationalism and Nation-State, 1840–1939' (Oxford, The University of Oxford, 2019).

102 See, for example, Wang Hui 汪晖, *Xiandai Zhongguo sixiang de xingqi 现代中国思想的兴起* [*The Rise of Modern Chinese Thought*] (Beijing: Sanlian shudian, 2004), 737–830.

103 Hsiao, *K'ang Yu-Wei*, 125.

104 Wei-Ming Tu, *Way, Learning, and Politics: Essays on the Confucian Intellectual* (Albany: State University of New York Press, 1993), 141–61; see also Jesús Solé-Farràs, *New Confucianism in Twenty-First Century China: The Construction of a Discourse* (Routledge, 2013), xxi, 105; Gan Chunsong 干春松, *Kang Youwei yu Ruxue de xinshi 康有为与儒学的新世* [*Kang Youwei and the New Epoch of Confucianism*] (Shanghai: Huadong shifan daxue chubanshe, 2015).

105 For a good discussion between Kang and New Confucianist schools, see Gan Chunsong 干春松, 'Kang Youwei yu xiandai Ruxue sichao de guanxi bianxi 康有为与现代儒学思潮的关系辨析 [On the Relationship between Kang Youwwei and Contemporary Schools of Confucianist Thought]', *Zhongguo renmin daxue xuebao 中国人民大学学报* [*Journal of China Renmin University*], no. 5 (2015): 22–30.

106 For an update of the collective proposals of New Confucianists based on the Chinese Mainland, see Jiang Qing 蒋庆 et al., *Zhongguo bixu zairuhua 中国必须再儒化* [*China Must Be Confucianized Again*] (Singapore: World Scientific, 2016).

107 For an informed, comprehensive critique of the New Confucianist 'political turn', see Ge Zhaoguang 葛兆光, 'Yixiangtiankai: jinnianlai dalu Xinruxue de zhengzhi suqiu 异想天开: 近年来大陆新儒学的政治诉求 [A Fantasy: On Recent Political Appeals of the New

In summary, Kang's body of work has returned to the centre of academic studies nearly a century after his death.

The publication in 2007 of his twelve-volume *Complete Works* (*Kang Youwei quanji* 康有為全集, hereafter cited as *KYWQJ*) has increased the convenience for academics who wish to approach Kang's life and work with renewed interest. An extremely prolific writer, Kang had been writing over the course of his whole life. A considerable portion of those writings, however, were not published during his lifetime, primarily due to the political and material restrictions caused by his exile and the loss of or damage to his manuscripts under the wartime conditions. It was not until the 1970s that an incomplete collection of his writings was printed posthumously in Taiwan.[108] The 2007 collection is the most comprehensive one so far, for it has included several seminal manuscripts Kang had composed in his youth (which were once thought to have been lost), as well as his various letters and cables collected from archives and private owners from across China and abroad. This collection, together with the relevant writings of his close disciples and associates, Liang Qichao 梁啟超, Tan Sitong 譚嗣同 and Chen Huanzhang 陳煥章 in particular, constitutes the primary sources of my study, which aims to enrich our understanding of Kang's thought through a rigorous (re)examination of its religious and scientific dimensions as gleaned from his writings.

4 Implications

This book seeks to contribute to three areas of scholarship. First, it constitutes another effort to add to the presentation of a more well-rounded picture of Kang's thought by focusing attention on the religious and scientific dimensions of his work. While Kang has traditionally been studied as a political figure, a utopian thinker and a philosopher, my study has established that he should also be regarded as a religious figure both in terms of his private belief and public involvement in religious activities, by reference to his early religious experience and through a consideration of the written evidence regarding his personal belief in supernatural forces, his construction of a Confucian

Confucianists on the Chinese Mainland]', www.aisixiang.com 爱思想, 1 July 2017, http://www.aisixiang.com/data/104951.html; see also Kiri Paramore, '"Civil Religion" and Confucianism: Japan's Past, China's Present, and the Current Boom in Scholarship on Confucianism', *Journal of Asian Studies* 72, no. 2 (2015): 269–82.

108 Jiang Guiling 蒋貴麟, ed., *Kang Nanhai xiansheng yizhu huikan 康南海先生遺著汇刊* [*Collected Works of the Late Mr. Kang of Nanhai*], 22 vols (Taipei: Hongye shuju, 1976).

theology involving the existence of *Shangdi* or 'God', the will of Heaven, divine retribution and the attributes of the human soul, and his endeavour to institutionalize Confucianism as a state religion. Similarly, as a pioneering advocate of Western science, Kang's conception of science, though rudimentary in certain aspects, has proven to have been more nuanced and balanced than the ideas maintained by the indiscriminate promoters of scientism among the younger generation of the New Culture intellectuals. In this connection, this book can be considered as a contribution to both the literature on the modern reformation or transformation of Confucianism[109] and the small yet growing literature that has challenged the conventional, secularist historiography of religion in late imperial and early republican China.[110]

Second, this study demonstrates the rich and complex nature of the relationship between science and religion during the transitional era of modern Chinese intellectual history. It proposes and defends the unconventional argument that one whole generation before the Conflict Thesis took root in Chinese thought in the late 1910s, the late Qing reformist thinkers, represented by Kang, had already reflected on the relationship between the domains which are now generally categorized as science and religion. Their conceptualization of that relationship, as we will see in the following chapters, extended beyond a simple conflict or harmony thesis. While maintaining that there was a fundamentally complementary relationship between religion and science, they also treated that relationship in consideration of individual religions and the overall classification of knowledge. Their acknowledgement of conflict or tension between science and certain religious doctrines did not make them reject religion in its entirety, nor did their recognition of the power and value of science convert them to scientism. This intellectually moderate and philosophically nuanced position is of considerable value to the increasing scholarly awareness of the need for renewed reflection on and a critical re-appreciation

109 See, for example, Jensen, *Manufacturing Confucianism*; John Makeham, *New Confucianism: A Critical Examination* (New York: Palgrave Macmillan, 2003); Chen, 'Confucianism Encounters Religion'; Sun, *Confucianism as a World Religion*; Sébastien Billioud, *Thinking Through Confucian Modernity* (Leiden; Boston: Brill, 2011); Wang, *Xiandai zhongguo sixiang*; Gan, *Xuxue xinshi*; Peng Chunling 彭春凌, *Ruxue zhuanxing yu wenhua xinming* 儒学转型与文化新命 [*The Transformation of Confucianism and the New Cultural Mission*] (Beijing: Bejing daxue chubanshe, 2014).

110 This includes Rebecca Nedostup, *Superstitious Regimes: Religion and the Politics of Chinese Modernity* (Cambridge, MA.: Harvard University Asia Center, 2009); Paul R. Katz, *Religion in China and Its Modern Fate* (Waltham, MA.: Brandeis University Press, 2014); Peter van der Veer, *The Modern Spirit of Asia: The Spiritual and the Secular in China and India* (Princeton: Princeton University Press, 2013); Goossaert and Palmer, *Religious Question*; and David Ownby's forthcoming book on the KMT official and religious figure Li Yujie.

of Chinese tradition and culture.[111] It also reminds us of the previously-mentioned cautionary and visionary statement by Whitehead, a thought-provoking idea which places upon the shoulders and minds of present-day scholars the duty to examine the science and religion question with fresh eyes.

Third, as this discussion is being carried out in the context of China's unique intellectual tradition, this book may also enrich the global discourse on the relationship between science and religion by presenting an alternative perspective, a non-Western, Confucian-inspired configuration of their relationship. The ancient Chinese concepts that underpinned Kang's personal embrace of science and religion, such as a holistic view of the universe, creation without a creator, law without a law-giver, and the idea of the natural law of constant change,[112] are capable of offering new insights into the current debates on the topic, including those between evolution and Intelligent Design.[113] Also admirable are his efforts to bridge the gap between what he sincerely believed to be a superior religious system, namely, a reformed or 'restored' Confucianism, and its complementary role as a moral force by which to harness the power of Western science and technology in order to serve the need to provide material prosperity to the people of China and the world. Kang's readiness to give up any doctrine when such a doctrine was contradicted or invalidated by verifiable scientific facts, and his more balanced approach in his differentiating treatment of individual religions in relation to science, all seem to have become increasingly consonant with today's academic predisposition to construct more global conceptualizations of the science and religion question.

Kang's message is clear – that there exists a deep complementary concord between true religious teachings and true scientific principles, because both conform to the nature of a cosmos that unites humanity, heaven and earth. A false dichotomy and an unnecessary division find no place in his mind, a mind that was trained in the holistic thinking of Chinese tradition and which was

111 See Fang Min 方敏, 'Kexue yu zongjiao guanxi de xueshu guanzhu 科学与宗教关系的学术关注 [Scholarly Attention to the Relationship between Science and Religion]', *Ziran bianzhengfa tongxun* 自然辩证法通讯 [*Journal of Dialectics of Nature*], no. 3 (2008).

112 For a general discussion on these concepts in ancient Chinese thought, see Yu-lan Fung, *A History of Chinese Philosophy*, trans. Derk Bodde, 2nd ed., vol. 1 (Princeton: Princeton University Press, 1983), 22–31; Laurence G. Thompson, *Chinese Religion: An Introduction*, 5th ed. (Belmont: Wadsworth Publishing Company, 1996), 1–3.

113 For a concise, updated account of the challenge to the theory of evolution from scientific creationism, of which Intelligent Design is a prominent variety, see Thomas Dixon, *Science and Religion: A Very Short Introduction* (Oxford: Oxford University Press, 2008), 81–103.

devoted to a search, which he engaged in throughout his adult life, for unity at various levels. From a contemporary perspective, Kang's holistic worldview, coupled with his firm belief in the eventual unification of the human race in a universal commonwealth, seems to fit in well with the emerging discourse on the Chinese-inspired concept of 'a community of shared future for humankind'.[114]

5 Chapter Organization

To bring out the salient features in the development of Kang's thought, the chapters of this book are organized according to the five distinctive phases of his eventful life: his early years (1858–1888); his career as a radical reformer (1889–1898); his exile abroad (1899–1912); the early Republican years, when he was back in China (1913–1917); and his later years (1918–1927).

The first chapter discusses Kang's early life and thought within the religious and scientific contexts of the late Qing. Born to a family steeped in Confucian tradition in a village in southeast Guangdong, the young Kang Youwei demonstrated strong religious leanings whose sources could be traced to the traditional belief in *Shangdi* or Heaven in classical Confucianism, the local folk beliefs with their admixture of Buddhist-Daoist concepts, and his mystical experience brought about by his exposure to Buddhist ideas and meditation. His knowledge from the Confucian and Buddhist sources was further complemented by his Western learning which he had acquired through reading translations of Western scientific books. His learning culminated in an 'enlightenment' in the mid-1880s, through which he was reinforced in his earlier image of himself as a sage for this age, with a bodhisattva-like mission to deliver from suffering not only his compatriots in China but the whole of humanity as well. Believing that he had arrived at an understanding of the great *Dao*, he worked out a set of what he regarded as universal principles based on his knowledge of scientific and religious sources, thereby laying the foundation for his design of

114 For the discussion of the concept of 'a community of shared future for humankind' from various perspectives, see, for example, Fengna Xu, 'Shaping "A Community of Shared Future for Mankind": New Elements of General Assembly Resolution 72/250 on Further Practical Measures for the PAROS', *Space Policy* 44–5 (2018): 57–62; Zhou Min 周敏, 'Zouxiang renlei mingyun gongtongti: yige bijiao wenhua de shijiao 走向人类命运共同体：一个比较文化的视角 [Towards a Community of Shared Future for Mankind: A Perspective from Comparative Cultural Studies]', *Shanghai jiaotong daxue xuebao 上海交通大学学报* [*Journal of Shanghai Jiaotong University* (*Philosophy and Social Sciences*)] 27, no. 126 (2019): 68–75.

an ideal future world community. With that vision, he was ready to enter the political stage and launch a religious reform.

The following chapter presents Kang's endeavour in two interconnected lines of action in the 1890s: as a revisionist Confucian educator who also promoted Western science, and as a political as well as religious reformer. The curriculum at his academy in Guangzhou was characterized by its teaching of Western learning and his radical revision of Confucianism based on the New Text exegesis. His recasting of Confucianism presented Confucius as both a divine founder of a religion and a visionary reformer who sanctioned political reforms. Equipped with the progressive Confucian doctrine of the Three Ages (*sanshi* 三世) that buttressed his political theory, Kang spearheaded the reform movement of the late 1890s. His reform proposal included the inclusion of modern science in the examination and education systems, and the establishment of a national Confucian Church in order to better counter the foreign religious challenge and preserve the Confucian faith. He concluded that to save China from the imminent danger of dismemberment and to make her wealthy and strong, science and technology were crucial, while a parliamentary polity and a common spiritual system were no less important. He believed that knowledge was one: ancient Chinese classics had already contained the origins of Western sciences, and principles of nature discovered through science could lead to the more profound teachings enshrined in Confucianism. Inspired by Kang, disciples such as Liang Qichao (1873–1929) and Tan Sitong (1865–1898) articulated a holistic system of knowledge, which was neither Western or Chinese, and which contained both religious and scientific elements interacting in harmony.

Chapter Three treats Kang's completion of his reinterpretation of the Confucian Classics in the early years of his exile abroad, and his new insights on science and religion obtained as a result of observations from his world travels. In his Confucian exegesis, Kang sought to refine or elaborate on many of his views, which had been expressed in his earlier lectures and works, on such topics as the 'divinity' of Confucius, the idea of serving Heaven, the concept of *jiao*/religion, the nature and immortality of the human soul, and the principle of Heaven-ordained laws applicable to all things. In doing so, he had actually established a Confucian 'theology'. His study of other religions during his extensive travels to Europe and America strengthened his belief in the doctrinal superiority of Confucianism, which he believed was rooted in Heaven but designed for humanity, encompassing and employing both the way of the spirits (*shendao*) and the human way (*rendao*). Kang also speculated on the eventual disappearance of Confucianism, Christianity and Islam, and prophesized the rise of a form of new religion in the ideal world of the Great Unity to come, a world that would be constructed on a high degree of scientific and technological development. While Liang Qichao embraced the Western conception of religion under the Japanese contemporaneous intellectual influence, analytically excluded Confucianism from the realm of religion and thereby indicated a relationship of conflict between religion and science, Kang remained firm in his view of science and religion as being compatible and complementary branches of one unified system of knowledge. This line of reasoning was most clearly demonstrated in his appropriation of scientific notions, in particular 'ether' and 'electricity', for his interpretation of such Confucian concepts as *ren* (benevolence) and *xing* (human nature). Convinced of the utmost importance of material development, he spoke in glowing terms of the value of science and technology and their practical and utilitarian aspects as being the foremost prerequisites for the reconstruction of China.

The next chapter is devoted to Kang's Confucian movement in the early Republican years in the 1910s. Convinced that the success of a country depended on the two complementary aspects of politics and religion, Kang proposed both a political and a religious solution for post-Revolution China – the adoption of a constitutional monarchy, which he now termed as a 'titular monarchical republic', and the institutionalization of Confucianism as a religion. While the former was rejected outright by his fellow countrymen, the latter achieved some limited progress by way of the 1912 establishment of the Confucian Religion Association with the support of his young and capable disciple, Columbia graduate Chen Huanzhang (1880–1933). Apart from defending the religious nature of Confucianism and elevating the station of Confucius to the status of the founder of a religion, the Association also launched an impressive campaign to have Confucianism enshrined as the state religion in the new Constitution of the Republic. This campaign (which failed), as well as the personality of Kang, was severely criticized by Chen Duxiu 陳獨秀 (1879–1942) and his fellow New Culture intellectuals, who drew on the authority of science to mount and launch their opposition to religion. These iconoclastic attacks on Confucianism in particular and on religious belief *per se* sensationally established the Conflict Thesis in the mainstream of Chinese thought.

The final chapter looks at the last decade of Kang's life. Despite his political setbacks, Kang kept abreast of both domestic and international affairs. He saw a silver lining in the otherwise dark clouds of the Paris Peace Conference – the proposal for the creation of the League of Nations, which he immediately connected with the concept of the Three Ages and which in his interpretation constituted the beginning of the realization of the final Age of the Great Unity. As an indirect response to the unprecedented debate among Chinese intellectuals on the question of science and metaphysics (including religion) in the early 1920s, Kang reaffirmed his position on the complementary relationship of Confucianism and Western science at both national and individual levels during his impressive lecture tour to the North. In doing so, he elaborated on the important role of scientific knowledge in understanding key Confucian concepts and the human soul as well as in shaping one's views on life, the world and even the universe. In his last book, which was a compilation of his lecture notes at his newly-established Academy of Celestial Peregrination, Kang blended knowledge of astronomy with his metaphysical speculations. While speaking of science in glowing terms, he also affirmed his firm belief in the existence of God, whose nature he claimed lay beyond the comprehension of the limited faculties of scientists. Kang's posthumous fame underwent dramatic changes during the following decades after his death in 1927, as exemplified by the fate that befell his skull.

CHAPTER 1

A New Sage

> I suddenly perceived heaven and earth and whatever is between them come into oneness with me, radiating bright light in great profusion. Believing myself to be a sage I laughed with joy; then thinking of the sufferings of the people in the world, I wept in sorrow.
>
> KANG YOUWEI[1]

∵

Kang Youwei was born in 1858 in Nanhai 南海 (literally, south sea)[2] in the southern Guangdong province, where the Kang family had lived for many generations since the Southern Song dynasty (1127–1279). A short distance to the west of the village of his birth, the Xijiang river, the fourth longest in China, flows southward through the Pearl River Delta before entering the South China Sea. Dozens of miles to the east of the village lies the city of Guangzhou (formerly known as Canton), the cultural and commercial centre of south China, which had been China's only 'open port' before the First Opium War (1939–1942). Its economic and cultural connection with the outside world had created favourable conditions for the emergence and acceptance of new ideas in the region. Growing up in close proximity to Guangzhou, which became the centre for the dissemination of Western learning in China in the mid-nineteenth century, would prove highly advantageous to the young Kang Youwei's intellectual development.[3]

Kang's birth took place in the context of two devastating wars being fought in China, on sea and on land, respectively. On the one hand, the Second Opium

1 Kang Youwei 康有为, 'Woshi 我史 [My History (1899)]', in *KYWQJ*, vol. 5, 2007, 62.

2 Nanhai had been designated a county since the third century BCE. It is now a district, still bearing the same name, of the Foshan prefecture, Guangdong. Following the tradition of the *diwang* 地望, according to which a person of rank was often respectfully referred to by the name of his birthplace, Kang was also known as Kang Nanhai or Mr. Kang of Nanhai.

3 Kang's family possessed a piece of property in Guangzhou, initially obtained by his great grandfather, who once held a provincial post. Later in 1890, Kang and his immediate family would move to the urban residence. The house has recently been rebuilt and converted into a museum honouring Kang. See Kang, 'Woshi', 58, 81.

 | DOI: 10.1163/9789004468221_003

War (1856–1860) was halfway through and would end with the capture of the Forbidden City by the Anglo-French troops, the Chinese Emperor's flight and the burning of the Summer Palaces. After the war, the opium trade was legalized in China, and so was missionary work, with the Christian religion, which had been banned in 1767, being granted 'full security' of its 'members, property and religious practices' on the Chinese land.[4] The humiliating defeat of the war gave rise to China's modernization campaign known as the Self-Strengthening Movement. On the other hand, the Taiping Rebellion against the Qing dynasty, which had started in the neighbouring Guangxi province in 1850, was still raging and would not be completely suppressed until 1865. The civil war brought both loss and honour to Kang's family, many of whose male elders, including his father, were involved in fighting against the rebels. Some of them had been killed in battle, and the returning survivors were promoted to civil service positions because of their military feats. One of Kang's granduncles, General Kang Guoqi 康國器, for example, was appointed Financial Commissioner and, subsequently, acting Governor of Guangxi before his retirement. Though not a scholar, he built a handsome family library that housed a collection of over 20,000 books. The library would become a storehouse of knowledge for the young Kang Youwei, for which he was ever grateful to the granduncle.[5]

Kang's family upheld a scholarly tradition rooted in Neo-Confucian orthodoxy, the imperially sanctioned school of Confucianism based on the interpretations of Zhu Xi 朱熹 (1130–1200) of Southern Song. Also known as the Learning of the Way (*daoxue* 道學) or the Learning of the Principle (*lixue* 理學), the 'rationalist' school held a well-defined genealogy of the Confucian Way: originally conceived by ancient Chinese Sages, it was passed down to the Duke of Western Zhou (*circa* 11th century–771 BCE), and then to Confucius, who compiled and edited its sacred Texts, the Five Classics, namely, the *Yi* 易 or *Changes*, the *Shu* 書 or *Documents*, the *Shi* 詩 or *Odes*, the *Li* 禮 or *Rites*, and the *Chunqiu* 春秋 or *Autumn and Spring Annals*. From Confucius it was transmitted through Mencius (372–289 BCE) and Xunzi 荀子 (313–238 BCE) to the Neo-Confucian masters including Zhu Xi. As a complement to the Five Classics, Zhu Xi further compiled, edited and commented on a compendium of the Four Books, including the *Lunyu* 論語 or *Analects*, the *Mengzi* 孟子 or *Mencius*, the *Zhongyong* 中庸 or *Doctrine of the Mean* and the *Daxue* 大學

4 'Treaty of Tientsin [Tianjin], 1858', 1860, http://www.chinaforeignrelations.net/node/162.

5 Kang, 'Woshi', 59, 60. After the death of this granduncle, Kang wrote him a moving tribute praising his military and civil accomplishments, see Kang Youwei 康有为, 'Kanggong shizhuang 康公事状 [A Eulogy for His Excellency Mr. Kang (1884)]', in *KYWQJ*, vol. 1, 2007, 5–13.

or *Great Learning* (the last two being two treatises taken out from the *Book of Rites*). Zhu's annotated edition of the Four Books had become the standard curriculum for the imperial civil service examinations (*keju* 科舉) during the following Yuan, Ming and Qing dynasties, and his philosophy, which stressed moral cultivation and rationalism, exerted a great influence upon the Chinese literati and society.[6]

Kang's great-great grandfather, Kang Hui 康輝, a teacher of local renown, was regarded as the founder of the scholarly tradition in the Kang family. He passed down to his descendants the Neo-Confucian philosophy, with emphasis on both personal cultivation and contribution to the state. His legacy was carried on by one of his grandsons, Zanxiu 贊修, Kang's grandfather, who became a provincial graduate (*juren* 舉人) and served as an educational officer in the region. As no member of the next generation of his family succeeded in passing the civil service examinations, Zanxiu had great expectations of Youwei, his first-born grandson. 'Long have I wished for a grandson till my eyes are weary,' he wrote on hearing the news of Youwei's birth, '[but now] I have restored the scent of books [literary pursuit] which you should perpetuate.'[7] Youwei, who lost his father at age ten, was very close to his grandfather, who took care of him and supervised his moral upbringing and literary education.

A diligent and avid learner, the young Kang proved an excellent pupil. At five, he started his formal education by learning to recite the Four Books. At eight, he was already reciting difficult passages from the *Documents*. By the age of nine, he had almost finished mastering the recitation, from memory, of such classics as the *Changes* and the *Rites*. His reading list also included titles on government as well as official documents, which suggested his early interest in politics and statecraft. The list quickly became long as he immersed himself in various books among the collections in the family library.[8] However, Kang was loath to practice writing the rigid 'eight-legged' essays in preparation for the civil service examinations. Under family pressure, though, he did make efforts in that direction, but in vain, as he repeatedly failed the local examinations.[9] It was this unpleasant experience that partly gave rise to his later attempts to

---

6 For the genealogy of the Confucian Way, see Fung, *Chinese Philosophy*, 1983; Yu-lan Fung, *A History of Chinese Philosophy*, trans. Derk Bodde, vol. 2 (Princeton: Princeton University Press, 1983).

7 Kang, 'Woshi', 58; translated in Jung-pang Lo, ed., 'Chronological Autobiography', in *K'ang Yu-Wei: A Biography and a Symposium*, trans. Jung-pang Lo (Tucson: University of Arizona Press, 1967), 23.

8 Kang, 'Woshi', 59–60.

9 Kang, 60–61, 64.

have the examination system abolished, an idea that would find its way into his future reform memorials to the Throne.

In 1876, Kang Youwei sought the tutelage of Zhu Ciqi 朱次琦 (1807–1881), a leading Confucian scholar in the region, who had been a friend of Kang's grandfather in addition to being his father's teacher. Rising above the popular partisan separation between the rival schools of Song Learning (which emphasized moral philosophy) and the Han Learning (which prioritized textual research), Zhu adopted an eclectic approach in instructing his students in 'the five fields of learning', namely, Classics (*jing* 經), histories (*shi* 史), government (*zhanggu* 掌故), moral philosophy (*xingli* 性理) and literary writing (*cizhang* 詞章). Zhu emphasized the moral praxis of Confucian teachings in improving the wellbeing of others and the society, and opposed the advocacy of intuition and meditation by the Learning of the Heart-and-Mind (*xinxue* 心學).[10] A rival to the Learning of the Principle, this 'idealist' school was based on Wang Yangming's 王陽明 (1472–1529) idealistic interpretation of Confucianism under the heavy influence of Chan (Zen) Buddhism. Kang received Zhu's teaching with great enthusiasm, like 'a blind man seeing the light' as he later put it, and was able to harmonize the various sources of knowledge he had previously acquired.[11] Kang's intensive study for the next three years under Zhu consolidated his foundation in Confucian learning, and made him more aware of the practical significance of Confucian scholarship. At the same time, his ambition to become a sage grew and intensified, an ambition that he had started to cherish in his childhood.

Two events, which took place during the period of Kang's studies under Zhu, are worthy of mention, for they were not only of personal importance, but also showed his strong and independent character. One was Kang's wedding in 1876, at which, to the disappointment of the guests, he refused to follow the custom of allowing them to tease his bride, because he considered it inappropriate. The other was a tragedy – the drowning of his beloved grandfather in a flood. Overcome by grief, he studied the mourning ceremonials as recorded in the *Rites*, and carried out the ancient rituals in every movement and motion. The villagers, who had scoffed at his extreme ritual obedience, came to respect him as he persisted in his ways.[12] His high regard for the ancient classics, as

10 For Zhu's scholarship, see Kang Youwei 康有为, 'Nanhai Zhu xiansheng mubiao 南海朱先生墓表 [Epitaph for Mr. Zhu of Nanhai (1882)]', in *KYWQJ*, vol. 1, 2007, 1; Jian Chaoliang 简朝亮, 'Nianpu 年谱 [A Chronological Biography]', in *Zhu Jiujiang (Ciqi) xiansheng ji* 朱九江（次琦）先生集 [*Collected Works of Mr. Zhu Ciqi*] (Taipei: Wenhai chubanshe, 1967), 9–91.

11 Kang, 'Woshi', 61.

12 Kang, 61–62.

evinced by his strict compliance with ritual prescriptions, and his nonconformist thinking were characteristic of many of his future endeavours.

Kang resumed his studies under Zhu after his grandfather's burial. In early 1879, however, he decided to leave Zhu, his last acknowledged teacher, as he had developed a revulsion against conventional book learning. For the next ten years or so, he was left to educate himself, exploring other avenues of knowledge that were not in accordance with the Neo-Confucian tradition. He resumed his early interest in Buddhism and practised spiritual self-cultivation through meditation, which led him to the exploration of several mystical experiences. Meanwhile, his trip to Hong Kong turned him into an eager student of Western learning through his reading and intensive study of Chinese translations of Western books. By the middle of the 1880s, Kang had achieved a seminal breakthrough in his learning; he achieved this, as attested in his own writing, by synthesizing various sources of his knowledge, Confucian, Buddhist and Western.[13]

1 Religious Leanings

To say Kang Youwei was a 'religious man' often raises the eyebrows of those who are familiar with the literature on Kang. Two questions they may readily ask in response are: Did Kang maintain a belief in God or in gods and spirits? Did he, like many religious figures in history, have personal religious experiences?[14] These are legitimate questions. For one thing, the issue of Kang's religious convictions has not received adequate treatment in the relevant literature, and the mystical states he had experienced in his youth have been overlooked or disregarded as trivial and sentimental.[15] For another, many religious figures in history did have personal religious experiences, which were considered of paramount importance for their faith and life. 'Personal religious experience', the American philosopher and psychologist William James wrote in his 1902 classic *Varieties of Religious Experience*, 'has its root and centre in mystical states of

13 Kang Youwei 康有为, 'Yu Shen Xingbu Zipei shu 与沈刑部子培书 [A Letter to Shen Zipei (1889)]', in *KYWQJ*, vol. 1, 2007, 237.

14 These questions, for example, occurred frequently during my discussions with several scholars at the Institute of Modern Chinese History, Chinese Academy of Social Sciences, when I was a visiting academic there in early 2015.

15 See for example the treatment in Ma, *Pingzhuan*; Huang Jing 黄晶, *Kang Youwei zhuan 康有为传* [*A Biography of Kang Youwei*] (Beijing: Beijing lianhe chuban gongsi, 2013).

consciousness.'[16] James proposed two essential criteria for marking out mystical states from others. The first is 'ineffability', meaning the mystical state must be directly experienced and cannot be imparted or transferred to others. The second criterion is 'noetic quality'. To those who experience them, mystical states seem to be 'states of insights into depths of truth', which most often carry with them 'a curious sense of authority for after-life'.[17]

In this section, I will attempt to give a 'yes' to both of the aforementioned questions by demonstrating that Kang was well informed concerning ancient Chinese religious ideas through his Confucian education, that he was exposed to the folk beliefs of his time, with their characteristic admixture of Buddhist and Daoist ideas, and that his attempts at Buddhist meditation had led him to vividly and intensely experience certain autobiographically recorded mystical states, which can be considered as being clearly among the 'varieties of religious experience' written about by William James.

1.1 *Classical and Folk Beliefs*

Through his training in Confucianism, Kang became informed of ancient Chinese religious thought as recorded in the 'Pentateuch' of the Confucian Canon, in particular the *Documents* and the *Odes*. Although the time of their composition has been the subject of controversy, at least some chapters of those texts are generally believed to have been written in pre-Confucian times.[18] In them was recorded a belief in the existence of a Supreme Ruler, called *Shangdi*, a designation that was used interchangeably with the more abstract *Tian* 天 or Heaven. As the Supreme Ruler, *Shangdi* gave decrees to man, as described in one of the earliest passages in the Documents (which Kang recited at a young age), a speech by King Tang, the founder of the rising Shang dynasty (circa 17th–11th century BCE).

> It is not I, the little child, who dare to undertake what may seem to be a rebellious enterprize; but for the many crimes of the sovereign of Hea

16 William James, *Varieties of Religious Experience* (London: Longmans, Green, and Co., 1902), 379.

17 James, 480–81. Pahnke and McNamara have built on James' work and expanded the list of features of religious experience derived from more recent studies. Richard C. DeBold and Russell C. Leaf, eds., *LSD, Man & Society* (Middletown, Ct.: Wesleyan University Press, 1967), 65–66; Patrick McNamara, *The Neuroscience of Religious Experience* (Cambridge University Press, 2009), 15–16.

18 For details see Michael Loewe, ed., *Early Chinese Texts: A Bibliographical Guide* (Berkeley: Institute of East Asian Studies, University of California, 1993), 377–80.

> [Xia] Heaven has given the charge to destroy him. [...] as I fear God [*Shangdi*], I dare not but punish him.[19]

After he overthrew the Xia dynasty and established his own dynastic reign, Tang gave an announcement to the people, in which he offered a prayer to the 'spiritual Sovereign of the high heavens' (*Shangtian shenhou* 上天神後).[20] Apart from being recorded in the *Documents*, a summary version of the prayer found its way into the *Analects*, to be committed to memory by generations of future Confucian students:

> I, the child Lü, presume to use a dark-coloured victim, and presume to announce to Thee, O most great and sovereign God [*Di*], that the sinner I dare not pardon, and thy ministers, O God, I do not keep in obscurity. The examination of them is by thy mind, O God.[21]

Like the *Documents*, the *Odes* also contains many references to *Shangdi* (or *Tian*). Sometimes he appeared to be an anthropomorphic Deity – 'God is with you, Have no doubts in your heart'; 'Great is God, Beholding the lower world with majesty.'[22] At other times, *Shangdi* or *Tian* was portrayed as an abstract Providence, who not only conferred blessings but also meted out punishments through natural phenomena, such as sending down rain for crops or causing natural disasters – 'God has reversed [His usual course of procedure], And the lower people are full of distress'; 'The doings of High Heaven, Have neither sound nor smell.'[23] As the prominent contemporary Chinese philosopher Feng Youlan 馮友蘭 has observed, the references to *Tian* in these two oldest Confucian Classics, except those referring to the physical sky, generally designate the ruling anthropomorphic type of *Tian*.[24]

---

19 *Shu* 書 (*Documents*) 4:1, in James Legge, trans., *The Chinese Classics III: The Shoo King, or The Book of Historical Documents* (Hong Kong: At The Author's, 1865), 173–74. Here and in other parts of this book, I generally follow the time-honoured translations of Chinese Classics by Legge, who, as a rule, translated *Shangdi* or *Di* by God. See also James Legge, *The Notions of the Chinese Concerning God and Spirits* (Hong Kong: The Hongkong Register Office, 1852).

20 *Shu* 4:3, in Legge, *Chinese Classics III*, 187–90.

21 *Lunyu* 20:1, in Legge, *Chinese Classics I*, 350.

22 *Shi* 詩 (*Odes*) 3:1:2, 3:1:7, in James Legge, trans., *The Chinese Classics IV: The She King, or The Book of Poetry* (Hong Kong: Lane, Crawford & Co., 1871), 436, 448.

23 *Shi* 3:2:10, 3:1:1, in Legge, 499, 431.

24 Fung, *Chinese Philosophy*, 1983, 1:31.

The belief in an anthropomorphic *Tian* or Heaven was maintained in the *Analects*. Confucius said, for example, that he knew the 'decrees of Heaven' at fifty, that 'there is Heaven – that knows me', that 'Heaven produced the virtue that was in me', and that 'he who offends against Heaven has none to whom he can pray'.[25] When his disciples asked leave to pray for his recovery, the Sage said that 'my praying has been for a long time'. He was described as 'a bell with its wooden tongue' used by Heaven; he agreed that 'Death and life have their determined appointment; riches and honours depend upon Heaven'.[26] On one occasion when he encountered a life-threatening danger at a place named Kuang, he spoke calmly to his disciples:

> After the death of King Wen, was not the cause of truth lodged here in me? If Heaven had wished to let this cause of truth perish, then I, a future mortal, should not have got such a relation to that cause. While Heaven does not let the cause of truth perish, what can the people of Kuang do to me?[27]

Passages such as those quoted above demonstrate Confucius' strong faith in Heaven, his certitude in his Heaven-endowed mission to carry forward the cause of truth, and in the providential arrangement of his life. It has been argued that *Tian* in the classical Confucian tradition functioned 'as a religious authority or absolute often theistic in its portrayal'.[28] Although Confucius did not speak much about life after death or spiritual beings,[29] he never denied their existence; instead, he stressed proper sacrifice to them. 'He sacrificed to the dead, as if they were present. He sacrificed to the spirits, as if the spirits were present.'[30] Unlike the Judaeo-Christian God, who alone should be worshipped to the exclusion of all others, *Shangdi* or *Tian* was believed to have associates or junior deities and various spirits, all deserving sacrifice from man. It was recorded in the *Documents*, for example, that the legendary Emperor Shun not only sacrificed to *Shangdi*, but he also 'offered appropriate sacrifices

---

25 *Lunyu* 2:4, 14:37, 7:22, 3:13, in Legge, *Chinese Classics I*, 146–47, 289, 202, 159.

26 *Lunyu* 7:34, 3:24, 12:5, in Legge, 206, 164, 253.

27 *Lunyu* 9:5, in Legge, 217–18.

28 Taylor, *Religious Dimensions*, 2.

29 It was famously said in the *Analects* that the subjects on which the Master did not talk were 'extraordinary things, feats of strength, disorder, and spiritual beings'. *Lunyu* 7:20, in Legge, *Chinese Classics I*, 201.

30 *Lunyu* 3:12, in Legge, 159. As we will see in this book, Kang would later argue that Confucius' teachings on God and spirits should primarily be sought in the Five Classics rather than the *Analects*.

to the hills and rivers; and extended his worship to the host of spirits'.[31] As will be seen later in this book, Kang shared the same beliefs as those expressed by Confucius, and on many occasions in his writings unabashedly likened himself to Confucius.

Compared with *Shangdi,* which modern scholars take to have originally been the tribal god of the Shang people or speculate might possibly have meant the high ancestor of the ruling house,[32] *Tian* carries naturalistic, fatalistic and ethical connotations in addition to its anthropomorphic meaning.[33] As *Tian* gradually and eventually replaced *Shangdi* and became the customary term, the Chinese concept of the ultimate Reality also came to assume an increasingly impersonal character. In the later Neo-Confucian tradition, *Tian*, or *Tianli* (天理, Principle of Heaven), had become an overwhelmingly abstract, monistic concept.[34] Nevertheless, *Shangdi* was still retained as part of the name of the Supreme Ruler in the Confucian Classics and the imperial rites of worship. Worship of *Shangdi* or Heaven, however, had become reserved for the emperor alone. As was the case in the Qing dynasty, the Supreme Ruler was represented by a wooden tablet inscribed with the words 'August Heaven the Supreme Ruler' (*Huangtian Shangdi* 皇天上帝) placed at the Temple of Heaven (*Tiantan* 天壇) in the capital, where prayers and ritual sacrifices were offered by the emperor at the Winter and Summer solstices.[35] The common people, on the other hand, could only offer sacrifices to local deities and spirits, including the spirits of their ancestors, at shrines and altars.

The arrival of the Jesuits had ignited new discussions on *Shangdi* among the Chinese literati. The early Jesuits in China had adopted both *Shangdi* and *Tian* for *Deus* before Pope Clement XI's 1704 order forbidding the use of those terms and declaring the proper Chinese equivalent for God to be *Tianzhu* (天主, literally, Lord of Heaven). That decree, together with the Papal interdiction of Chinese rites such as worship of Heaven, Confucius and ancestors among Chinese converts, was refuted by the angered Emperor Kangxi 康熙 (1662–1722), who subsequently banned the Catholic missionary work in China. This series of

31 *Shu* II 1:3; Legge, *Chinese Classics III*, 33–34.

32 Thompson, *Chinese Religion*, 2–3.

33 Fung, *Chinese Philosophy*, 1983, 1:31.

34 For more on this observation, see Taylor, *Religious Dimensions*, 2.

35 The imperial worship of the Qing dynasty at the Temple of Heaven started on the first day of the tenth lunar month, 1644, when the Manchu Emperor Shunzhi offered a prayer and announced his mounting the throne of China. For a description of the solemn ceremony and a translation of his prayer, see Joseph Edkins, *Religion in China; Containing a Brief Account of the Three Religions of the Chinese*, 2nd ed. (London: Trübner & Co. Ltd, 1893), 18–20.

FIGURE 1.1 Tablet of August Heaven the Supreme Ruler (*Huangtian Shangdi*, in both Manchu and Mandarin), placed in the Hall of Prayer for Good Harvests, Temple of Heaven, Beijing. Photo Date: 1932.
IMAGE COURTESY OF HISTORICAL PHOTOGRAPHS OF CHINA, UNIVERSITY OF BRISTOL (https://www.hpcbristol.net/visual/fh01-174)

disputes subsequently came to be known as the Rites and Term Controversy.[36] In response, some of the Confucian scholars had offered their own interpretations of the concept of *Shangdi*. Wang Qiyuan 王啟元 (1559–?), for example, criticized the Jesuits' attempt to equate the Confucian *Shangdi* with the Christian God or *Tianzhu*, listing various differences between the two concepts. In particular, he argued that the latter, described as being born of a human being in a particular country, could not be reconciled with the former that was characterized by utmost power and impartiality and was understood to have existed with the world from the very beginning. Similarly, Zhuang Cunyu 莊存與 (1719–1788), the pioneering advocate for reviving New Text Confucianism during the Qing dynasty (more on this revival in the next chapter), re-examined the multiple facets of the notion of *Shangdi* in Confucian Classics.[37] These can be regarded as efforts by Confucianists to maintain and defend their right to interpret the seminal concept of *Shangdi* in their encounter with the Christian religion. As will be seen later in this book, Kang would draw on the writings and views of both his Confucian predecessors and the Christian missionaries in his reflections and discussions on the concept of *Shangdi*.

Kang was also exposed to the local beliefs and practices of his childhood home. The ancestral shrine of the Kang family was built by Kang's great-great grandfather, and was later expanded by his granduncle, the general.[38] As for the Confucianists, they had set up Temples of the Literati (*wenmiao* 文廟, also known as Temples of Confucius), at county levels and above, to honour Confucius and his associates, and had built Shrines of the Worthies (*xianliangci* 賢良祠) at various localities. One of Kang's granduncles (not the general), for example, was honoured at the local shrine for his literary accomplishments as well as his contribution to the local welfare.[39]

---

36 See Minamiki, *The Chinese Rites Controversy*, 25–76. See also the 'Translators' Introduction' in Matteo Ricci, *The True Meaning of the Lord of Heaven* [*Tianzhu Shiyi*], trans. Douglas Lancashire and Peter Kuo-chen Hu (St. Louis: Ricci Institute, 1985), 13–55.

37 Wang Qiyuan 王啟元, *Qingshu jingtan 清署经谈* [*Discussions on Scripture at Qingshu*] (Shanghai: Shanghai guji chubanshe, 2017), 331–33; Zhuang Cunyu 莊存與, *Weijingzhai yishu: Xici zhuan lun 味經宅遺書: 繫辭傳论·* [*The Unpublished Writings of the Weijingzhai: A Treatise on the Xici of the Book of Changes*] (Yanghu zhuangshi, 1882), 94–95; see also Zhang, 'Ruxueshi xushu de fenduan'.

38 Kang Youwei 康有为, 'Kangshi jiamiao bei 康氏家庙碑 [An Inscription on the Stone Tablet of the Kang Shrine (1917)]', in *KYWQJ*, vol. 10, 2007, 438–40. The family shrine, together with the ancestral tombs, was destroyed by imperial order after Kang's reform was cut short by the Coup of 1898. During a recent visit to Kang's restored old house, I noticed an altar built into the inner side of a wall in the yard, possibly for the purpose of honouring certain household deities.

39 Kang, 438.

To honour spirits by offering sacrifice to them was fine, but to engage in communication with them was too much for a Neo-Confucian family like the Kangs. The practice of interacting with the spirit world by way of a trance medium was popular at the grassroots level in traditional China, but a strict Confucianist was supposed to follow the example of Confucius and avoid even talking about it, let alone practising it. That is why it was considered a serious problem if a family member started the practice. In his eulogy for his granduncle, the general, Kang praised him for divorcing his wife because 'she was talkative and fond of the art of *wu* 巫 and *xi* 覡'.[40] The character of *wu* means a female spirit medium, while *xi,* a male trance medium, both commonly rendered into English as 'shaman'.[41] The practitioners were believed to have the power to interact with spirits for the purpose of healing, divination and supplicatory prayers.[42] The ancient shamanic art of *wu* and *xi* continued to be practised during Kang's times, but it was generally looked down upon by Confucianists.

With the rise of Daoism and the coming of Buddhism, more deities were added to the pantheon of the Chinese people, and accordingly more temples and monasteries were built. Thus at the local level, the communal belief system was often a mixture of the Three Teachings, namely, Daoism, Buddhism and Confucianism, as well as some forms of the *wu*. This was particularly true with regard to Kang's hometown. About a dozen miles to the south of his old house was the mountain of Xiqiao (西樵), renowned for its natural scenery and spiritual heritage. An ancient, extinct volcano, the mountain has many peaks, cliffs and caves, adorned with waterfalls and small woods. Since the seventh century, Buddhist and Daoist monks had often visited; some had built monasteries there over the course of the next millennium. Neo-Confucian scholars came during the Ming dynasty (1368–1644) and set up Confucian academies. While many of the old monasteries and academies had been destroyed in wars, some had survived or been restored, and new ones were to be built. The reconstruction of the Celestial Cloud-Spring Abbey (*Yunquan xianguan* 雲泉仙舘), for example, was completed in the very year of Kang's birth.[43] By the mid-nineteenth century, the Xiqiao mountain had become known as a resort of the Three Teachings.

---

40 Kang, 'Kanggong shizhuang', 12.

41 L. C. Hopkins, 'The Shaman or Chinese Wu: His Inspired Dancing and Versatile Character', *The Journal of the Royal Asiatic Society of Great Britain and Ireland*, no. 1 (1945): 3–16.

42 For an observation on the Chinese *Wu-ism* by a contemporary Western scholar, see J. J. M. De Groot, *The Religious System of China*, vol. VI (Leyden: E. J. Brill, 1910), 1212.

43 Guan Xiang 关祥, ed., *Xiqiaoshan zhi* 西樵山志 [*A Gazette of the Xiqiao Mountain*] (Guangzhou: Guangdong renmin chubanshe, 1982), 86.

1.2 *Buddhist Inspiration*

Kang developed an interest in Buddhism in his early years. This religious predilection, or a 'natural liking for religion' as his disciples and early biographers observed,[44] was first nourished by his visits to religious sites and reading Buddhist texts. In his boyhood, Kang was often brought by his grandfather to visit local scenic spots including Buddhist temples and Daoist monasteries, on which occasions the grandfather would speak of great names associated with those sites. That created in the young Kang a strong desire to emulate and associate himself with such illustrious men as Huineng 慧能 (638–713, the famous Sixth Patriarch of Chan Buddhism), and Qiu Chuji 丘處機 (1148–1227, the renowned Daoist Master).[45] Kang's first reading of Buddhism had taken place before he went to study under Zhu Ciqi, who was known for his distaste for Buddhism. Kang later recalled that he had perused the *Huayan jing* 華嚴經 or *Garland Sutra* as early as age sixteen, and found it 'comprehensive, profound and inexplicable'.[46] The sutra is one of the most influential Mahayana works and constitutes the basic text for the creation of the Chinese Huayan School. The text describes a cosmos consisting of infinite realms that mutually contain each other; in particular, it offers a detailed account of the various stages of the practice of the bodhisattva, a Buddhist practitioner who out of compassion delays entering nirvana in order to remain behind and save other beings.[47] This bodhisattva concept of compassion would have an important bearing on the development of Kang's thought.

Kang did not actively pursue his interest in Buddhism until towards the end of his study under Zhu, when he decided to discontinue his book-learning based training in orthodox Neo-Confucianism by reason of his realization that it would only 'smother his spiritual faculties'. Instead, he opted to seek ways to 'achieve peace of mind' and to 'fulfil my mission in life'.[48] Privately, and against his teacher's counsels, he turned his attention to the rival Learning of the Heart and Mind.[49] He converted to the belief that the Chan idea of sudden

---

44 Lu Naixiang 陆乃翔 and Lu Dunkui 陆敦骙, 'Nanhai xiansheng zhuan 南海先生传 [A Biography of Mr. Kang of Nanhai]', in *KYWQJ*, vol. 12, 2007, 457.

45 Kang, 'Woshi', 60.

46 Kang Youwei 康有为, 'Chang'an yanjianglu 长安演讲录 [Recorded Talks (by Kang) in Xi'an (1923)]', in *KYWQJ*, vol. 11, 2007, 293; see also Kang, 'Woshi', 67–68.

47 See 'Flower Ornament Sutra華嚴經', Digital Dictionary of Buddhism 電子佛教辭典 (University of Tokyo), http://www.buddhism-dict.net/ddb/).

48 Kang, 'Woshi', 62.

49 Mao Haijian 茅海建, *Cong jiawu dao wuxu: Kang Youwei Woshi jianzhu 从甲午到戊戌：康有为我史鉴注* [*Kang's 'My History' Verified and Annotated: 1894–1998*] (Beijing: Sanlian shudian, 2009), 15. Mao's examination of Kang's manuscript of *My History* shows

enlightenment through self-reflection might facilitate rather than obstruct his pursuit of sagehood, a belief that could be traced back to the beginning of Neo-Confucianism.[50] As a result, he abandoned his studies and started to practise Chan meditation by secluding himself in his room. Through this practice he had an extraordinary experience as described in his own words which are quoted at the beginning of this chapter. He had written of having experienced the feeling of being at one with heaven and earth while suffused in supernal light.[51] Kang's fellow students, who had heard his unnatural laughter and cries, thought he had become insane or had suffered a mental breakdown.[52] Kang, however, believed that he was in a state of limbo, eagerly and intensely seeking truth but not yet obtaining it, a state of mind described as 'a demon flying into one's mind' in the Buddhist texts.[53] This mystical experience gave Kang the conviction that he was born to be a sage with a bodhisattva-like mission to deliver mankind from suffering. His 'sage complex', something comparable with a 'Messianic complex', which he had cherished since boyhood, now received a mystical endorsement, and from then on would assume a strong presence as a relentless driving force in his life.[54]

After leaving his teacher and returning home, Kang went into semi-seclusion by residing in the White Cloud Cavern (*Baiyundong* 白雲洞) at the northern foothills of Xiqiao. Regarded as the jewel of the scenic attractions in the mountain, the cavern was a sizable natural cave leading into a cliff, a few meters from the ground and facing an attractive waterfall. Since the late sixteenth century when stone steps were built leading to its entrance, it had become a famous resort for Confucianists to engage in scholarly reflections and meditative pursuits. In its vicinity lay the recently-refurnished Three Lakes Confucian Academy (*Sanhu shuyuan* 三湖書院), the Ancient White Cloud Monastery (*Baiyun gusi* 白雲古寺), where the Buddha was worshipped, and the grand, newly-restored Celestial Cloud-Spring Abbey, where the famous

---

that Kang had first written and then crossed out this sentence: "Privately I was fond of [the learning] of [Wang] Yangming."

50 For a case study of Chan influences on Neo-Confucianism, see Ari Borrell, 'Ko-Wu or Kung-an? Practice, Realization, and Teaching in the Thought of Chang Chiu-Ch'eng', in *Buddhism in the Sung*, ed. Peter Gregory and Daniel A. Getz (Honolulu: University of Hawai'i Press, 1999), 62–108.

51 Kang, 'Woshi', 62.

52 This has been described by some scholars as one of Kang's 'traumatic life experiences'. See Tay, 'Kang Youwei', 59.

53 Kang, 'Woshi', 62.

54 Geng, 'Kang Youwei de shengren qingjie jiqi yi Kongjiao wei guojiao shuo 康有为的"圣人"情结及其以孔教为国教说 [Kang Youwei's Sage Complex and His Views on Confucianism as a State Religion]'.

FIGURE 1.2 The White Cloud Cavern (*Baiyundong*), now a tourist attraction.
PHOTO COURTESY OF JIA XIAOYE, 2017

Daoist Immortal Lü Dongbin 呂洞賓 was worshipped. Hence the Cavern had long been considered a blessed spot for Buddhism, Daoism as well as Confucianism. The academy, the monastery and the abbey housed a good collection of books of the Three Teachings, to which Kang was given access, as well as a group of resident monks and scholars, whose intellectual company Kang may have sought.[55]

There Kang immersed himself in reading Buddhist and Daoist books so as to 'nourish his spirit and purify his mind from impure thoughts'.[56] As a result, he had another extraordinary experience, or a 'great enlightening' (*dache dawu* 大澈大悟) in the words of his disciple and biographer Liang Qichao.[57] According to his autobiography penned ten years later, Kang stayed up all night for a whole month and let his mind wander freely, reflecting on the heavens and the human world, experiencing alternative states of ecstasy and of profound sorrow. At first, his mind was full of 'wild, demon-like thoughts', but as his practice progressed he felt 'spiritually transcendent and joyfully at ease'.[58] In addition, he practised the *Wushengdao* (五勝道), probably a Daoist art,[59] through which he said he was able to 'see' his 'spiritual self' (soul), a self that existed outside his body but which could also enter his body when he so wished. In other words, he realized that his physical body was not indispensable to his spiritual self. As a result, he began to view his own body as nothing but a carcass and those of others as no better than swine.[60] It was also during this period that Kang attempted to write a commentary on the *Laozi* (老子), one of the primary Daoist texts. But eventually, he found the work 'abhorrent' and abandoned his effort.[61] Indeed, that seemed to be the beginning of the end of Kang's affection for Daoism; rarely would he revisit Daoist texts or Daoist practice in his later life.

His seclusion in the mountain ended in autumn under pressure from family elders (mother, uncles and great uncles) who wanted him to prepare for the civil service examinations: otherwise they would cut off financial support for him. As an aspiring scholar without a source of income, Kang depended on them for his living, and had to consider their demands and hopes for him to bring honour to the family by passing the exams.[62] However, the profound

---

55 Wenshizu 文史组, ed., *Nanhai wenshi ziliao: Xiqiaoshan zhuanji 南海文史资料: 西樵山专辑* [*Historical Materials of Nanhai: On Xiqiao Mountain*] (Nanhai, 1983), 47, 48, 51–59.

56 Kang, 'Woshi', 62.

57 Liang, 'Nanhai Kang', 486.

58 Kang, 'Woshi', 62.

59 What Kang meant by *wushengdao* is not exactly clear. Jung-pang Lo, Kang's grandson who rendered Kang's *Woshi* into English, interpreted it as 'the theory of the successive ascendency of the five elements in their cycle'. See Lo, 'Autobiography', 34, 145. Some biographers of Kang, however, have argued that it was a particular Daoist spiritual practice. See, for example, Huang, *Kang Youwei zhuan*, 16.

60 Kang, 'Woshi', 62; see also Liang, 'Nanhai Kang', 486.

61 Kang, 'Woshi', 62.

62 Kang, 62. It was an established tradition in the Kang family, a practice common among many families at that time, that one branch of the family specialized in Confucian learning

insights he gained from the experience during his seclusion, namely, the archetypal Buddhist bodhisattva's compassion for mankind, would have a lasting effect on his future life and thought. For some time during his youth, as he recalled in his evening years, he had even harboured the idea of 'casting aside worldly cares' (*yishi* 遺世) and 'becoming a Buddhist monk' (*chujia* 出家), but was held back by his compassion for the sufferings of his mother and his widowed sister and the sorrows of the people around him. So, he did not give up Confucianism to convert to Buddhism, for he thought Confucianism would better prepare him for his self-imposed mission to lighten the sufferings in the human world.[63]

His interest in Buddhism, however, remained throughout his life, and Buddhist ideas can be easily discerned in his later thought. Historians have rightly pointed out that Kang was formally introduced to Buddhism by the byway of Neo-Confucianist works, which were heavily influenced by Buddhism.[64] In this he was not without predecessors and comrades. Quite a few prominent Confucian scholars before him, like Gong Zizhen 龔自珍 (1792–1841) and Wei Yuan 魏源 (1794–1857), were deeply attracted to Buddhism, and some were even completely converted. It has been argued that there was a 'Buddhist revival' among Chinese intellectuals in the second half of the nineteenth century. The central figure among them was the Confucian-turned-Buddhist Yang Wenhui 楊文會 (1837–1911), whom Kang later befriended. Apart from printing and circulating Buddhist texts in China, Yang was believed to be the first Chinese Buddhist to have visited Europe where, probably under the influence of Max Müller (1823–1900), he began to conceive of Buddhism as a world religion in a modern age of science.[65]

In a sense, the Chinese Buddhist revival was a response to the impact from Christianity following China's defeat in the Opium Wars. The increasing need

---

in order to pass the civil service examinations and obtain official positions, and the other branch engaged in economic activities to provide financial support for the entire family. See Richard C. Howard, 'K'ang Yu-Wei (1858–1927): His Intellectual Background and Early Thought', in *Confucian Personalities*, by Arthur F. Wright and Denis Twitchett (Stanford: Stanford University Press, 1962), 297; see also Richard C. Howard, 'The Early Life and Thought of K'ang Yu-Wei, 1858–1895' (Ph.D., Columbia University, 1972), 13–17.

63 Kang Youwei 康有为, 'Wuxu lunzhou zhong juebishu ji Wuwu bahou 戊戌轮舟中绝笔书及戊午跋后 [My "Last Note" Penned in a Ship in 1898, with Postscripts Added in 1918]', in *KYWQJ*, vol. 5, 2007, 4–5.

64 See Liang, 'Nanhai Kang', 487; Hsiao, *K'ang Yu-Wei*, 108. But that view holds less weight now that we know Kang had already become interested in Buddhism and read Buddhist literature in his boyhood.

65 Chan, *Buddhism in Late Ching*, 30–49; see also Holmes Welch, *The Buddhist Revival in China* (Harvard University Press, 1968).

for national salvation and the preservation of Chinese culture gave rise to an 'intellectual eclecticism' which was intent upon searching for alternatives or new perspectives from Buddhism in addition to Confucianism.[66] Some of the contemporary intellectuals regarded Buddhism as a better countervailing force to Christianity than Confucianism. '[H]ad it not been for the presence of Christianity,' one scholar has observed, 'Buddhism would not have been so enthusiastically studied' by late Qing Confucian scholars.[67]

1.3 *Knowledge of Christianity*

Unlike many of his educated contemporaries who resented or ignored Christianity, the young Kang paid serious attention to the 'Western' religion. His motivation, however, was more for the purpose of learning about it so as to mitigate its impact, than to seek truth from it. Around the time of Kang's birth, Catholicism and Protestantism had obtained legal status in China under the treaty system established after the Opium Wars, and during Kang's youth, Christian missionary work became increasingly active in the coastal areas including his home province, Guangdong. As the missionaries intensified their attacks on Chinese morality and such folkloric practices as zoolatry and ancestor worship, the resentment of the Chinese gentry-scholar class against Christianity continued to grow.[68] Interestingly, as I will discuss in the next chapter, Kang would adopt some of the concepts and rhetoric of the missionary discourse in his later advocacy of reforming Chinese morality and beliefs.

Another source of Kang's childhood impressions of Christianity might have come from the distorted lens of the ongoing, and ultimately devastating, Taiping Rebellion. As stated at the beginning of this chapter, many male members of Kang's family had joined the imperial forces in the fight against and the eventual suppression of the uprising. The millenarian rebels, who called themselves God-worshippers, were led by Hong Xiuquan 洪秀全 (1814–1864), who claimed to be another son of God the Father and the younger brother of Jesus Christ. Although its true nature has remained a subject of scholarly controversy, there is no doubt that the Taiping ideology carried a strong religious character, with Anglo-American Protestant inspiration combined with traditional Chinese religious elements. A recent study based on a close reading of the Taiping corpus has argued that the Taiping appeal to the common people

66 Chan, *Buddhism in Late Ching*, 48.

67 Chan, 49.

68 See Paul A. Cohen, 'Christian Missions and Their Impact to 1900', in *The Cambridge History of China*, ed. John K. Fairbank and Kwang-Ching Liu, vol. 11 (Cambridge: Cambridge University Press, 1980), 560–65.

came from Hong's representation of his Taiping Christianity as 'a revival and a restoration of the ancient classical faith in *Shangdi*'.[69] As will be shown later in this book, Kang would also attempt to restore the ancient Chinese belief in *Shangdi* in his endeavour to reform Confucianism.

Kang's first reading of Christianity, as he recalled it in 1904, had occurred twenty-five years earlier, that is, in 1879 when he had stopped his Confucian training under Zhu and started to explore other avenues of learning.[70] That year, he read Wei Yuan's *Haiguo tuzhi* 海國圖志 (An Illustrated Gazetteer of Maritime Countries), which contained a good deal of general introduction to Christianity as practised in the West. The multi-volume *Illustrated Gazetteer*, first published in 1843, was regarded as the first significant Chinese work on the West, whose composition was stimulated by the Anglo-Chinese First Opium War. Kang also read some Christian texts by missionaries, although it is not easy to ascertain the exact titles. As an eager reader of the influential and informative Chinese weekly, *The Globe Magazine* (1874–1883, Wan-kuo Kung-pao 萬國公報), edited by American missionary Young J. Allen (1836–1907), Kang would probably have read the religious treatises and articles included in its collected issues, which he purchased in 1883.[71] It can also be inferred that he possessed the Chinese translation of the Bible and at least read some of it. As indicated in his later correspondence, he had lent an eleven-volume copy of the *Old Testament* to the Confucian scholar and provincial censor Zhu Yixin 朱一新, with whom he discussed the relationship between the 'ancient' doctrines of Moses and the later teachings of Jesus.[72] In his early writings in 1880s, he also made references to Biblical names, such as aligning the time of Fuxi 伏羲, the legendary ancestor of the Chinese people, with that of Adam and Eve.[73] Included in his reading list might also have been *Tianzhu shiyi* 天主實義 (The True Meaning of the Lord of Heaven), a catechism written in classical

---

69 Thomas H. Reilly, *The Taiping Heavenly Kingdom: Rebellion and the Blasphemy of Empire* (Seattle: University of Washington Press, 2004), 4. For another recent work reaffirming the religious nature of the Taiping movement, see Zhou Weichi 周伟驰, *Taiping tianguo yu Qishilu* 天平天国与启示录 [*Taiping Heavenly Kingdom and Revelation*] (Beijing: Zhongguo shehui kexue chubanshe, 2013).

70 Kang Youwei 康有为, 'Yidali youji 意大利游记 [An Account of My Travels in Italy (1904)]', in *KYWQJ*, vol. 7, 2007, 397.

71 Kang, 'Woshi', 63.

72 Kang Youwei 康有为, 'Yu Zhu Yixin lunxue shudu 与朱一新论学书牍 [Letters to Zhu Yixin on Learning (1891)]', in *KYWQJ*, vol. 1, 2007, 334. Zhu did not seem to have been convinced by Kang's arguments. Kang, however, alleged that Zhu had been convinced and 'enlightened', suggesting that Zhu had expressed contrary opinions in his letters and essays was 'only for the sake of appearance'. See Kang, 'Woshi', 81.

73 Naiwaipian, p. 111.

Chinese by Matteo Ricci. First published in 1603, the work was probably the most influential religious text produced by the Jesuit mission to China. Its title and summary were included in the 1795 imperial collection *Siku quanshu zongmu* 四庫全書總目 (Bibliography of the Complete Library in Four Sections). Reprinted in 1868, the work was spoken of highly by many, such as the protestant missionary Richard Timothy, and Kang's future disciple Liang Qichao.[74]

In his study of Christianity, Kang consciously compared it with Buddhism, a comparison that led him to the surprising conclusion that the Christian Faith had originally 'stemmed' from Buddhism.[75] 'Each of the Christian doctrines,' he later wrote, 'be it on the soul, love, miracles, or on confession, salvation, hell and heaven, speaks directly to the heart and is in no way different from Buddhism.'[76] Whether Kang was justified in this claim is not as important (given the objective of this book) as what his claim implied: he had already begun to make efforts to synthesize the Western religion with Chinese religious traditions, a direction that he would continue to pursue in his later works. This process of his intellectual quest would eventually culminate in his third mystical experience or enlightenment, which occurred towards the end of 1884. Although Kang's understanding of Christianity, as Hsiao has observed, might have been fragmentary,[77] Christian practices exerted a strong impact on his thinking and his later proposal to turn Confucianism into a state religion.

2 Scientific Pursuits

In addition to the avenue of religious exploration, Western science was another source of knowledge for Kang to pursue outside the Neo-Confucian tradition. As his early education was strictly focused on the Confucian Classics,

74 For a discussion on reactions towards the book, see the 'Translators' Introduction' in Ricci, *True Meaning*, 38–47.

75 Kang Youwei 康有为, 'Kangzi neiwai pian 康子内外篇 [The Esoteric and Exoteric Essays of Master Kang (1886)]', in *KYWQJ*, vol. 1, 2007, 103.

76 Kang, 'Yidali youji', 397. On Buddhist influences on Christianity, the most important (if fraudulent) contributor to the debate was Nicolas Notovitch, see Nicolas Notovitch, *The Unknown Life of Jesus Christ*, trans. Alexina Loranger (Chicago and New York: Rand, McNally & Company, 1894), 13–62. Modoern scholars, who acknowledge 'many parallels concerning the births, lives, doctrines, and deaths of the Buddha and Jesus', have remained divided on the issue, see Jerry H. Bentley, *Old World Encounters: Cross-Cultural Contacts and Exchanges in Pre-Modern Times* (Oxford: Oxford University Press, 1993).

77 Hsiao, *K'ang Yu-Wei*, 111.

the young Kang had had little knowledge of the outside world. His teacher Zhu Ciqi's curriculum paid little attention to natural studies and Western learning. At that time, although decades-long educational efforts, missionary or government-sponsored, had achieved some success in disseminating Western knowledge in China, they did not produce any major impact on the mind of the generality of the Chinese literati concerning the importance of modern science. A case in point was the controversy over a scientific education programme offered by the Beijing Tongwenguan (京師同文館, Academy of Combined Learning, also known as Interpreters' College in English), established by the Zongli Yamen (Board of Ministers for Foreign Affairs) in 1862 during the initial stage of the Self-Strengthening Movement. In 1866 the college proposed to recruit students from those who had passed the civil service examinations to study mathematics and astronomy with Western tutors, but it was fiercely opposed by the conservative officialdom, who were imbued with uninformed, negative notions respecting foreigners. Their accusations were severe – that Confucian scholars treading 'the right path' (*zhengtu* 正途, i.e. passing civil service examinations) should not engage in those technical pursuits that would corrupt their minds, and if they were to learn technical knowledge from Western tutors, they would be put in the danger of being induced to convert to Christianity in the same manner 'as half of the ignorant masses have done'.[78] So learning modern science from Westerners was considered likely to have dire moral consequences for Confucian scholars. Although the programme was commenced and continued for years by virtue of the support of the Zongli Yamen, the academic quality of the students enrolled in the programme remained low largely due to the controversy.[79] The introduction and acquisition of Western science was often met with opposition or indifference from the influential conservative officialdom and the mainstream Confucian scholars, who would ignore anything but the Five Classics and Four Books as their passports to securing state offices.

2.1 *The Window of Geography*

It was only in 1874 when Kang was sixteen that he saw world maps for the first time and laid his hands on a popular contemporary geography, Xu Jiyu's

78 Quoted in Xiong Yuezhi 熊月之, *Xixue dongjian yu Wanqing shehui 西学东渐与晚清社会* [*The Eastward Dissemination of Western Learning in the Late Qing Dynasty*], 2nd ed. (Beijing: Zhongguo renmin daxue chubanshe, 2011), 257.

79 For a good discussion on the controversy, see Xiong, 256–62; Timothy B. Weston, *The Power of Position: Beijing University, Intellectuals, and Chinese Political Culture, 1898–1929* (Berkeley: University of California Press, 2004), 15–17.

徐繼畬 (1795–1873) *Yinghuan zhilue* 瀛寰志略 (A Compendious Description of the World). This work became the first window through which Kang began to view the world, and by which he learned the basics of world geography and the history of other countries.[80] It seems, however, he read the work with a condescending attitude towards foreigners, for he still shared the prevailing perception of them as 'barbarians'. Kang did not question this view until five years later, when his brief yet intellectually significant visit to Hong Kong in 1879 made him realize that 'the Westerners have the art of good government and should not be regarded as barbarians as depicted in the old and outdated perspective'.[81] With this fresh understanding as well as a newly-ignited passion for Western knowledge, he perused for a second time *A Compendious Description of the World*, along with Wei's *Illustrated Gazetteer*.[82]

Kang's earliest knowledge of Western science must have come from those geographies which he specifically named and perused. Therefore, an examination of their 'scientific content' would help us ascertain Kang's 'scientific intake' around 1880. Xu's *Compendious Description* demonstrated a scientific attitude: 'It is gratifying to observe,' wrote a contemporary European reviewer, 'how readily our author admits the truths of western science, even when [these were] opposed to the favourite hereditary opinions of his country.'[83] What was referred to by the reviewer was the fact that Xu decided to accept, after verifying with his European informants, the existence of the 'Frozen Ocean at the South Pole', which had not been shown in the earlier Jesuits' maps in China and whose existence was contrary to the traditional Chinese view.[84] In addition, unlike the earlier Jesuits, Xu placed China in East Asia on the map in his book, demonstrating that China, far from being the centre of the world, was only one of the countries located in Asia. His neutral stance in his accounts of Western people and political systems was another sign of his objectivity, but it was also probably one of the reasons for his eventual dismissal from his post.[85] In reference to this, the *New York Times* observed in 1868, 'Till within a quarter of a century, the study of barbarian annals has been one of the most

80 Kang, 'Woshi', 61.

81 Kang, 63.

82 Kang, 63.

83 'Review: A Compendious Description of the World (Lit. the Oceanic Circle) by Seu Sung Lung in 6 Vols, 1849', *The North-China Herald*, 28 September 1850.

84 'Review: A Compendious Description'.

85 See also Yuezhi Xiong, *The Eastward Dissemination of Western Learning in the Late Qing Dynasty*, trans. Yanxin Chan, Jiao Li, and Junchao Wang, vol. 1 (Singapore: Enrich Professional Publishing, 2013), 172.

dangerous branches of learning a Chinese could pursue; and a candid geographer has dared the fate of Galileo.'[86] In addition, the work also covered the history of various countries, and in doing so it touched upon the development of Western technology, though little was said about the exact sciences. Kang must have appreciated Xu's objectivity when he picked up the work a second time. Fifteen years later, he still spoke highly of it for its accuracy and clarity and recommended it to his students and other scholars as the best primer on world geography.[87]

Unlike Xu's *Compendious Description*, Wei's *Illustrated Gazetteer* in its final edition of 1852 devoted a few sections to Western astronomy. With the understanding that to better understand geography one was required to possess a knowledge of astronomy, the compiler included excerpts from a recent work on geography written in Chinese by Martinho José Marques 瑪吉士 (1810–1867), a Macau-born Portuguese sinologist.[88] Marques posited that the earth was spherical rather than flat as previously thought. This was followed by an accurate presentation of the heliocentric theory, presenting the Sun at the centre with a rotating Earth and other planets revolving around. This was no longer a theory, Marques argued, but a truth proved by generations of Western astronomers.[89] That was not the first time that a heliocentric theory was introduced into China, for nearly a century earlier, the French Jesuit Michel Benoist had already presented the Copernican system to Chinese astronomers.[90] However, because the theory ran contrary to the traditional geocentric conception, it was ridiculed by such leading scholars as Ruan Yuan 阮元 (1764–1849), the author of the influential *Chourenzhuan* 疇人傳 (Biographies of Mathematical Astronomers). Although he included the theory in his collection, Wei Yuan did not consider it to be true: 'If the earth did rotate and revolve continually [as the Westerners claimed],' he commented, 'then the Pole Star should also be moving rather than remain in its constant position.'[91]

86 'America in China.', *New York Times*, 29 March 1868.

87 Kang Youwei 康有为, 'Guixue wenda 桂学问答 [Answers to Questions on Learning in Guiling (1894)]', in *KYWQJ*, vol. 2, 2007, 23, 19. As the *Illustrated Gazetteer* was basically a compilation of various excerpts from various works, it contained some inconsistence or inaccuracies, which, in Kang's view, made it inferior to Xu's *Compendious Description*.

88 Ma Jishi 瑪吉士 [Martinho José Marques], *Xinshi dili beikao quanshu 新釋地理備考全書* (Guangzhou: Haishan xianguan, 1847).

89 Wei Yuan 魏源, *Haiguo tuzhi 海國圖志* [*An Illustrated Gazetteer of Maritime Countries*] (Yangzhou: Guweitang, 1852), 2a, 3a-b.

90 A. Udias, 'Jesuit Astronomers in Beijing, 1601–1805', *Quarterly Journal of The Royal Astronomical Society* 35, no. 4 (1994): 463–78.

91 Wei, *Haiguo tuzhi*, 4a.

At the time Kang first read the *Illustrated Gazetteer*, the heliocentric theory was still not widely accepted among Chinese intellectuals. 'Of ten Chinese who heard the theory,' commented Li Gui 李圭 (1842–1903), who is on record as one of the first Chinese nationals to have circled the globe, 'eight or nine would not believe it.'[92] Li himself was not convinced until he made the trip in 1876–7, travelling all the way eastward and finally arriving in Shanghai where he had started. We have no idea if Kang accepted the theory upon reading the *Illustrated Gazetteer*; but even if he did not, he was likely to have been convinced by Li's account. In any case, as was indicated in his later correspondence with Zhu Yixin, Kang had by 1891 already accepted and, indeed, had even begun to advocate, the heliocentric theory, commenting that the theory did not start with Copernicus, but had been originally invented by Pythagoras.[93]

2.2 *Book Purchase at Shanghai*

Not content with reading works on the West by native Chinese authors, Kang decided to collect and peruse translations of Western books. He did not seem to have a speedy start in that direction until the end of 1882, when he stopped at Shanghai on his way back home from Beijing, where he had sat for, but failed, the triennial provincial examination. Designated a treaty port after the Opium War and with an increasingly large International Settlement, Shanghai had developed into a prosperous international city. Kang's visit to Shanghai intensified his desire to learn from Westerners, a desire that was first ignited during his visit to Hong Kong three years earlier. 'The prosperity of the Shanghai [International Settlement]', he observed, 'further convinced me that Westerners' art of government must have had its own source of learning.'[94] For Kang, who did not read in any foreign language, the only viable way to learn from the West was through reading translations of Western books, and nowhere could he find a better place than Shanghai to purchase them, especially scientific titles. Since the 1860s, Shanghai had replaced Guangzhou and become the centre for the translation and publication of Western books as well as for the dissemination of Western learning, or to be specific, Western science.[95]

The central figure of the scientific translation enterprise in Shanghai was the British scholar John Fryer (1839–1928傅蘭雅), the chief translator of the Translation Department at the Jiangnan Arsenal. The department was set up in 1868

92 Li Gui 李圭, *Huanyou diqiu xinlu* 环游地球新录 [*A New Account of a Trip Around the Globe* (*1878*)], New edn. (Changsha: Hunan renmin chubanshe, 1985), 158.

93 Kang, 'Yu Zhu Yixin', 334.

94 Kang, 'Woshi', 63.

95 Xiong, *Xixue dongjian*, 338.

with the sponsorship of the open-minded Zeng Guofan 曾國藩 (1811–1872), the chief Confucian general who suppressed the Taiping Rebellion. It aimed at 'the translation and publication of books relating to the Arts and Sciences of the West'. By the year 1880, Fryer and his staff of native and foreign translators had translated an impressive number of 143 titles, two thirds of which had already been printed by the Arsenal, as is shown by the table below.

TABLE 1 Translations by the Jiangnan Arsenal translation department, 1880.

Subjects	List 1, published	List 2, unpublished	List 3, in course of translation
Mathematics, Surveying, &c.	22	2	3
Engineering, &c.	7	3	1
Chemistry, &c.	5	1	1
Geography, &c.	8	0	2
Geology, Mining, &c.	5	0	0
Astronomy, Navigation, &c.	9	3	0
Physical Science	6	4	1
Medicine	2	1	3
Arts and Manufactures	13	9	0
Naval & Military Science	15	9	2
Chronology, News summaries, &c.	6	1	0
Naval Architecture	0	3	1
History	0	5	0
International Law	0	2	0
Miscellaneous	0	2	0
Total	98	45	13

Note: The number of the published translations would double by the time the Translation Department stopped functioning in the early 1900s. See also Xiong, *Xixue dongjian*, 396.

SOURCE: REPRODUCED (WITH ADAPTATIONS) FROM JOHN FRYER, 'AN ACCOUNT OF THE DEPARTMENT FOR THE TRANSLATION OF FOREIGN BOOKS AT THE KIANGNAN ARSENAL, SHANGHAI', *THE NORTH-CHINA HERALD*, 29 JANUARY 1880.

The translations covered a wide range of subjects, but as might be expected, a large share of attention had been given to the military and naval sciences as well as engineering and surveying. Apart from their military utility, these translations were for sale, and copies of them were found in mission schools as well as personal libraries of some eminent native scholars. They were also on display in the Chinese Reading Room in Shanghai, which was set up in connection with the Shanghai Polytechnic Institution (*Gezhi shuyuan* 格致書院). According to Fryer, as many as 31,111 copies of the works had been sold up to mid-1879.[96] That was an impressive number given that there had been no regular means of communication, no postal or railway delivery, and few means of advertising or distributing them to the public.

As Fryer acknowledged, most of the translations of the department were either too specific or too profound in their content to be understood by the general popular readership. In order to remedy this defect, Fryer, who professed a firm belief in the role of translation in 'bringing about the intellectual and moral regeneration' of China, started in 1876 as an individual initiative, *The Chinese Scientific Magazine* (*Gezhi huibian* 格致彙編). The monthly continued publication for four years in two periods of 1876–78 and 1880–82. The focus of the magazine was the introduction of elementary knowledge of modern science, and its Q&A column proved to be especially popular among its readers. By the time of Kang's visit, its past issues had been collected and reprinted, and were made available for sale at the Polytechnic Institution. As the 'first scientific magazine' in China, a Chinese historian of science has observed, Fryer's magazine played an important role in the 'scientific enlightenment' of late Qing intellectuals.[97]

Concurrently, Fryer and the Translation Department were behind another endeavour of a similar nature – the production of introductory textbooks for Chinese students. For that purpose, Fryer chose the science primers published by Macmillan & Company in England in 1872. Compiled by Henry Roscoe (1833–1915) in collaboration with several other top British scholars including Thomas Huxley, the series was considered to have represented the state of knowledge in the scientific fields in Great Britain.[98] The Chinese translation was completed with the collaboration of Young J. Allen before being published by the Jiangnan Arsenal in 1879–80 under the title of *Gezhi qimeng* 格致啟蒙 (Primers on Science Education). That would presage a more ambitious project, an expanding textbook series, the *Gezhi xuzhi* 格致須知 (Science Outline Series), which was to be accompanied and complemented by an illustrated series, *Gewu tushuo* 格致圖說 (Science Handbook Series). Conceived as a

96 Fryer, 'An Account of Translation of Foreign Books'.

97 Xiong, *Xixue dongjian*, 328–29.

98 Elman, *Science in China*, 321.

multivolume set of texts that would keep growing, the series was published in 40 volumes from 1882 to 1898.[99]

Another motivation for scientific translation was more missionary in nature. There were various missionary-related printing presses in Shanghai. Although their primary focus was on religious publications, they also printed various scientific texts, with the intention of using science as a medium for conveying religious truth. The tradition of employing science to assist missionary work had been started by the Jesuits in late sixteenth century China, and was inherited by mid-nineteenth century Protestant missionaries in China. As may well be expected, these scientific titles were informed by natural theology. A typical example was the Scottish Missionary Alexander Williamson's *Gezhi tanyuan* 格致探原 (literally, Inquiry into the Origins of Things Based on the Investigation of Them), which was known in English as *Natural Theology and the Method of Divine Government*.[100] Parts of the work had been published as a series of articles in the *Globe Magazine* before being published by the Presbyterian Mission Press in 1876 and used as a textbook in the Tongwenguan and mission schools.[101] Williamson tried to demonstrate the design of the Creator through science and stressed the harmony between science and religion. '[God] has consciously and purposefully embodied His thoughts in the laws and works of nature;' he wrote in another work, 'and therefore in investigating the phenomena around us we are in reality reading an autograph of the Almighty.'[102]

In summary, Shanghai had the largest collection of copies of scientific books for purchase. Kang seized that opportunity. 'I purchased a large number of translated titles,' he wrote, 'and took them home while travelling by boat and cart.'[103] Although we have no idea what those titles were, it was possible that he might have obtained many of the titles mentioned above. He did not stop there, as evidence shows that he would continue to collect books on Western learning. The next year, for example, he purchased the collected issues of the then recently-discontinued *Globe Magazine*. There has been an unsubstantiated estimation that by the Reform of 1898, Kang may have collected over

---

99 Elman, 321, 323 and appendix 3.

100 As its English name indicates, the work was compiled by drawing on William Paley's *Natural Theology*, the Bridgewater Treatises and James McCosh's *Method of the Divine Government*.

101 Wei Lianchen 韋廉臣 [Williamson], *Gewu tanyuan* 格物探原 [*Natural Theology and the Method of Divine Government*], 2nd ed., 6 vols (Shanghai: Christian Literature Society, 1880); see also Elman, *Science in China*, 321.

102 Wei Lianchen 韋廉臣 [Williamson], *Jidu shilu* 基督實錄 [*The Light and Life of the World*] (Shanghai: Meihua shuguan [The Presbyterian Press], 1879), Preface.

103 Kang, 'Woshi', 63. His purchase of these books, in all probability, was funded by his uncles.

3,000 copies of books on Western learning.[104] These books set the stage for Kang to explore another source of knowledge, Western science.

2.3 *Gleaning from Translations*

'Upon my return in the eleventh month [of 1882],' Kang wrote, 'I embarked on a thorough study of Western learning.'[105] According to his autobiography, his reading list for that year covered such subjects as acoustics, optics, chemistry, electricity and mechanics, as well as histories and travelogues. He even had the ambition to compile *Wanguo wenxian tongkao* 萬國文獻通考 (A General Study of Documentary Records of the Countries of the World), in which he said he would include discussions on such subjects as music, prosody, and cartography as well as those mentioned above.[106] For the next five years up to 1888 when he made his debut on the political stage, he continued his reading of translations of Western scientific texts. In his diary of 1887, for example, he recorded that on the twenty-second day of the fifth lunar month, he read the ninth volume of the 1877 collection of Fryer's *Chinese Scientific Magazine*.[107] In particular, the first twelve issues ran a serialized, illustrated *Introduction to Modern Sciences*. The series contained 300 items carefully selected by John Fryer from English science primers, which were organized according to such major fields or topics of modern science as astronomy, gravitational laws of matter and motion, geology, geography, heat, light, electricity, chemistry, botany, and physical anthropology. These introductory texts provided Kang with a good knowledge of the basics in the sciences. Kang considered the magazine to be one of the best works on Western science, and he would later recommend it to his students.

Kang was particularly fascinated by astronomy. As was mentioned earlier, his encounter with modern astronomy had first occurred during his reading of the two geographies of the world. His astronomical knowledge was expanded by reading the *Chinese Scientific Magazine,* which often discussed astronomy. Its first issue, for example, covered such subjects as the vastness and infinity of the universe, the stars, the solar system, the comets and the phenomena of eclipses.[108] These were among the favoured topics often found in Kang's later talks and writings. His reading list on the subject also included the *Tantian* 談天, a translation of John F. Herschel's *Outlines of Astronomy* (1858). Kang

104 Jiang Guangxue 蒋广学, *Liang Qichao pingzhuan* [*A Critical Biography of Liang Qichao*] (Nanjing: Nanjing University Press, 2005), 21.

105 Kang, 'Woshi', 63.

106 Kang, 63.

107 Kang, 68; see also Elman, *A Cultural History*, 124.

108 Fu Lanya 傅蘭雅 [John Fryer], 'Gezhi luelun 格致略論 [Sketches of Science]', *Gezhi huibian* 格致彙編 [*The Chinese Scientific Magazine*] I, no. 1 (1876): 1–6.

tried to apply astronomical knowledge to practical use as well. He recalled that in 1886 he had 'studied astronomy at night', in an attempt to work out 'a natural calendar' in order to solve the confusion caused by simultaneously using the Muslim, the solar and the lunar calendars.[109] On another occasion, he recollected that he had 'used a telescope to observe the heavens day and night' along with his reading of the *Lixiang kaocheng houbian* 歷象考成後編, a Chinese compendium of observational and computational astronomy.[110] Kang found new astronomical knowledge very inspirational and thought-provoking. It motivated him to reflect on 'the realms of the heavens' and 'the realms of the stars', speculating on 'the orbital movements of the heavens' and 'communications with the moon and the stars' in the far-off future.[111] He believed that contemplations on the heavens would help reorient one's views of life and the world. His interest in astronomy would be maintained throughout his life and it would reach a climax in his last years, during which he lectured on astronomy (as will be discussed in Chapter Five).

Reading of Western sciences not only forced Kang to shed his old views, but it also infused him with new insights. This was evidently shown in his study of mathematics. Although mathematics was not part of his Confucian curriculum, Kang made an effort to study it on his own. Euclid's *Elements* would have been among the books he was studying, as he quoted terms from this work in his contemporary writings. His reading list also included mathematical treatises by native scholars and other Western mathematical texts, such as those titles translated by Fryer, many of which he had purchased in Shanghai. But his perceptions of a decline in the power of his memory as a result of a severe headache in 1885 caused him to decide to discontinue that mathematics study.[112] He did not, however, abandon it altogether. A diary entry in mid-1887, for example, recorded that he studied the method of *tianyuan* 天元, the Chinese version of algebra.[113] He was reading from the *Baifutang suanshu* 白芙堂算書, a recent collection of over twenty mathematical treatises by native scholars from the thirteenth century onward. With these readings, he acquired a fair knowledge of mathematics.

Around 1885 when he was contemplating the future world of the Great Unity, he proposed that carefully-selected books should be collected and

109 Kang, 'Woshi', 65.
110 Kang Youwei 康有为, 'Zhutian jiang 诸天讲 [Lectures on the Heavens (1927)]', in *KYWQJ*, vol. 12, 2007, 12.
111 Kang, 'Woshi', 64.
112 Kang, 65.
113 Kang, 68.

compiled, for the universal edification of mankind. For the genre of specialized learning, he suggested such subjects as music, mathematics, chemistry, medicine, astronomy, geography, and physics.[114] From this it can be inferred that he had probably studied books in those fields. It also shows that he had, by that time, begun to accept the modern classification of Western sciences. He further suggested that for every five years, the essentials of each subject should be selected from various books and published as a compilation, so that 'students would not lose direction in their pursuit of those specialized sciences'.[115] That might be an indirect indication that he may have found himself lost in his extensive reading of various translations of Western books. This explains why he stressed the importance of such selected collections of Western books as the *Xixue dacheng* 西學大成 (Great Anthology of Western Learning), an 1888 collection that included many new editions of works originally produced by Fryer. Kang also appreciated scientific experiments, and would try them out whenever possible. For example, he did an experiment on the killing of bacteria with soap water, apparently in relation to the Pasteur's theory of germs. In an entry in his 1887 journal, he recorded: 'I observed things through a microscope and saw many tiny animals and plants [bacteria], which can only be killed by soap water.'[116]

It can be observed that, without formal training in science, Kang basically taught himself by reading translations of Western scientific books. His grasp of Western scientific knowledge, though sometimes understandably shallow and fragmentary, was rather remarkable among his cohort of educated contemporaries in the late nineteenth century. It is safe to conclude that his intensive and extensive study had turned him into one of the most learned Confucian scholars of Western sciences of his time.

3 Enlightenment

By the mid-1880s, Kang achieved an intellectual breakthrough, in which he attempted to integrate all the sources of knowledge he had been pursuing. In early 1884, he read Buddhist texts extensively along with the literature of Brahmanism at two Buddhist monasteries in Guangzhou. He later recalled that he

114 Kang Youwei 康有为, 'Shili gongfa quanshu 实理公法全书 [A Compendium of True Principles and Universal Laws (c. 1885)]', in *KYWQJ*, vol. 1, 2007, 160.

115 Kang, 160.

116 Kang, 'Woshi', 69.

had also written a treatise on Brahmanism,[117] a work which is no longer extant. In the meantime, he continued his reading of translations of Western science. During the autumn and winter of that year, he spent time in semi-seclusion, this time in his family library named the House of Calm and Contentment (*Zhanrulou* 詹如樓). According to his own account penned at a later time, he speculated about the mystery and infinity of time and space, explored deep and abstruse statements in the classics and in other philosophic works, examined the hidden meanings of Confucianism and Buddhism, and compared the tenets of various religions.[118] One of his reading titles was likely to have been the Chinese translation of Alexander Williamson's *Gujiao huican* 古教彙參 (Ancient Religions and Philosophies).[119] Published in 1882, it had soon become the primary work in Chinese that discussed the various doctrines of ancient religions around the world, including those of Babylonia, Assyria and Egypt as well as Brahmanism, Buddhism and Zoroastrianism. As a result of this process of reading and reflection, Kang said later, he had arrived at an understanding of the profound truths and principles in the universe and conceived of himself as a sage with the mission to deliver the world from its sufferings:

> As I realized that the spirit is immortal, I made my principal work the explanation of the phenomenal world; as I realized that coarseness and refinement, purity and filth are all relative, I directed myself to intuitive thought; …I made benevolence the centre of my philosophy, which was to follow the will of Heaven in bringing about unity to the earth, the unity of nations, the unity of races, and the unity of religions.[120]

He was convinced that to bring about unity of religions on the earth was the will of Heaven and a manifestation of the virtue of benevolence. In that context, he saw his mission in clearer terms. 'I was born in this world', he wrote, 'to save the masses of living things. For their sake I would be happy go to purgatory instead of residing in heaven.'[121] In the terminology of Buddhist thought, it could be surmised that he had, by this declaration of intent, stated his desire to serve humanity as a 'bodhisattva'. Thus, he seems to have assigned himself

117 Kang, 81.

118 Kang, 64.

119 Wei Lianchen 韋廉臣 [Williamson], *Gujiao huican* 古教彙參 [*Ancient Religions and Philosophies*], trans. Dong Shutong 董樹堂, 3 vols (Shanghai: Yizhu shuhui, 1882).

120 Kang, 'Woshi', 64; translated in Lo, 'Autobiography', 40–41.

121 Kang, 'Woshi', 64.

the role of a messianic saviour, and privately, he even pictured himself as Confucius reborn for this age.[122]

Kang lost no time in elaborately organizing his insights from those experiences and attempted to develop a philosophy based on them, a common practice, according to William James, for professional mystics at the height of their development.[123] Clearly, his mystical experiences had furnished him with spiritual insights and fortified him with great confidence and fearless tenacity. Now that he believed that he had found the truth (*dao*), he felt contented and fearless even in face of death.[124] Within a couple of years, he produced a number of works in manuscript, most notably the *Kangzi neiwai pian* 康子內外篇 (Interior and Exterior Chapters of Master Kang)[125] and the *Shili gongfa quanshu* 實理公法全書 (A Compendium of True Principles and Universal Laws).[126] The former, portions of which were published in 1899, expressed his general philosophical and religious thought, while the latter, the manuscript of which was not published until the late 1970s, encapsulated his early ideas about a world utopia.[127] An examination of these works would enable us to see how Kang attempted to syncretize Chinese and Western knowledge, to harmonize Western science and the Confucian Classics, and to bring about the unity of religions.

3.1 *Universal Laws*

Nowhere was Kang's syncretism of Chinese and Western systems of knowledge more evident than in the *Compendium*. It is written in a style similar to that of a treatise on geometry, in a manner reminiscent of Euclid's *Elements*. The first six books of *Elements* were translated into Chinese in 1607 by Ricci in collaboration with his Chinese convert, Xu Guangqi. Since then the translation had become arguably the most famous and popular translated Western text, and remained influential in the late Qing during which the remaining nine books

122 Indeed, as is indicated by Kang's corrections on his manuscripts, he had initially considered himself to be 'Confucius' before replacing the word 'Confucius' with 'Sage'. See Mao, *Woshi jianzhu*, 15.

123 James, *Varieties of Religious Experience*, 382.

124 Kang, 'Woshi', 65.

125 Kang, 'Neiwaipian'.

126 Kang, 'Shili gongfa'. Because this work contains historical information that was not available until the early 1890s, some scholars have argued that it could not have been written, as Kang claimed, in the mid-1880s. In my view, it was more likely that the manuscript was written earlier but was revised by Kang at a later time, with pieces of new information inserted.

127 Provisional translations of these two works can be found in San-pao Li, 'K'ang Yu-Wei's Iconoclasm: Interpretation and Translation of His Earliest Writings, 1884–1887' (Ph.D., California, University of California, Davis, 1978).

of *Elements* were also translated. Like Newton who had adopted the structure of *Elements* to codify his new mechanics, or Spinoza who formulated his *Ethics* in the formal apparatus of geometry, Kang also attempted, as he himself pointed out, to use the *Elements* as a template to organize his 'axioms for mankind'. While *Elements* consists of a set of definitions, axioms, theorems, propositions, lemmas, and scholia, Kang's *Compendium* contains definitions, true principles, universal laws, lemmas and commentaries.

Apart from its formal structure, Kang's *Compendium* contains various religious, scientific, Chinese and Western elements. Its aim was to set forth principles and institutions for the use of mankind based on an extensive examination of those principles in various religions and cultures. 'True principles' were defined as those being in agreement with verified scientific truths, time-tested traditions, or geometrical axioms. Once true principles were understood, the universal laws could then be determined. The work consists of ten sections on key aspects of human life, including the relations between husband and wife, parents and children, sovereign and subjects, friend and friend, as well as rites, religion, law, and the art of government.[128]

Now that he had written down the Truth he believed that he had obtained, Kang felt he could die without regret. When he suffered from the above-mentioned severe headache that defied Chinese doctors, for example, he 'calmly' awaited death while checking over his notes and essays.[129] He did not die, however, for he cured himself by taking some Western medicine (it is a pity he did not specify the medicine). That experience further strengthened his confidence in Western science and buttressed his own conviction in his Heaven-endowed mission, something he had been spared to pursue, a mission that remained yet to be accomplished.

3.2 *Unity of* Shangdi (*God*)

Unification of religion, as was mentioned above, was an essential part of Kang's vision of a future world of the Great Unity. In his discussion of religions around the world, he employed the Chinese term *jiao*. In Confucian tradition, *jiao* was understood as cultivation of the way for the development of the Heaven-endowed human nature. Following the traditional *yin-yang* philosophy, Kang classified all religions into two groups:

> I say there are [in reality] only two religions. All those that [teach] the establishment of states and governments; recognize the relationships between ruler and subject, father and son, husband and wife, elder and

128 Kang, 'Shili gongfa'.

129 Kang, 'Woshi', 65.

> younger brothers; authorize the occupations of the scholar, farmer, artisan, and trader; maintain the customary rituals practised by shamans and priests in connection with [the belief in deities and spirits]; and approve of the eating of vegetables, fruits, fish, and meat, [are essentially identical with] Confucianism....On the other hand, there are those that prohibit meat-eating; require celibacy, worshipping their founders by day and by night, and refraining from carrying on the four occupations; reject the four tenets; follow the rules of supernatural beings; and transcend human feelings and desires, are [essentially identical with] Buddhism from which stemmed Christianity, Islam and other [similar] religions.[130]

Yang religions, represented by Confucianism, were more oriented towards meeting human needs, while *yin* religions, as typified by Buddhism, more oriented towards suppressing human passions and sensual desires. Which of these two would win and which would lose? In Kang's view, both kinds of religions were needed, and their alternative approaches and contrary spiritual practices would complement each other and guide mankind in progression to the future world of the Great Unity. His classification of religions into two basic categories was often repeated in his later writings or speeches, sometimes from a different perspective, as we will further discuss in the following chapters. It is worth noting that at this stage, Kang did not categorically assert the superiority of Confucianism, a stance that he would later adopt.

Meanwhile, Kang reflected on the concept of *Shangdi* or 'God'. As discussed earlier, in ancient Chinese literature, there were two main terms – *Shangdi* and *Tian* – that were used to denote the ultimate Reality – *Shangdi* and *Tian*. The former carries the strong sense of a personal God, who lives in heaven but comes into contact with human beings through a system of rewards and punishments. The latter is a more abstract and vague concept that is usually associated with destiny, calamities and blessings. *Shangdi* had been used in ancient times before it was gradually replaced with *Tian*, which then became the dominant term for the abstract, mysterious Being in an increasingly polytheistic Chinese culture. Herbert Giles, Professor of Chinese at Cambridge, whom Kang would meet later, observed that the Chinese use of the two terms could be compared with that of 'God' and 'Jehovah' in the Bible.[131]

As mentioned previously, the early Jesuits had adopted both *Shangdi* and *Tian* for *Deus* before switching to *Tianzhu*. The 'Term Question' also plagued

130 Kang, 'Neiwaipian', 103; translated in Hsiao, *K'ang Yu-Wei*, 165 with minor adaptations.

131 Herbert A. Giles, *Religions of Ancient China* (London: Archibald Constable & Co Ltd, 1905), 17.

later Protestants in nineteenth-century China. There was a protracted debate on whether *Shangdi* or another Chinese term such as *Shen* (神 literally, Deity), should be adopted as the proper name for 'God'. Despite heated discussions on the question among the missionaries in China, Europe and America, especially those that were involved in Chinese translation of the Bible, the exact connotations of these Chinese terms remained controversial. Eventually, a compromise was made by agreeing to disagree: English missionaries were in favour of *Shangdi*, while American missionaries in general preferred *Shen*. The impact of this difference would linger for the following century, as shown in the coexistence of two editions of the Chinese Bible, one using *Shangdi*, while the other *Shen*.[132] The unsolved Term Question was neither a merely linguistic issue nor was it solely a theological matter, for it was intrinsically related to questions on the nature and content of Chinese religion. For this reason, it lay at the heart of the encounter and comparison between Confucianism and Christianity, and in that sense, between Chinese and Western cultures as well.[133]

The debate on the Term Question went beyond the missionary circles and attracted the attention of native Christians as well. During the years 1877–8, for example, the popular and influential *Globe Magazine* had published a series of articles debating on the proper Chinese terms for God by a number of native Protestant converts as well as a couple of missionaries. The debate was initiated by the native pastor Huang Pinsan 黄品三, who, in his article on the 'sacred appellation', argued against the current options of *Shangdi* and *Shen* and proposed an innovative term, *Zaohuazhu* 造化主, literally, the Lord of Creation. The American missionary John Roberts (Lu Pei 陸佩), who disagreed with Huang's proposal and favoured the term *Shen*, followed up by inviting contributions from other native Christians. As a result, nearly 50 articles appeared in the magazine within less than two years, debating the Term Question between Roberts and Chinese Christians and sometimes among the latter themselves. Three years later, Young Allen initiated a prize essay contest, inviting non-Christian Confucian scholars to elaborate on the origins and connotations of *Shangdi*, *Tian* and *Shen*, and in due course two essays were selected as winners and were published in the *Globe Magazine*. Despite these efforts, the Term Question remained unresolved.[134]

---

132 For a summary account of the controversy among the Protestant missionaries over three decades, see S. Wells Williams, *Controversy among the Protestant Missionaries on the Proper Translation of the Words God and Spirit into Chinese* (Andover: Warren F. Draper, 1878).

133 For a recent review of the controversy over the Term Question, see Irene Eber, *The Jewish Bishop and the Chinese Bible: S.I.J. Schereschewsky (1831–1906)* (Leiden: Brill, 2016), 199–233.

134 For a good introduction, see Man-Kong Timothy Wong, 'The Rendering of God in Chinese by the Chinese: Chinese Responses to the Term Question in the Wanguo Gongbao.', in

A passionate reader of the *Globe Magazine*, Kang must have been motivated by the debate over the Term Question to reflect on the issue. With a firm belief in the unitary nature of the concept of *Shangdi*, Kang attempted to find a more suitable designation for it. 'In the world there are many names which fail to reflect realities,' he wrote in his *True Principles and Universal Laws*, 'but because *Shangdi* is the origin of all things, his designations must be carefully chosen.'[135] He listed as many as fourteen existing names from various religions, including the Chinese terms *Shangdi* and *Tian*, Christian terms like Creator, God, Jehovah, Deus and Elohim, Ahura Mazda in Zoroastrianism, and Brahma and Uma in Hinduism. None of them, however, were appropriate in Kang's view. Some of them could be rendered as the Everlasting, but that, Kang argued, suffered from being too vague. If they were to be translated as Heaven (*Tian*), they would suffer from being too shallow. Accordingly, he proposed three terms as proper designations for God, namely *Qihua* 氣化 (Qi-based Transformation), *Yuanzhi* 原質 (Primordial Element) and *Dazhuzai* 大主宰 (Great Sovereign).[136] The first term was a Neo-Confucian concept, through which it was believed the first members of each species of creature, including man, originated, something like spontaneous generation. The second, *Yuanzhi*, was originally a Chinese neologism created to translate the chemical elements. The third was one of the terms that was often employed by missionaries to refer to their God, such as in Williamson's *Natural Theology* in Chinese.[137]

These names, of which Kang favoured the first, represented different attributes of his concept of God. On the one hand, he maintained that although called by different names in different cultures, God was one. It may be worth highlighting here that this concept of the unitary nature of God is in agreement with the teachings of the Abrahamic line of religions, Judaism, Christianity, Islam and the Baha'i Faith. On the other hand, Kang believed that God as the origin of everything existed in constant flux and resided everywhere and within everything. This, however, differs from the Judeo-Christian conception of God as a Creator existing outside his Creation and entirely independent of it. In his later years, as will be shown in Chapter Five, Kang would pick up the subject again, taking great pains to demonstrate that belief in God (however named) was common to every nation and every religions.

Mapping Meanings: The Field of New Learning in Late Qing China, ed. Michael Lackner and Natascha Vittinghoff (Leiden; Boston: BRILL, 2004), 589–614; for a collection of those articles, see Li Chichang 李炽昌, *Shenghao lunheng* 圣号论衡 [*Discussions on the Chinese Translations of the Divine Name*] (Shanghai: Shanghai guji chubanshe, 2008).

135 Kang, 'Shili gongfa', 154.

136 Kang, 154.

137 Wei, *Gewu tanyuan*, Preface.

Towards the end of the 1880s, the always self-assured Kang felt more 'learned' and 'confident' than ever before, with a sense of 'being exalted above mundane things'.[138] His 'enlightenment' had not only helped him integrate the various components of knowledge he had acquired, Confucian, Buddhist, and Western, but it had also reinforced his self-image as a new sage, with a bodhisattva-like mission to deliver the whole of humanity from suffering, but with the priority given to his Chinese compatriots. His immediate ambition was to become a top advisor to the Throne, in which position he would be able to assist the Emperor in reordering and restoring the country in terms of economy, politics, diplomacy and culture.[139] To achieve that, however, he would first have to pass the civil service examinations.

138 Kang, 'Woshi', 31.

139 Kang, 'Neiwaipian', 97.

CHAPTER 2

The Confucian Luther

1888 was a watershed in Kang Youwei's life, the year in which he made his debut on the political stage. In June of that year, Kang, aged 30, left Guangzhou for Beijing to sit for the triennial provincial civil service examination. Six years before, he had travelled to the capital for the same purpose, but had failed the exam. This time, however, he felt much better prepared, as he had not merely consolidated his old (Chinese) learning, but had also studied the new (Western) learning. To his great disappointment, he again failed.[1]

Perhaps to help himself recover from that devastating setback, Kang visited the nearby Western Hills. Looking down from the mountain heights at the capital city, he could not help feeling deeply concerned for the future of China, whose position had further declined since the defeat by the French at the *Majiang* battle in 1884. He was convinced that imminent reforms must be carried out in order to save China from the encroachment of foreign powers. In this he had envisaged a central role for himself, a calling that was motivated by his sense of having been himself the recipient of a Heaven-endowed mission to save China. However, failure in the civil service examinations had hindered and handicapped him. Worse still, his political mentor in the capital, Zhang Yanqiu 張延秋, an imperial censor and academician who had encountered and befriended Kang during a tour to the Xiqiao mountain, had recently died of illness. Kang could no longer count on Zhang's recommendation letters and his network of highly-placed connections in order to become acquainted with the metropolitan officials.[2]

Kang did not give up easily, however, and he made friends with some of the intermediate level officials, such as the Ministry of Punishments secretary Shen Zengzhi 沈曾植, who sympathized with his reform ideas (I will discuss his reform proposals later in this chapter). Kang also approached and sought interviews with high officials, and, in many instances, he was customarily and cursorily ignored. His friend-turned-critic, Liang Dingfen 梁鼎芬, later accused Kang of trying to seek personal wealth and position by writing flattering letters

1 He seems to have even tried to cover this up in his autobiography by confusing it with the next examination he sat. For more details, see Zhang Yong 张勇, 'Kang Youwei de "zuowei" jiqi xiandu 康有为的"作伪"及其限度 [On Kang Youwei's "Forgery" and Its limits]', *Lishi yanjiu* 历史研究 [*History Studies*], no. 6 (2005).

2 Kang, 'Woshi', 72.

 | DOI:10.1163/9789004468221_004

to high officials.[3] That accusation was not entirely justified, for Kang's letters also contained his reform proposals, and even stern terms reminding the high officials of their responsibility. His reproachful tone even angered some of the recipients. The Minister of Personnel Xu Tong 徐桐, for example, had Kang's letter returned the very next day with a rebuke.[4] As a result, Kang's letters created a stir among the metropolitan officialdom, a consternation that was probably due more to their tone than their contents.

1 A Memorial to the Throne

Failing to enlist the support of high officials, Kang conceived an even bolder act, and he decided to petition the Throne directly. That was a phenomenally audacious move, for common scholars did not have the privilege of communicating with the 'Son of Heaven'. By taking this unusual initiative, Kang hoped that his voice could be heard by the young Emperor Guangxu 光緒, who was positioned to take over full power from Empress Dowager Cixi 慈禧 the following year. At the same time, however, he was prepared for the likely consequences, including imprisonment.[5] For his memorial to reach the Throne, Kang had to find a high official to submit it on his behalf. Of those he requested, only the President of the Censorate Qi Shichang 祁世长, whom Kang praised for his courage and integrity, consented to help. On the morning of 10 December 1888 when Kang was about to leave his hotel room to hand in his memorial, he learned that a public execution was being carried out nearby and that traffic was so heavy that his carriage could not traverse the street. He considered the event an ominous sign that he could lose his life as a result of his imminent action. He recalled his hesitation in his autobiography: 'I had an old mother at home – how could I die and leave her?' That thought, however, was immediately pushed out of his mind by the recurring sense of a bodhisattva-like mission: 'If I was born to save the world, how can I withdraw now?'[6] Fortified by a firm belief that one's life and death were predestined by Heaven,

3 Liang Dingfen 梁鼎芬, 'Bo panfan Kang Youwei nishu 駁叛犯康有為逆書 [Refuting the Blasphemous Letter from the Fugitive Kang Youwei]', *Shen Pao* 申報 [*Shanghai News*], no. 9173 (27 October 1898): 3.

4 Zhu Weizheng 朱维铮, *Qiusuo zhenwenming* 求索真文明 [*In Search of True Civilization*] (Shanghai: Guji chubanshe, 1996), 189, 210; Kang, 'Woshi', 72.

5 Kang Youwei 康有为, 'Yu Youbo shu 与幼博书 [A Letter to Youbo (1888)]', in KYWQJ, vol. 1, 2007, 175.

6 Kang, 'Woshi', 72.

he went to the Censorate's office by a circuitous route. The appointment with Qi, however, did not take place due to the latter's sudden illness.

Although his memorial failed to reach the Throne, its contents were made known, for he had copies of it circulated at the same time in the capital and even within his home province.[7] Like his letters to those high officials, it contained a tone of reproach. He took advantage of the recent occurrence of a landslide at the Imperial Mausoleum and interpreted it as a sign of Heaven's warning to the Throne, a warning that he thought showed 'the utmost love from the heart of Heaven'.[8] To show respect to Heaven and return His kindness, the Empress Dowager and the Emperor should, he wrote, in accordance with the ancient tradition, reflect upon their own behaviour, and the ministers should apologize for negligence in the discharge of their duties. Kang deplored the fact that he could not see any evidence that these proprieties were being observed. Instead, construction work continued in the Summer Places and 'exotic techniques and improper technologies' like electric bulbs and model trains were being introduced for the entertainment of the imperial household.[9] Therefore, he urged the Throne to heed Heaven's warning and make efforts for the reformation of China.[10]

Unlike the technology-oriented proponents of the Self-Strengthening Movement, Kang emphasized the importance of political reforms, calling for a change in the outmoded customary ways of governance.[11] In his view, the fact that the introduction of Western-style factories and such technological achievements as electricity lines, mining machines and warships had not strengthened China as had been expected, was largely due to the lack of effective means of communication between the Throne and the populace. He harboured the idea of a parliament similar to the British model, arguing that it had had a precedent in ancient China and could be set up in China, with local officials forming the House of Lords and elected degree-holders the House of Commons.[12] Oblivious to Parliament's legislative power, however, Kang did not seem, at that time, to have grasped the nature of the British parliamentary system. He would revisit the concept of parliamentarism in his future political endeavours.

---

7 Kang, 'Yu Youbo'. He had sent copies of the memorial to his brother at home.

8 Kang Youwei 康有为, 'Shang Qingdi diyishu 上清帝第一书 [My First Memorial to the Throne (1888)]', in *KYWQJ*, vol. 1, 2007, 182.

9 From another perspective, this is a sign of their interest in Western technology.

10 Kang, 'Shang Qingdi diyishu', 182.

11 Kang, 182–84.

12 Kang Youwei 康有为, 'Lun shiwu 论时务 [On Current Affairs (1888)]', in *KYWQJ*, vol. 1, 2007, 165–66.

1.1 *Countering Christian Threats*

In his memorial to the Throne, Kang also stressed the factor of religious threats from outside. He cited the fate of Vietnam as an object lesson for China. After taking control of Vietnam, a former vassal state of Qing, the French had 'gradually won the hearts of the Vietnamese', Kang wrote, 'by setting up churches to lure them to Catholicism as well as by building railways to promote commerce'.[13] The French missionaries had also increased their activities in China's peripheral provinces of Guangxi and Yunnan. According to Kang's estimation (or exaggeration), the majority of the local people in those provinces had been coaxed to convert to Catholicism. Recounting the recent rumours that the Vietnamese, together with the people of Laos and Siam (now Thailand), had accepted a French priest as their General Chief,[14] Kang warned that in a few years the French might lead their Vietnamese followers in invading South China, with Chinese Catholic converts collaborating from inside. Bearing in mind China's recent defeat by the French, Kang predicted that if the French warships could also make their way up the Yangtze river, the Catholics in inner China would also rise and help them.[15] Kang's recognition that Catholicism and Protestantism were Western imperialist tools, and his corresponding distrust of the Chinese converts to a foreign religion, were not surprising, for such views were widely shared among his contemporaries. Those sentiments had been formed and spread since the Opium Wars, and were successively intensified by various religious disputes (*jiaoan* 教案), in which foreign Christians and Chinese converts were often treated more favourably in trials than non-believer locals, since the former group were invested with extraterritorial privileges.

Kang's solution for countering the Christian threat was a vigorous programme of education of the populace. He observed that the main reason for the massive conversion to Catholicism in those peripheral provinces was lack of education. The provincial folk were uneducated and most of them were illiterate. They would often turn to local *wu* (shamans) for help in making decisions and for guidance during significant occasions in their lives. Kang declared that they were illiterate, ignorant, impoverished and superstitious, and these were conditions which provided the French missionaries a good opportunity

13 Kang, 'Shang Qingdi diyishu', 181.

14 Kang was probably referring to the formation of French Indochina in 1887, which consisted of Annam, Tonkin, Cochinchina (which together form modern Vietnam) and Cambodia; Siam would be added in five years' time after the Franco-Siamese War in 1893, and the leased Chinese territory of Guangzhouwan would be added in 1898.

15 Kang, 'Shang Qingdi diyishu', 181.

to convert them.[16] Here Kang seems to have held an unstated impression that Catholicism was similar to or could be treated like a form of *wujiao* or shamanism, which was considered by Chinese Confucianists to be something disdainful and a product of a lack of education.

At that time, the Chinese equivalent of the Western type of religion was a subject of scholarly discussion. A typical view was expressed by Peng Guangyu 彭光譽, Qing's Commissioner to the 1893 World's Columbian Exposition, at the World's Parliament of Religions held in Chicago in connection with the Exposition. He argued that 'religion' in the West had its proper Chinese equivalent in *wu*, the form of religion practised in ancient China long before the appearance of the Confucian school. However, he added that in later times both practitioners and books of *wu* had always been placed under a legal ban to prevent them from corrupting the people, for 'China was never found wanting in men who are versed in supernatural things'.[17] I will revisit the controversy on the Chinese equivalent of 'religion' in later chapters. Suffice it to say here that Kang's implied assertion that education could help get rid of *wujiao* or that sort of 'religion' would become the prevailing idea among intellectuals in twentieth-century China.

Kang observed that in addition to the general populace, many of the officials and students of Confucianism across the country were not properly educated in Confucianism either.[18] He considered this fact to be one of the grievous problems faced by the dynasty. Although he did not elaborate on this particular issue in his memorial, he discussed in detail the ways and means by which to restore Confucian learning in localities across the entire country in his correspondence with his friend Pan Yiqin 潘峄琴, who had just been promoted to Educational Officer of eastern Zhejiang province. One of his proposals was to have the most morally upright and learned among the deceased local Confucian scholars worshipped in shrines and temples so that they could serve as exemplars and provide inspiration for the student population in China. For the funds needed to build the shrines and temples, Kang's solution was daring: to turn some of the Buddhist or Daoist temples, which existed in each locality, into Confucian temples, simply by replacing the old images or statues with the tablets of Confucian scholars. As for those temples known as 'improper

16 Kang, 181; see also Kang Youwei 康有为, 'Yu Pan Wenqin shu 与潘文勤书 [A Letter to Pan Wenqin (1888)]', in *KYWQJ*, vol. 1, 2007, 168.

17 John Barrows, *The World's Parliament of Religions*, vol. 1 (Chicago: The Parliament Publishing, 1893), 376. Peng's address was translated into English under the title of 'Confucianism'; its Chinese original, *Shuojiao* 說教, was published in China in 1896 by an imperial order, demonstrating that it represented the orthodox views on Confucianism and religion.

18 Kang, 'Shang Qingdi diyishu', 182.

shrines' (*yinci* 淫祠), that is, those not sanctioned by the state (and often honouring animal spirits), Kang showed no leniency to them, and believed they should be utterly destroyed.[19] His proposal, if implemented, would have met with great opposition from the general populace, and would have come under strong criticism from the local authorities for inviting disorder. Perhaps that was why Pan did not follow it. Kang remained unsympathetic to the popular forms of worship and considered them as a distraction from, or even a threat to the Confucian faith. In this he may well have been inspired by missionary critiques of Chinese popular beliefs and by his impression that in Europe the Christian faith was practised to the exclusion of other forms of worship. He would revisit the idea in his memorials to the Emperor during the later Reform of 1898.

It should not be concluded, however, that Kang's idea to turn some of the Buddhist temples into Confucian shrines meant that he was opposed to Buddhism. As discussed in the previous chapter, not only had he read Buddhist literature and practised Buddhist meditation himself in his youth, but he had also emphasized the supplementary role of Buddhism to Confucian training. In a letter to the Imperial Academy compiler Huang Shaoji 黄绍箕, whom he had also befriended, Kang observed that the Buddhist practice of self-discipline and self-restraint was conducive to the cultivation of determination and perseverance, which were essential qualities for accomplishing any worthy cause. The Confucian training, on the other hand, was broad in scope yet less strict in self-discipline; in Kang's view, the disciplinary fortitude of Buddhist practice would benefit Confucianism. For this reason, he compared Buddhism to medicine, which one needed to take when ill, while Confucianism was like staple food that was indispensable to one's daily life.[20] He explained that for a Confucian scholar, excessive desire for material benefits were like his internal ailments, while habitual commitments to essay writing and textual criticism could be likened to skin ulcers.[21] Drawing on his own experience, he urged Huang to read Buddhist books and try some Buddhist practice instead of confining himself to Confucianism. Kang recounted that even the great Confucian master Zhu Xi was fond of Buddhism up to the age of 50 and benefited a lot from it. Kang wrote that if he himself had spent more time exercising his mind through Buddhist practices in the mountains, he would have probably

19 Kang Youwei 康有为, 'Yu Yiqin xueshi shu 与峄琴学士书 [A Letter to Pan Yiqin, a Member of the Hanlin (1888)]', in *KYWQJ*, vol. 1, 2007, 190.

20 Kang Youwei 康有为, 'Yu Huang Zhongtao bianxiu shu 与黄仲弢编修书 [A Letter to Huang Zhongtao, the Hanlin Compiler (1888)]', in *KYWQJ*, vol. 1, 2007, 188.

21 Kang, 'Yu Shen Xingbu'.

achieved more by the time they were engaged in correspondence.[22] As we will see later in this book, this view of a complementary relationship between Buddhism and Confucianism would remain a constant element in Kang's religious thought.

Towards the end of 1888, Kang was frustrated. He had achieved a measure of fame because of his outspoken letters to high officials and his failed memorial to the Throne, but he had also received many criticisms from metropolitan scholar-officials. Meanwhile, the situation in the capital became such that it was no longer considered appropriate for him to make critical comments on state affairs, as the imperial Court was busy preparing for the wedding of the young Emperor. To celebrate the imperial event, it was announced that an extra provincial examination was to be held during the following autumn. Kang decided to remain in the capital in order to have another go at it. To amuse himself, Kang started to study stele inscriptions collected in the capital city. An accomplished calligrapher himself, he soon turned a pastime into a serious scholarly pursuit. Studying stylistic changes in the development of Chinese calligraphy energized his mind for political reform. In one year's time, Kang was able to produce a treatise on the history of calligraphy, *Guang yizhou shuangji* 廣藝舟雙楫 (Extended Paired Oars for the Boat of Art), which would immediately become a best seller following its publication in 1891.[23] What is pertinent to this book is that his study on ancient inscriptions helped him revisit the Old Text versus New Text controversy (which will be discussed later in this chapter).

In Autumn 1889, Kang sat the extra examinations but repeated the same failure as at his previous attempts. Beyond disappointment, he was furious. He accused the chief examiner, Xu Tong (who had been offended by the reproachful tone of his letter a year before) of failing him after presumably recognizing his essay by its style.[24] This failure had deprived him of the opportunity, at least for the next three years, to become an official, which in turn meant he would not be able to help save China in the role of an imperial official. Kang then began to consider what profession he should pursue. He reckoned that he could not become a scientist, for he did not possess the funds to purchase scientific instruments which he considered indispensable to pursuing science; neither did he have a community of fellow scientists with whom to collaborate. The only viable option for him was to become a teacher and educator,

22 Kang, 'Yu Huang Zhongtao'.

23 See Kang, 'Woshi', 73. For a recent study of Kang's calligraphic art in the English language, see Wong, *The Other Kang Youwei*.

24 See Kang, 'Woshi', 73.

for which he believed he was well-prepared by virtue of his integration, in the mid-1880s, of various branches of knowledge, religious and scientific, Chinese and Western.[25]

2 An Unconventional Teacher

Returning to his home province, Guangdong, he was warmly received as a celebrity scholar. Although he had failed the civil service examination, he had achieved fame for his bold attempt to send a memorial to the Throne as well as for his outspoken call for reforms. After he moved his residence to Guangzhou in 1890, some young local scholars came to seek his tutelage, among whom was Liang Qichao, who would become his most renowned disciple and eventually aid him in his reform cause.

Like Kang, Liang had received a good Confucian education during his childhood at home. In 1887 he entered the prestigious *Xuehaitang* 學海堂 (Academy of the Sea of Learning) in Guangzhou, where he, as a distinguished student, devoted himself to the philological study of the classics and refined styles of writing. At sixteen (in 1889), he sat and passed the provincial examinations with high honours and had thereby received a provincial degree (*juren*).[26] That, however, marked the summit of his success at the civil service examinations, for he was never to pass the metropolitan examinations, despite his several later attempts.

The year 1890 was a watershed in Liang's intellectual development. His eyes were opened up and his intellectual horizons broadened by his contact with Kang. Their first meeting lasted a whole day, during which Liang was as much awed by Kang's insights and learning as he was depressed by the discovery of the 'uselessness' of his own previous learning.[27] Liang thereupon decided to leave his old school and come to study under Kang's tutelage. For a provincial graduate to become a student of someone like Kang, who had not yet succeeded in earning the same degree, was extraordinary, and Liang's 'conversion'

25 Kang, 'Yu Shen Xingbu'. It appears that he did not reveal his new doctrines to his associates until the end of his sojourn in the capital, when he shared a copy of his *Interior and Exterior Chapters* with his friend Shen Zengzhi.

26 Ding Wenjiang 丁文江 and Zhao Fengtian 赵丰田, eds., *Liang Qichao nianpu changbian* 梁启超年谱长编 [*A Chronological Biography of Liang Qichao*] (Shanghai: Shanghai renmin chubanshe, 1983), 1–22.

27 Liang Qichao 梁启超, 'Sanshi zishu 三十自述 [An Account of Myself at 30 (1902)]', in *LQCQJ*, vol. 1, 1999, 958.

created quite a stir among local scholars. This episode illustrates Kang's charismatic influence over the minds of the young scholars of his time.

As the number of his students increased, Kang decided to set up a private academy in Changxing Lane 長興里 in Guangzhou in 1891. He later moved the academy to a larger residence and renamed it as the *Wanmu caotang* 万木草堂 (Thatched Hall among Ten Thousand Trees). The academy thrived, and by the mid-1890s the number of its students had exceeded a hundred.[28] Kang's curriculum was reminiscent of his teacher Zhu Ciqi's in its eclectic approach, but it differed in a very important aspect: it included unconventional subjects such as mathematics and Western learning. According to Liang's summary, Kang's teaching programme was founded on Confucianism and Buddhism and was complemented by history and Western learning.[29] As a generic or protean term, Western learning at that time included knowledge of the Western political system, science and technology.

2.1 *Western Sciences*

The negative attitude of the literati towards Western learning (see Chapter One) remained largely unchanged up to the early 1890s. As a result, modern science was hardly ever included among the curriculum of regular Chinese academies, nor was Western learning given due attention in their lectures. Being an avid student and beneficiary of Western science, Kang found it natural and necessary to include the pursuit of scientific knowledge as a theme in his curriculum. He followed a progressive approach by starting with an emphasis on the ancient concept of *liuyi* 六藝 or six arts, including rituals (*li* 禮), music (*yue* 樂), archery (*she* 射), charioteering (*yu* 御), writing (*shu* 書) and arithmetic (*shu* 數). Kang advocated the restoration of those long-forgotten arts. As archery and charioteering were no longer useful in today's world, he suggested that they be replaced with cartography (*tu* 圖) and shooting (*qiang* 槍).[30] The message Kang was trying to convey was clear: Confucian students should follow the ancients in paying due attention to practical knowledge and useful skills, and to do so was both legitimate and justified.

Then, in another published treatise on the ways of acquiring knowledge, Kang urged young Confucianists to make efforts to acquire Western sciences by reading translations of scientific texts. The titles he recommended covered

---

28 Kang Youwei 康有为, 'Kang Nanhai xiansheng jiangxue ji 康南海先生讲学记 [Notes of Mr. Kang's Lectures (1896)]', in *KYWQJ*, vol. 2, 2007, 105.

29 Liang, 'Nanhai Kang', 483.

30 Kang Youwei 康有为, 'Changxi xueji 长兴学记 [The Curriculum for Changxing Academy (1891)]', in *KYWQJ*, vol. 1, 2007, 345–46.

such fields as astronomy, geography, botany, zoology, optics, acoustics, electricity, mechanics, chemistry, most of which had been included in the 1888 collection of *Great Anthology of Western Learning* (*Xixue dacheng*). He stressed that it would be ideal if one could purchase the translations by the Jiangnan Arsenal Translation Department 'in their entirety'.[31] He also listed an earlier title on anatomy, Benjamin Hobson's *Quanti xinlun* 全體新論 (Human Anatomy, 1851). It was no surprise that Kang singled out his favourite, the collected issues of John Fryer's *Chinese Scientific Magazine*, which he still considered as the best reading for its comprehensive coverage of nearly every natural science. As for mathematics, Kang recognized its study as being of fundamental significance to the pursuit of other subjects such as astronomy and geometry. He suggested that learners should start with Euclid's *Elements* before embarking on subjects such as algebra and calculus.[32] The way he discussed these titles indicates that he had read them himself and obtained a fair knowledge of each of the subjects.

Although he did not give lectures on any particular science, Kang often incorporated scientific knowledge in his lectures on general topics. He established a library at the academy with the books he had collected over the past decade, and encouraged students to browse through them and to turn to him for instruction when necessary.[33] Appointed by Kang as assistant lecturer, Liang had designed a monthly reading schedule for the younger students. Like Kang, Liang considered acquisition of Western learning as a must for modern scholars. After obtaining a fair knowledge of the history of Western countries, their current state, and the source of their power and wealth, Liang wrote, that one should proceed to read primers on natural sciences. For this purpose, he recommended the multi-volume *Science Outline Series* (*Gezhi xuzhi*), *Outlines of Astronomy* (*Tantian*), and the *Dixue qianshi* 地學淺釋, an 1871 translation of the Scottish geologist Charles Lyell's *Elements of Geology* (1865) which became a geological textbook in the early 1870s in China.[34]

2.2 *Chinese Origins*

While many scientific books translated or written by missionaries were informed by a natural theology, which used scientific knowledge to glorify the Christian God, Kang considered Western science as a field separate

31 Kang, 'Guixue wenda', 23.
32 Kang, 23.
33 Liang, 'Sanshi zishu', 958.
34 Liang Qichao 梁启超, 'Dushu fenyue kecheng 读书分月课程 [A Monthly Reading Schedule (1892)]', in *LQCQJ*, vol. 1, 1999, 8.

and distinct from Christianity, as knowledge with a universal character. This understanding resonated with the Confucian concept that everything has an ultimate organising principle and this principle can be ascertained through investigation of it (*gezhi*). On this basis, Kang upheld the view that Western sciences had their origins in Chinese traditions (*xixue zhongyuan*). This was not a new idea, though. Since the Jesuit mission to China, the notion – whose weak version implied that modern science had been anticipated by Chinese classical scholarship – had been developed among the Chinese literati, including the late Emperor Kangxi, who himself had studied Western mathematics under the tutelage of missionaries.[35] The view was also shared and promoted by some of the missionaries in China. William A. P. Martin, for example, argued for the notion of 'Oriental origins of Occidental sciences' in his Chinese review of Western learning.[36] Martin further asserted that ancient Chinese arts must have been spread to Greece through the intermediaries of Egypt, Persia and India.[37]

Kang elaborated on Martin's claim: all of the branches of learning had once been highly developed in the times of the legendary ruler Huangdi 黄帝 about four thousand years before, but they had not been passed down to posterity in China. Their remnants were somehow spread to the foreign lands and were picked up and further developed by the Westerners.[38] For Kang, it was easy to find supporting evidence in mathematics, astronomy and geography, which were ancient and traditional fields of learning in China. Echoing the view of Emperor Kangxi, Kang claimed that the modern methods of algebra and calculus had originally been introduced from the East (India), and cautioned his students not to disregard them as the 'heterodox learning from the Westerners'.[39] The idea of a spherical earth, for another example, could be found in the *Zhoubi suanjing* 周髀算经, one of the oldest Chinese mathematical texts.[40] When it came to the modern exact sciences, Kang argued that their principles could also be seen in Chinese classics. 'In the times of Mozi 墨子 (*circa* 479–81 BCE),' he wrote, 'the principles of optics and mechanics were already known.'[41]

35 See Elman, *Science in China*, 155–56; Han, *Tongtian zhixue*, 106–26.

36 Ding Weiliang 丁韪良 [William A. P. Martin], *Xixue kaolue 西學考略* [*A Brief Review of Western Learning*] (Beijing: Zongli yamen, 1883), II, 55a.

37 Ding, I, 52a.

38 Kang Youwei 康有为, 'Mingong pian 民功篇 [On Civil Feats (1886)]', in *KYWQJ*, vol. 5, 2007, 75.

39 Kang, 'Changxing xueji', 246.

40 Kang, 'Neiwaipian', 106.

41 Kang Youwei 康有为, 'Yu Hong Youcheng jijian lun zhongxi yixue shu 与洪右臣给谏论中西异学书 [A Discussion with Hong Youchen, the Supervising Censor, on the Differences between Chinese and Western Learning (1891)]', in *KYWQJ*, vol. 1, 2007, 337.

In addition, Kang claimed that the Chinese had made ingenious inventions long before the Westerners created the same objects, inventions such as Zhang Heng's water-powered armillary sphere and Zu Chongzhi's 祖冲之 (429–500) south-pointing chariot. This demonstrated that the ancient Chinese, Kang wrote, were actually more intelligent than the Westerners. The fact that natural sciences had been developed into finer and higher stages in the modern West rather than in China, was not because of difference in intellect between Westerners and Chinese. Rather, it was due to difference in attitudes and practices. 'In the West,' Kang wrote, 'programmes have been designed to teach natural sciences at schools, both the sovereign and the ministers sponsored the sciences, and scholars jointly participate in the pursuit of the sciences.' In China, Kang observed, the contrary was true, since the sciences were not valued, sponsored or pursued in a serious manner. Consequently, China had fallen behind in her material and technological development. But if China were to make reforms to systematically pursue scientific learning in the Western manner, Kang reasoned, it would be merely a matter of time before China would catch up with the West.[42]

Kang's adoption of the notion of the Chinese origins of Western sciences would, he hoped, serve a threefold purpose. First, it would help offset the prevalent resistance to the teaching and learning of Western sciences. Second, it would ease the resistance to the proposed reforms: for if to study Western sciences was in fact to recover something that was originally and rightfully Chinese, to reform society according to the Western model was also to recover a lost heritage in China. Thirdly, it would counter the alleged Western mental superiority, as the gap in scientific development between the West and China did not imply any intrinsic deficiency or inferiority in Chinese intellectual capacities.

2.3 *Cosmic Evolution*

Cosmic evolution was another area for which Kang claimed a Chinese historical priority, and by which he had been fascinated since his youthhood. Kang recollected in 1904, after a visit to the Museum of Natural History in London where his eyes fell upon the statues of Darwin and Huxley, who he said 'were like old friends' to him, that the evolution theories of the two were regarded as 'new inventions' in the land of Europe, but he had already developed similar, though less elaborate, ideas before he had ever read 'a word of text in any Western book'; for the idea of biological evolution, he asserted, had been enshrined in ancient Chinese thought and in the Chinese classics.[43]

---

42 Kang, 337.

43 Kang Youwei 康有为, 'Yingguo youji 英国游记 [An Account of My Travels in Great Britain (1904)]', in *KYWQJ*, vol. 8, 2007, 23.

The Darwinian theory of evolution, along with its social implications, was formally introduced into China in the late 1890s through Yan Fu's 嚴復 paraphrased translation, known as *Tianyan lun* 天演論 (On Cosmic Evolution), of Thomas Huxley's *Evolution and Ethics,* in which the translator gave a fair account of Darwin and his theory in the form of notes. In 1896/7, when Kang read Yan's manuscript translation, which was already circulating before its appearance in newspaper instalments and eventual publication in 1898,[44] Kang was deeply impressed, but he appeared to be more confirmed than inspired, as he had developed similar ideas and had discoursed on them in his lectures. According to Liang Qichao, who was among the earliest readers of Yan's manuscript, he and his fellow students had learned about similar ideas on evolution from Kang, though these ideas had not been explored by Kang to the same depth as Darwin had. 'Don't think,' Kang cautioned, 'that these are original principles, for there might be many Westerners who have studied them for many years.'[45]

According to his autobiographical remarks, Kang seriously reflected on the 'evolution of Heaven and Earth' and the 'origin of living things' at the age of 26, that is, in 1884 when he renewed his basic worldviews.[46] By that time, European evolutionary notions, including fragmentary information of Darwin and his new theory, had already been introduced into China through the 1871 Chinese translation of Lyell's *Elements of Geology* as well as John Fryer's 1877 essay 'On Chaos' published in his *Chinese Scientific and Industrial Magazine*.[47] As an avid reader of translations of Western texts and an admirer of Fryer's magazine, he could not have missed those European notions on evolution, traces of which can be spotted in his early manuscript writings in the mid-1880s, in particular, in his treatise on 'civil feats' (*Mingong pian* 民功篇). Drawing on the Confucian concept of 'change' and his fragmentary knowledge of Western geology and astronomy, Kang developed a comprehensive conception of evolution that embraces the successive development of the physical universe, the biological organisms and then the human world. After the physical world came

44 Hexuli 赫胥黎 [Huxley], *Tianyan lun* 天演論 [*On Cosmic Evolution* (*Evolution and Ethics*)], trans. Yan Fu 嚴復, 2 vols (Gaiyang: Shenshijizhai 慎始基斋, 1898).

45 Liang Qichao 梁启超, 'Yu Yan Youling shu 与严又陵书 [A Letter to Yan Fu (1897)]', in *LQCQJ*, vol. 1, 1999, 73.

46 Kang, 'Woshi', 64.

47 Leixiaer 雷侠兒 [Lyell], *Dixue qianshi* 地學淺釋 [*Elements of Geology*], trans. Magaowen 瑪高溫 [MacGowan] and Hua Hengfang 華衡芳, new ed., vol. 13 (Shanghai: Kiangnan Arsenal, 1873), 3–16; Fu Lanya 傅蘭雅 [John Fryer], 'Hundun shuo 混沌說 [On Chaos]', *Gezhi huibian* 格致彙編 [*The Chinese Scientific Magazine*], no. 7 (1877): 6–8. See also Jin, 'The Evolution of Evolutionism in China, 1870–1930', 49–50.

into being by virtue of the primal substance *qi*, according to Kang, biological organisms appeared and eventually gave rise to animals. From animals humans evolved and are still evolving today, with different races at different stages.[48]

About six years after his contemplations on 'evolution' and 'creation', Kang confided to one of his most talented students, Chen Qianqiu 陳千秋, that 'humans could produce the horse and vice versa', and that 'man had evolved from apes', which Chen believed and for which he was said to have found supporting evidence.[49] We do not know what the evidence was, as Kang did not elaborate on it, and Chen died young leaving few written works of his own. The idea of humans producing the horse or vice versa can be found in several pieces of ancient Chinese literature, such as, the Daoist text *Liezi* 列子, whose first chapter dwells on the processes of evolution or devolution between species.

> All species develop in their own ways: …When withered and decayed, plantains turn back into dirty soil and become dark-footed grass whose roots in turn are transformed into bombyx and whose leaves into butterflies. … The insects in turn produce leopards. Leopards produce horses. Horses produce humans who in the end return to the cycle of change in the great Nature. [1.4][50]

The other idea that 'man had evolved from apes', however, can hardly be found in ancient Chinese texts, and may therefore have come from a European source, such as Fryer's 'On Chaos', in which the idea of 'apes transforming into humans' was listed as one of rival theories on the origin of man in late-nineteenth century Europe.[51] By juxtaposing the two sources, but with the Chinese followed by the European, Kang may have implied that the Chinese could claim historical priority on the discovery of the concept of the animal ancestry of man.

To conclude this section, Kang's advocacy of the teaching and learning of Western sciences was not limited to his labours in his academy or within his region. His vision was to have them included in the state examination and education systems. He supported the establishment of the special examinations on political, economic, military, and natural sciences (*jingjiteke* 經濟特科), for which he was recommended as a candidate in 1898. In his memorials to the Throne, he suggested that the practice of the special examinations could be

---

48 Kang, 'Mingong pian', 65–79.

49 Kang, 'Woshi', 81.

50 Liezi, *Liezi* [*in Chinese-English Version*], trans. Liang Xiaopeng (Beijing: Zhonghua shuju, 2005), 7–8.

51 Fu, 'Hundun shuo', 6.

emulated in the reform of the civil service examinations, arguing that the rigid, book-based 'eight-legged' essays should be replaced by policy papers (*celun* 策論) on specialized subjects, including astronomy, geography, chemistry, optics, electricity, mechanics, cartography, arithmetic, mining and law as well as the Confucian Classics. He further proposed that the natural sciences should be included in the school curricula across the whole county, so that young students would spend their precious time and energy pursuing those useful subjects instead of confining themselves to empty talk and the Four Books.[52] As will be seen in the next section, Kang's proposals were taken seriously by the Emperor, who issued an edict on reforming the civil service examinations during the Reform of 1898.

3 Recasting Confucianism

What was most unusual about Kang's teaching, however, was that rather than toeing the line of orthodox Neo-Confucianism, it was based on a radical revision of Confucianism, which he had only recently developed, drawing on the New Text exegesis. He had cherished the desire to become an independent sage or a founder of a religion (*jiaozhu* 教主),[53] but now he may have come to see it as being more realistic if he were to bow to the authority of Confucius and become, in the words of Liang Qichao, 'the Martin Luther of Confucianism',[54] by restoring Confucianism to what he believed was its original sacred and pristine state.

To better understand Kang's intellectual reorientation away from the Old Text school, which had underpinned orthodox Confucianism, to the New Text school, which had once been the orthodox ideology of the Former Han dynasty (206 BCE–25 CE), an account of the origin and history of New Text Confucianism is required here. According to the traditional account, the Confucian Classics had been destroyed in the Burning of Books and Burying of

52 Kang Youwei 康有为, 'Qingfei bagushitie kaifa shishi gaiyong celun zhe 请废八股试贴楷法试士改用策论折 [A Memorial to the Throne Requesting the Replacement of Eight-Legged Essays with Policy Papers in the Civil Service Examinations (1898)]', in *KYWQJ*, vol. 4, 2007, 78–79; Kang Youwei 康有为, 'Qingzhao jingjiteke li tuixing shengtong suikeshi pian 请照经济特科例推行生童岁科试片 [A Request for the Application of the Practices of the Special Examinations to the Civil Service Examinations at Various Levels (1898)]', in *KYWQJ*, vol. 4, 2007, 50–51.

53 Kang Youwei 康有为, 'Canyiyuan tiyi liguo zhi jingshen yishu hou 参议院提议立国之精神议书后 [On the Senate's Proposal to Establish the Moral Spirit to Sustain China (1914)]', in *KYWQJ*, vol. 10, 2007, 206.

54 Liang, 'Nanhai Kang', 486.

Scholars (*fenshu kengru* 焚書坑儒) during the Qin dynasty (221–206 BCE). In early Former Han, surviving Confucians wrote down from memory the Five Classics, and for their instruction and dissemination, erudite scholars (*boshi* 博士) were appointed in the Imperial Academy under state sponsorship. As these texts were recorded in contemporary-style script (*jinwen* 今文), they were later dubbed as 'New Texts'. Then, New Text Confucianism was exalted as the officially sanctioned doctrine, with the promotion of Dong Zhongshu 董仲舒 (179–104 BCE), the leading Confucian scholar of the period. Of the Classics, the *Annals* by Confucius as interpreted in the *Gongyang Commentary* (*Gongyangzhuan* 公羊傳) came to be regarded as the most valued text; this portrayed Confucius as a charismatic prophet or 'Uncrowned King' (*Suwang* 素王) who was accorded the Mandate of Heaven but did not attain kingship. The commentary provided not only the name but also key doctrines to the Gongyang school, which was regarded as the heart of New Text Confucianism.

Around the same time, however, alternative versions of the Confucian Classics, recorded in 'old' or pre-Han script (*guwen* 古文, hence 'Old Text'), were allegedly discovered at places such as Confucius' old residence. Among them was the *Zuozhuan* 左傳, another commentary on the *Annals*. In contrast to the *Gongyang*, the *Zuo Commentary* portrayed Confucius as a respected teacher and a mere transmitter of classical learning, a learning that was believed to have been passed down from the ancient sagely rulers, in particular the Duke of Zhou 周公 of the eleventh century BCE. Subsequently, some scholars claimed superior authenticity for Old Text Classics over the New Text version on the ground that the former were written in more ancient script. Liu Xin 刘歆 (*circa* 50 BCE–23 CE), the Imperial Librarian, vigorously called for the incorporation of Old Text Classics into the officially-sanctioned texts. Despite opposition from New Text scholars, Liu's proposal received the support of his patron, Wang Mang 王莽 (*circa* 45 BCE–23 CE), the powerful regent who eventually seized the throne and established the transitory Xin dynasty (9–23 CE). Apart from the relative authenticity of their respective versions, the two schools differed ideologically: New Text Confucianism endorsed various superstitious and supernatural beliefs as advocated in the contemporary *Apocryphal Texts* (*Weishu* 緯書), while the more rationalistic Old Text Confucianism tended to shun those tendencies, a difference that was most evidently manifested in their portrayals of Confucius. In the following Later Han dynasty (25–220 CE), the competition between the New Text and Old Text schools continued, but eventually the latter got the upper hand and the former fell into obscurity. As a result, the Confucian orthodoxies in later dynasties, including Neo-Confucianism in the Ming and Qing periods, were primarily based on Old Text sources. It should be noted, however, that over a period of time the two versions of texts became more or less mixed up during the ensuing discourse and the distinctions between them were no longer clear.[55]

New Text Confucianism, on the other hand, had basically been forgotten for fourteen hundred years until the mid-eighteenth century, when the Qing scholars reopened the New–Old Text controversy. The New Text advocates, including the previously mentioned Zhuang Cunyu, claimed that some of the Old Text sources might well have been fabricated during the Xin dynasty of Wang Mang. In their search for authenticity, they revisited the *Gongyang Commentary* and revived the Gongyang school. One of the leading New Text scholars of Kang's time was Liao Ping 廖平 (1852–1932), who pinpointed Liu Xin as the forger of Old Text Classics and upheld Confucius as a charismatic visionary and institutional reformer.

On his way back home after his unsuccessful candidature in Beijing, Kang met and conversed with Liao in Guangzhou in early 1890. It was widely believed that Liao was the inspiration behind Kang's conversion to New Text Confucianism. Liang Qichao, for example, stressed the great contribution Kang had made to the New Text scholarship but at the same time admitted the 'undeniable influence' of Liao on Kang's thinking.[56] Kang, however, never acknowledged his debt to Liao; instead, he claimed that he had discovered the truth of the New Text version independently in 1888/1889 (that is, before he had met and conversed with Liao), when he was working on stele inscriptions and reviewing the Confucian Classics.[57] According to his account, he had already considered the Gongyang scholarship even before that. As early as in 1880, he had studied the *Gongyang Commentary* and the writings of earlier Gongyang scholars. Initially, he did not seem to have been convinced by the Gongyang doctrine. He even wrote a treatise, *Heshi jiumiu* 何氏糾謬 (He's Errors Exposed) to refute He Xiu (何休 129–83), the leading Gongyang scholar in Later Han, who took an interest in supernaturalism and was inclined to deify Confucius, assigning to him a messianic role. However, Kang said he soon came to realize that he was wrong and consequently had burned his critique.[58] The incident suggests that Kang began to see some validity in the Gongyang tradition. Following his 'enlightenment' in the mid-1880s, although he remained in the camp of orthodox Confucianism, he expressed such views as would lend support to the New

55 For a historical treatment of the Old–New Text controversy, see Fung, *Chinese Philosophy*, 1983, 1:17–18, 65–66; Fung, *Chinese Philosophy*, 1983, 2:133–36; see also Liang Qichao 梁启超, 'Qingdai xueshu gailun 清代学术概论 [Intellectual Trends in the Qing Period (1920)]', in *LQCQJ*, vol. 10, 1999, 3095.

56 Liang, 'Qingdai xueshu', 3097; see also Liang, *Intellectual Trends*, 92.

57 Kang, 'Woshi', 73.

58 Kang, 63.

Text school in his *Jiaoxue tongyi* 教學通義 (The General Import of Teaching and Learning).[59]

3.1 *Restoring the Religion of Confucius*

Kang's recasting of Confucianism started with his *Forged Classics of the Xin Dynasty* (*Xinxue weijing kao* 新學偽經考, 1891), a work that was described by Liang Qichao as a 'hurricane' that hit the Chinese intellectual world.[60] In this work, Kang upheld New Text Confucianism and attacked the Old Text Classics in the course of reinterpreting Confucianism in a radical fashion. Kang made a number of bold assertions. First, contrary to the traditional account, the Confucian Classics had not been totally destroyed in the book-burning in the Qin dynasty, for, he argued, those copies kept in the imperial library had survived. It followed that New Text Classics as transmitted in the Han period were authentic and complete texts. Second, the Old Texts had, in their entirety, been forged by Liu Xin, who had also tried to cover up the traces of his forgery by interpolating glosses into ancient works. The reason for his forgery, according to Kang, was to help Wang Mang usurp the Han throne and legitimize the Xin dynasty. Among other things, Liu had compiled two influential works: one was the *Zhouli* 周禮 (Official System of Zhou) under the name of the Duke of Zhou, making the Duke a rival to Confucius; the other was the *Zuo Commentary*, whose fact-finding statements went against the mystical, religious interpretations of the *Gongyang Commentary*. Liu's spurious works, Kang claimed, became widely diffused and accepted through the influential commentator Zheng Xuan 鄭玄 (678–751), whose popular edition and annotation of the Confucian Classics was based primarily on the Old Texts. As a result, many books of the New Texts became mixed up with the Old Text version and it remained difficult to distinguish the true classics from the corrupted texts. Third, the extant Five Classics had been authored by Confucius himself, rather than merely edited by him as had been traditionally supposed.[61] The *Annals*, in particular,

59 See Kang Youwei 康有为, 'Jiaoxue tongyi 教学通义 [The General Import of Teaching and Learning (1885)]', in *KYWQJ*, vol. 1, 2007, 17–58. For a good discussion on the Gongyang influence on Kang's early thought as expressed in the work, see Liu Wei 刘巍, 'Jiaoxue tongyi yu Kang Youwei de zaoqi jingxue luxiang jiqi zhuanbian 教学通义与康有为的早期经学路向及其转变 [The Jiaoxue tongyi and Kang's Approach to Classical Studies and Its Ensuing Change]', 历史研究 [*Historical Research*], no. 4 (2005): 190–49.

60 Kang Youwei 康有为, 'Xinxue weijing kao 新学伪经考 [An Investigation into the Forged Classics of the Xin Dynasty (1891)]', in *KYWQJ*, vol. 1, 2007, 353–558; Liang, 'Qingdai xueshu', 3097.

61 In his later writings, he would take a more lenient view on Confucius' authorship of all the Classics, acknowledging that Confucius 'edited' the *Documents* and 'trimmed' the *Odes*.

enshrined 'great principles hidden in esoteric language' (*weiyan dayi* 微言大義), through which Confucius had advanced the ideas of *tuogu gaizhi* 托古改制 or institutional changes proposed on the basis of false attribution to earlier sage kings.[62]

Having made these claims, Kang attempted to substantiate them by painstakingly going through various ancient works and selecting an array of evidence in his favour. Whether Kang's claims were entirely correct or not is beyond the scope of this book,[63] but his reinterpretation shook the foundation of imperial Confucianism, for its very sources, the Old Text Classics, had now been called into question. No doubt this new interpretation would pave the way for his reform proposals, but at the same time, Kang was, as might be expected, accused of distorting the orthodox teachings by the mainstream Confucianists, causing such a furore that copies of *Forged Classics* were ordered to be destroyed by an imperial order. Predictably, the critiques and the issuance of the imperial order only served to bolster its fame and popularity.

Kang sought to defend the station of Confucius upheld in the New Text school: Confucius was not a mere transmitter of ancient teachings as had been traditionally regarded, but an originator of new teachings and hence, conceivably, a founder of a religion (*jiaozhu*). In the *Apocryphal Texts* popular in the Han period, such as the *Chunqiu wei* 春秋緯 (Apocrypha on the *Annals*), Confucius was presented as a supernatural being. According to one account, his mother was once taking a walk on a hill when she fell asleep. In her dream, she was invited to the 'Black God' (*Heidi* 黑帝) and had intercourse with him. When she awoke she felt pregnant and eventually gave birth to Confucius within a hollow mulberry tree. For that reason Confucius was called the Black Sage (*Xuansheng* 玄聖). In other accounts, the capture of a unicorn (*huolin* 獲麟, allegedly in 481) was interpreted as an omen that Heaven had conferred its Mandate on Confucius to found a new dynasty, and a small red bird was believed to have brought an inscription from Heaven, announcing that Confucius had created institutions and laws following the Mandate of Heaven.[64]

See, for example, Kang Youwei 康有为, 'Zhongyong zhu 中庸注 [The Doctrine of the Mean Annotated (1901)]', in *KYWQJ*, vol. 5, 2007, 389.

62 As Liang Qichao stressed, by institutional change Kang meant a kind of political and social reform. Liang, 'Qingdai xueshu', 3098.

63 For a thorough critique of Kang's claims, see Qian Mu 钱穆, 'Liu Xiang Xin fuzi nianpu 刘向歆父子年谱 [Chronicles of Liu Xiang and Liu Xin, Father and Son]', *Yenching Journal*, no. 7 (1930).

64 See Fung, *Chinese Philosophy*, 1983, 2:129–30.

Kang did not hesitate to follow these *Apocryphal Texts* in deifying Confucius. In a following work, *A Treatise on Confucius as a Reformer* (*Kongzi gaizhi kao* 孔子改制考, 1897), he wrote:

> Heaven, having pity for the many afflictions suffered by men who live on this great earth, (caused) the Black God to send down his semen so as to create a being who would rescue the people from their troubles – a being of divine intelligence, who would be a sage-king, a teacher for all ages, a bulwark for all men, and a religious leader [*jiaozhu*] for the whole world.[65]

By virtue of his miraculous conception, Divine authority was bestowed upon Confucius, and Kang portrayed him as a Heaven-ordained teacher for all ages. Liang Qichao later observed that because Kang had 'mistakenly' considered Christian worship as the basis of good government and state power in Europe, he had attempted to exalt Confucius to the rank of Christ as a founder of a religion.[66] There is no doubt that the role of Christ served as a reference for Kang in his religious endeavour, but considering his early religious experience, Kang almost certainly had a personal belief in the 'divinity' of Confucius. Kang stressed that, although divinely-originated, the religion of Confucius was human-oriented. This is a key theme that, as we shall see, he would reiterate frequently. The central mission entrusted to Confucius was to lay down infallible laws and formulate institutions valid for all ages, a system that would eventually bind together all people.[67] Kang believed that the teachings of Confucius were directed toward the regulation and improvement of human life, with the ultimate aim of bringing unity and harmony to China and to the whole earth.

Kang claimed that Confucius' essential teachings were encoded in the Classics and were primarily transmitted by oral tradition. As time passed, however, misinterpretations crept into the texts and some key doctrines of Confucius were forgotten. One may ask: what were exactly those doctrines and how did they get lost? Kang's discussions on this issue were scattered among his various writings and lectures to his students. A coherent summary was offered by Liang in his 1899 speech on 'Religious Reformation in China' at the Japanese Society of Philosophy.[68] According to Liang, just as Buddhism was comprised

65 Kang Youwei 康有为, 'Kongzi gaizhi kao 孔子改制考 [A Treatise on Confucius as a Reformer (1897)]', in KYWQJ, vol. 3, 2007, 3; translated in Fung, *Chinese Philosophy*, 1983, 2:675, with minor adaptations.

66 Liang, 'Qingdai xueshu', 3098.

67 Kang, 'Gaizhi kao', 3.

68 Liang Qichao 梁启超, 'Lun Zhina zongjiao gaige 论支那宗教改革 [On the Religious Reformation in China (1899)]', in LQCQJ, vol. 2, 1999, 263–66.

of two major schools, Mahayana and Hinayana, Confucianism also consisted of two schools: *xiaokang* 小康 (Lesser Prosperity) and *datong* 大同 (Great Unity). The doctrine of *xiaokang* represented Confucius' mundane teachings, which were taught to all his disciples through the *Odes*, the *Documents*, the *Music* and the *Rites*, which had been edited by Confucius.[69] The doctrine of *datong* represented Confucius's supramundane teachings, which were only transmitted through the *Changes* and the *Annals* to disciples of superior intellect. Being written by Confucius himself, these two books were the chief repositories of the essence of his teachings. In the former were enshrined Confucius' teachings on the 'supramundane dharma', while the latter 'encoded' his subtle principles on statecraft (*jingshi* 經世, literally, setting the world in order), which needed decipherment in order to fathom their inner meanings.[70]

During the ensuing Warring States period (403–221 BCE), Liang continued, the *xiaokang* school was primarily inherited by Xunzi, while the *datong* school was inherited by Mencius through the *Annals* and by Zhuangzi 莊子 (*circa* 369–286 BCE) through the *Changes*. During the Han period, however, the doctrine of *datong* was discontinued, while that of *xiaokang* prospered. That happened because, for one thing, the rulers of various dynasties preferred the *xiaokang* school, which tended to support despotism, to the *datong* school, which upheld democratic ideals. For another, the subtle yet profound teachings enshrined in the *Annals* were obscured, as people were increasingly sidetracked by Old Text Confucianism from the more reliable interpretation of the *Annals*, namely, the *Gongyang Commentary*. That meant, Liang stressed, that what had been practised in the name of Confucianism for the past two thousand years was only the *xiaokang* school transmitted through Xunzi, while the comprehensive, authentic *datong* school had been forgotten.[71]

Kang and Liang both considered religion to be a profoundly powerful source for the development and strengthening of a country. The fact that the modern

69 This account differed from Kang's claim in *Confucius as a Reformer* that all the Six Classics were authored by Confucius in the name of the ancient Sages. Liang offered a weak version of the authorship of the Six Classics to soften Kang's strong version, a position that Kang occasionally endorsed. See Kang, 'Zhongyong zhu', 389. A note is needed here on the term 'Six Classics'. The Confucian tradition holds that there had been originally 'Six Classics', but one, the *Music* (*Yue* 樂), became lost, hence the more commonly used term 'Five Classics'.

70 For a detailed discussion on the protean idea of *jingshi* [*chingshi*], see Hao Chang, 'The Intellectual Heritage of the Confucian Ideal of Ching-Shih', in *Confucian Traditions in East Asian Modernity*, ed. Wei-ming Tu (Cambridge, Mass.: Harvard University Press, 1996), 72–91.

71 Liang, 'Lun Zhina zongjiao', 263–64.

West enjoyed its present level of civilization was precisely due to the 'religious revolution' and the 'revival of classical learning'.[72] By contrast, China's present weakness and its inferiority to the West were due to the fact that the original teachings of Confucius had been lost. The pressing task for China, therefore, was to initiate a 'religious reformation' in order to bring the true teachings of Confucius out of obscurity, a mission that fell, Liang said, on the shoulders of none other than Kang himself. In summary, Liang declared that the teachings of Confucius as expounded by Kang conformed to the needs of modern society, as they had six qualities: they were progressive rather than conservative; they advocated equality instead of despotism; they were inclusive rather than exclusive; they were devoted to the greater social good rather than just the interest of individual persons, they encouraged independence instead of weakness; and they valued the soul above the body.[73] Liang then elaborated on each of the six qualities except the last one. One possible reason why he did not talk about the soul could be that Kang had not yet written much on the topic at that time.

Convinced that Confucius had originally been a founder of a *jiao* or religion whose essential teachings had been lost, Kang endeavoured to restore Confucianism to its 'pristine' state by recasting it as a religion, to be called *Kongjiao* 孔教. In doing so, he was consciously following the example of Martin Luther who had initiated the Protestant Reformation, and for that reason, Kang was often referred to by his disciples as the 'Martin Luther of the Confucian Religion'.[74]

3.2 *Reform and a Confucian Church*

Kang adopted the Gongyang concept of the Three Ages (*sanshi* 三世), which in his view had been inaccurately understood as merely a description of the past, and developed it into a progressive view of history that would furnish the basis of his political theory. As a society developed from the Age of Disorder (*juluan* 據亂) through the Age of Approaching Peace (*shengping* 升平) to the Age of Universal Peace (*taiping* 太平), the system of government would accordingly evolve historically from the rule by a despot through an intermediate stage of joint-rule by a despot and (the representatives of) the people to a third stage of rule by the people themselves. Late-nineteenth-century Qing, in his

72 Liang was probably referring to the Reformation and the Renaissance. Liang, 263.

73 Liang, 263.

74 Liang, 'Nanhai Kang', 486. It is interesting to note that Martin Luther had also had a mystical experience through which he suddenly 'saw the Scripture in an entirely new light' and straightway he felt as if he were 'born anew' and he had found 'the door of paradise thrown wide open'. Quoted in James, *Varieties of Religious Experience*, 382.

judgment, was in the Age of Approaching Peace, so its autocratic government ought to undergo institutional reforms towards the establishment of a constitutional monarchy.[75] Convinced of the truth of this Confucian doctrine, in the latter 1890s, Kang initiated a campaign for reform, which was triggered by China's debacle in the Sino-Japanese War (1894–95). China's defeat proved to be a watershed in the course of Chinese intellectual history in several ways. To begin with, the event signified the ultimate failure of the technology-oriented Self-Strengthening Movement: the clear policy implications were that technological improvement alone was inadequate to save and strengthen China. Secondly, shocked by the power of a Japan transformed through Westernization, many of the Chinese literati and officials mitigated their earlier opposition to reforms and acknowledged the Japanese experience as a more successful model for development from which China could learn useful lessons for achieving its goal of becoming wealthy and powerful. Thirdly, the result of the war had completely destroyed the Sino-centric view of the world, causing the reigning Confucianism to be seriously undermined, and attempts had to be undertaken by the Chinese intelligentsia to work out a new view of the world.[76]

When the news of China's defeat spread to Beijing, Kang was there for his metropolitan examinations – he had passed the provincial examinations two years before – and succeeded in becoming a *jinshi* 進士 (metropolitan graduate) before being appointed a second-class secretary in the Board of Works. He acted immediately to lead his fellow examination candidates gathered at the capital in sending a petition (known as *gongzhe shangshu* 公車上書, candidates' memorial) to the Throne urging the rejection of the humiliating Treaty of Shimonoseki, by which Taiwan was ceded to Japan; however, that effort was all in vain. Nevertheless, Kang and Liang (who also took the metropolitan examinations but failed) decided to remain in the capital in order to make new efforts to generate public awareness of the need for political reforms. They started a reform newspaper and established a reform association, *Qiangxuehui* 強學會 (Society for National Strengthening). Initially, the society was enthusiastically supported by the elite in the capital, whose sponsors and members included high officials and missionaries, as well as Kang's associates. Because of its critical review of state affairs, however, the society, as well as the newspaper,

---

75 Liang, 'Lun Zhina zongjiao', 264–65.

76 See Charlotte Furth, 'Intellectual Change: From the Reform Movement to the May Fourth Movement, 1895–1920', in *An Intellectual History of Modern China*, ed. Merle Goldman and Leo Ou-fan Lee (Cambridge: Cambridge University Press, 2002), 13; John K. Fairbank, 'Introduction: The Old Order', in *The Cambridge History of China*, ed. John K. Fairbank, vol. 10 (Cambridge: Cambridge University Press, 1978), 1–34.

was suppressed by a decree of the imperial Court early in 1896. Subsequently, a string of reform associations and newspapers were set up in several provinces, and particularly in Hunan, where Tan Sitong, a young scholar who had been inspired by Kang's reform calls, was playing an instrumental role (I discuss Tan in the next section).[77]

Because of the mounting hostility among the powerful conservatives to his activities, Kang left for Guangdong, where he resumed his teaching at the *Wanmu caotang* and made lecture tours to Guilin in the neighbouring Guanxi province. With a hard-earned *jinshi* degree and the newly-garnered fame as leader of the 1895 petition initiative, Kang enjoyed considerable respect from and commanded some influence over the local scholars and officials. He observed that scholars in Guangxi were very open-minded, like those from Guangdong and Hunan, and he conceived the idea of expanding the base of his political and religious cause to Guangxi. Therefore, he proposed and supervised the organization of *Shengxuehui* 圣學會 (Society of Sacred Learning) in Guilin in 1897. The society was set up, in Kang's words, to venerate Confucius alone, as the divine founder of a religion.[78] It should be noted, in this regard, that Kang's efforts to idolize or deify Confucius had started from the early 1890s when he had attempted to revive and practise ancient religious rituals for the worship of Confucius at his academy in Guangzhou.[79]

In taking this initiative, Kang held up Christianity as a reference. 'Western countries have promoted their religion across the planet and penetrated its influence inland in China,' he wrote. 'The common worship of Jesus, and Jesus alone, has united and empowered the missionaries. It is the Chinese scholars' fault, then, to have failed to revere and worship Confucius alone, and promulgate his sacred religion.'[80] Emulating Christianity, Kang introduced the practice of worshipping Confucius and having lectures every ten days. In an effort to restore ancient Chinese practices, he established a school affiliated to the society, teaching subjects in traditional Chinese learning and Western learning. The society also planned to sponsor translation of Western books so as to educate and enlighten the general populace. 'Learning and scholarship,' Kang observed, 'are needed if the Confucian doctrine is to be propagated.'[81] In

77 For a concise yet informative introduction to the reform movement in Beijing and Hunan, see Chang, 'Intellectual Change', 291–318.

78 Kang Youwei 康有为, 'Shengxuehui houxu 圣学会后序 [An Afterword to the Society of Sacred Learning (1897)]', in KYWQJ, vol. 2, 2007, 265–66.

79 See Kang, 'Woshi', 81.

80 Kang Youwei 康有为, 'Shengxuehui yuanqi 圣学会缘起 [An Introduction to the Society of Sacred Learning (1897)]', in KYWQJ, vol. 2, 2007, 268.

81 Kang, 268.

addition, the society started a weekly (which later became a monthly), the first private newspaper in the province. The society, together with its school and newspaper, thrived for over a year until it was banned as a consequence of the failed Reform of 1898. Liang considered the society to be the southern counterpart of the *Qiangxuehui* in Beijing as an important source of new learning and new thought.[82] Though short-lived, the society would serve, as I will show later, as a model for the future endeavours of Kang and his followers in exile, and as a forerunner to the Confucian Religion Association eventually established in China in the early 1910s (see Chapter Four).

Kang returned to the capital again in late 1897, when China was faced with another crisis. It started with the forced lease of Jiaozhou (Kiautschou) Bay to Germany, which was followed by the seizure of Port Arthur and Dalian by Russia, and the demands for territorial concessions made by France, Japan and Great Britain. These events and the poor performance of the Qing government led Kang and many of his contemporaries to believe that China was in imminent danger of partition. Conscious of, and building upon the widespread sense of crisis, Kang organized in April 1898 the short-lived *Baoguohui* 保國會 (Society for Preserving China), calling for prompt reforms in order to preserve 'the country, the Chinese race and the religion/teaching' (*baoguo* 保國, *baozhong* 保種, *baojiao* 保教). By *baojiao* Kang meant the protection of the Confucian tradition, 'the sacred teaching', from imminent cessation in the face of the foreign military and religious encroachment.[83]

The best way to preserve Confucianism, in Kang's view, was to reform and institutionalize it by modelling it after the then evidently successful Christian practices in the West. His proposal was motivated by at least three factors: his conviction that Confucianism was originally and essentially a religion; the perceived superiority of the church as an organizational form of Christianity in the West; and the alarming spread of Christianity in China. As early as in 1895, he had observed that the Chinese people were prone to be seduced by Christianity precisely because the true doctrines of Confucius had not been properly and duly promulgated. He lamented the fact that while Christian churches were seen in many places, there was only one Confucian temple in each locality. To counter the Christian influence, he proposed that the state should support Confucian scholars in their efforts to instruct the common people in the authentic doctrines of the Confucian Classics. Building on his

82 See Zhong Wendian 钟文典, *Guangxi tongshi* 广西通史 [*A Comprehensive History of Guangxi*], vol. 2 (Nanning: Guangxi renmin chubanshe, 1899), 407, 382–410.

83 Kang Youwei 康有为, 'Baoguohui zhangcheng 保国会章程 [The Charter of the Society for Preserving China (1898)]', in *KYWQJ*, vol. 4, 2007, 54–55.

1888 proposal to Pan Yiqin, he suggested that all improper shrines across the country be transformed into Confucian temples; and that Confucian teachers should even be sent abroad as missionaries to spread Confucian teachings and establish Confucian temples.[84]

All of Kang's efforts culminated in the launch on 21 June 1898 of a reform programme by Emperor Guangxu, who heeded his calls for reforms. In a following memorial to the Emperor, which accompanied the presentation of his *Confucius as a Reformer*, Kang proposed the establishment of a national Confucian Church (*Kongjiaohui* 孔教會). The church would, Kang suggested, have as chairman the Duke Yansheng (衍聖公), a title of nobility conferred to the eldest male in the main line of descent of Confucius.[85] Membership was to be open to everyone, high and low, educated and illiterate, in the empire. It would have two practical functions. First, it would relieve the government of the unpleasant task of having to deal with cases of religious disputes involving Christian converts (the Confucian Church would agree on and abide by codified rules with its Catholic and Protestant counterparts). As was well-known, religious disputes in Late Qing often led to diplomatic crises or even military conflicts with foreign powers. The second function of the Confucian Church was to revive the morality of the people by dispatching instructors across the country to diffuse Confucian teachings. For that purpose, Kang requested that an imperial decree be issued to have all improper temples, such as those erected to worship animal spirits, converted to local Confucian temples,[86] at which all people, including women and otherwise marginalized villagers who had previously been denied access to Confucian temples, should be allowed to worship.[87]

In a revised version of the memorial,[88] Kang set forth more details on the structure of a Confucian religion based on the Christian model. He did not

---

84 Kang Youwei 康有为, 'Shang Qingdi disanshu 上清帝第三书 [My Third Memorial to the Throne (1895)]', in *KYWQJ*, vol. 2, 2007, 75.

85 The title was conferred in accordance with a millennium-long tradition. The then Duke Yansheng was Kong Lingyi 孔令貽 (1872–1919), the 76th-generatoin descendent of Confucius.

86 Kang used the two words interchangeably, Confucian temples (*Kongzi miao* 孔子廟) and Confucian churches (*Kongzi jiaohui* 孔子教會).

87 Kang Youwei 康有为, 'Qing shangding jiaoan falv ... yi zun shengshi er bao dajiao zhe 请商定教案法律……以尊圣师而保大教折 [A Memorial to the Throne Requesting the Making of Laws on Religious Disputes ... so as to Revere the Sacred Teacher and Preserve Our Great Faith (1898)]', in *KYWQJ*, vol. 4, 2007, 93; see also Kang, 'Woshi', 94.

88 When he published it following the failure of the Reform, Kang did not reveal that it was a revised version rather than the original version. The discrepancies between the two were not proved until the 1980s when the original memorial was brought to light. For this

conceal his admiration for what he believed to be the role of Christianity as a unifying force for many Western countries.[89] In place of the Confucian Church he had proposed earlier, he now proposed that a ministry for religion should be created at the national level with an elected head; Confucian churches or temples should be set up in each locality; Confucian pastors should be elected and professionalized, with at least one assigned to each church; both men and women should gather in their local church every seven days to receive instruction in the Confucian Classics; Confucius, along with Heaven, should be worshipped as the divine founder of a religion for China; and, last but not least, all improper temples should be converted into Confucian temples by imperial order and with the use of force if necessary.[90] As a result, what he proposed was very similar to the Christian practice, with Christ replaced by Confucius, the Bible by the Confucian Classics, and God by Heaven. His reference to Christianity, however, was limited to the institutional aspect; at the doctrinal level, he had, as demonstrated in the previous section, turned to ancient Chinese texts. As Goossaert and Palmer have observed, Kang's proposal was a 'wild hybridization' of Confucian fundamentalism and Christianity as conveyed by missionaries in China.[91]

By suggesting the forcible conversion of all improper temples into Confucian ones, Kang put forward a precise plan to eradicate Chinese folk religion so as to establish a Confucian religion as the exclusive state religion. In both the 1898 memorial and its revised version, Kang criticized Chinese folk religion, lashing out at various polytheistic practices, which in his view were responsible for the degradation of the Chinese people. He listed the popular yet 'improper' cults of worshipping gods of fertility, wealth, carpentry and literature. 'It's a crying shame,' he lamented, 'that some Confucian scholars, heedless of the import of their learning, pray to and worship the so-called god of literature!'

Kang was accused of forgery himself, see Kong Xiangji 孔祥吉, *Kang Youwei bianfa zouzhang jikao* 康有为变法奏章辑考 [*A Critical Collection of Kang Youwei's Memorials on Reforms*] (Beijing: Beijing tushuguan chubanshe, 2008). This issue is not crucial insofar as this research is concerned, because the two versions differed in detail but not in principle. Kang's revision and editing of the original memorial also allows future scholars to follow the development of his thought.

89 See Kang, 'Woshi', 94.

90 Kang Youwei 康有为, 'Qingzun Kongsheng wei guojiao li jiaohui jiaobu yi Kongzi jinian er fei yinci zhe 请尊孔圣为国教立教会教部以孔子纪年而废淫祠折 [A Memorial Requesting that Confucius Be Worshiped as the Founder of the State Religion, that a Ministry of Religion and a Church Be Established, that Calendar Years be Reckoned from the Birth of Confucius, and that Improper Temples Be Suppressed (1898)]', in *KYWQJ*, vol. 4, 2007, 98.

91 Goossaert and Palmer, *Religious Question*, 46.

Even worship of the god of the town *Chenghuang* 城隍, a practice endorsed by the state, was, in Kang's view, incompatible with the Confucian doctrines. What appeared most absurd to Kang was for humans to worship such animal gods as the Dragon King, the Bull King and the Monkey King, not to mention the worship of a tree trunk or a rock.[92] Beyond that, it pained him to notice that Confucius was not widely worshipped among the general populace, while Guan Yu 关羽 was worshipped in every household, although the latter was no more than a distant follower of the former.[93] Kang's rhetoric was reminiscent of contemporary missionary attacks on Chinese polytheistic beliefs and superstitious practices.

Kang's proposal for creating a Confucian religion on the Christian model found little support among the wide circle of Chinese intellectuals and officials. Not only were the more conservative opposed to his proposal; even many of the more reform-minded, such as Yan Fu, found his ideas too radical.[94] Indeed, after the publication of his memorial to turn improper temples into Confucian churches, a rumour started to circulate that Kang may have converted the Emperor to Christianity, and a subsequent imperial edict protecting the missionaries only reinforced the impression that Christianity was out to eradicate Chinese religion.[95] In response to these circumstances, Kang changed his strategy. In a later memorial to the Emperor, he suggested that all temples in towns and villages across the country be turned into primary schools and that temple revenues be appropriated for funding those schools. Kang remained unswerving in his attacks on what he called immoral cults and improper forms of worship. The Emperor was so pleased with the proposal that he immediately promulgated an edict of the same nature, an edict that was widely understood as a call for the destruction of all temples. Although the edict was later recalled, Kang's proposal foreshadowed the large-scale confiscation and destruction of temples, and the anti-superstition campaigns of early twentieth-century China.[96]

92 Kang, 'Qingzun Kongsheng', 96.

93 Kang, 97.

94 Yan Fu 严复, *Yan Fu ji* 严复集 [*Collected Works of Yan Fu*], ed. Wang Shi 王栻, vol. 1 (Zhonghua shuju, 1986), 82, 85.

95 Goossaert and Palmer, *Religious Question*, 46.

96 Kang Youwei 康有为, 'Qinggai zhisheng shuyuan wei zhongxuetang xiangyi yinci wei xiaoxuetang 请改直省书院为中学堂乡邑淫祠为小学堂 ... [A Memorial to the Throne Requesting the Transformation of Provincial Academies into Middle Schools and Improper Temples in Towns and Villages into Primary Schools ...(1898)]', in *KYWQJ*, vol. 4, 2007, 317–19; see also Goossaert and Palmer, *Religious Question*, 46–47.

Kang's effort to institutionalize Confucianism and reform Chinese religion, however, soon came to an end when the whole reform enterprise was cut short, in less than 100 days since its start (hence the name 'Hundred Days Reform'), in a coup by the Empress Dowager, who remained the ultimate decision-maker of the imperial court. The Emperor was put under house arrest and his reform measures were cancelled; Kang and Liang escaped death and fled to Japan, with a reward placed on their heads (more on this in the next chapter), but six of their fellow reformers were beheaded, among whom was Tan Sitong, son of a provincial governor. Given his fame as an advocate of reform in Hunan, Tan had been, during the Reform of 1898, summoned to Beijing for an audience with the Emperor and then appointed to a low-ranking post in the Grand Council, assisting in reform measures and facilitating contacts between the Emperor and Kang. On the eve of the coup, Tan disdained flight and chose to be executed, fulfilling his expressed wish to become the first to shed blood for the cause of reform in modern Chinese history.[97] Tan's death brought great sorrow to Kang, who later penned these verses in his memory, 'A rare and heroic soul, he has been invited to reside beside God (*Di*).'[98]

As Kang's admirer and 'disciple in spirit', Tan's monist philosophy had been considered to have been inspired by Kang's thought; at the same time, Tan's thinking, especially his appropriation of Western scientific notions in his religious reflections, would exert a traceable impact on Kang's thought. Therefore, I will devote a large portion of the next section to Tan in relation to the reformists' efforts to construct a new, unified system of knowledge.

4 Knowledge Is One

Encouraged by the reformist ferment in the mid-1890s, there was a renewed and expanded quest for knowledge among the more open-minded Chinese intellectuals. As has been shown in the previous sections, a key feature of Kang's reform campaign was the advocacy of Western learning, in which he saw both a power to emancipate the Chinese mind and a means of strengthening

97 For a concise sketch of Tan's life, see Arthur W. Hummel, ed., *Eminent Chinese of the Ch'ing Period, 1644–1912* (Folkestone: Global Oriental, 2010), 792–95; for a detailed, comprehensive account of Tan's intellectual development, see the introduction to Sin-wai Chan, trans., *An Exposition of Benevolence: The Jen-Hsüeh of T'an Ssu-t'ung* (Hong Kong: The Chinese University Press, 1984).

98 Kang Youwei 康有为, 'Kang Nanhai xiansheng shiji 康南海先生诗集 [Collected Poems of Mr. Kang of Nanhai (1888–1927)]', in *KYWQJ*, vol. 12, 2007, 189.

China. To meet the increasing demand for new knowledge, Kang proposed that in addition to translating directly from Western texts, translation should be carried out from Japanese sources as well. His reasoning was that translating from Japanese would be more efficient because of the affinity between the two languages, and that Japanese translations of Western books represented a selection by the Japanese scholars of the best original titles.[99] For this purpose, by reviewing his own collection of Japanese books and referring to certain Japanese catalogues, Kang had compiled a lengthy, annotated bibliography of Japanese works. It listed more than 7,000 titles grouped under fifteen subjects, namely, physiology (medicine included), sciences (including philosophy, psychology and ethics also included), geography and cartography, politics, law, agriculture, industry, commerce, education, literature, languages and scripts, fine arts, fiction, and military arts.[100] It is worth pointing out that Kang used the Japanese neologism '*kexue*' (science) in his bibliophagy. For that reason, he was sometimes given the credit for being the first Chinese scholar to use the term,[101] but that was more the result of copying Japanese titles than an intentional, informed borrowing of a new term. In fact, he had serious reservation about that term, which he rarely used in his other writings.

Although Kang's recommendations, as stated in the list of books which in his view deserved translation, were not followed by the contemporary translators, Kang was right in his prescription, which, as it happened, turned out to be a good prediction – by the early 1900s, more books had been translated into Chinese from Japanese than directly Western sources. His bibliography contributed to his countrymen's awareness of the various branches of new knowledge and gave them an idea of the different possible ways of classifying knowledge. Traditionally, books in China were grouped into four categories: *jing* 經 (Confucian Classics), *shi* 史 (historical works), *zi* 子 (works of other schools of thought, works devoted to agriculture, medicine, mathematics and astronomy, and the miscellaneous works) and *ji* 集 (literary works). As they strove to acquire Western learning, the Chinese intellectuals, Kang's disciples, followers and associates in particular, started to reflect on the classification of knowledge as well as on the relationship between Chinese and Western learning.

---

99 Kang Youwei 康有为, 'Riben shumu zhi 日本书目志 [An Annotated Bibliography of Japanese Titles (1898)]', in *KYWQJ*, vol. 3, 2007, 264.

100 Kang, 'Riben shumu zhi'. For a discussion on the possible sources of Kang's bibliography, see Mao, *Woshi jianzhu*, 382–83.

101 See Fan, 'Cong gezhi dao kexue'.

4.1 *A Three-Tiered System*

A notable attempt at classification was the 'triumvirate of knowledge' proposed by Liang in his 1896 *Xixue shumu biao* 西學書目表 (A Bibliographical Table of Books on Western Learning). The table included over three hundred titles deemed 'worthy', selected from Western works translated into Chinese over the previous half century, many of which were accompanied by his concise annotations.[102] Breaking from convention, Liang classified Western learning into three categories: *xue* 學 (science), *zheng* 政 (management) and *jiao* 教 (religion). Under *xue* were listed such subcategories as mathematics, mechanics, electricity, chemistry, acoustics, optics, thermodynamics, astronomy, geology, zoology and botany, physiology, medicine and graphics. To the category of *zheng*, he assigned historiography, law, bureaucracy, education, agriculture, mining, commerce, shipping, and the art of war. He listed *jiao* as the third category, although he omitted books on religion in his bibliographical table. Liang acknowledged the difficulty involved in classifying Western books, pointing out that as the art of management was based on the sciences, there was no clear demarcation of boundaries between books on the sciences and books on management.[103] He further indicated that scientific discoveries or principles were to conform or testify to the validity of the subtle, profound principles of religion. Therefore, he reasoned, understanding of the former was conducive to a surer grasp of the latter. A basic knowledge of mathematics, astronomy and geography, for example, should make it easier for pupils to comprehend the subtle teachings of the Confucian Classics (although Liang did not elaborate on how).[104]

Liang's triumvirate of knowledge was embraced by Tan Sitong, when the two met and conversed in Beijing in early 1896. By that time, Tan had already become an earnest student of Western learning and a convert to the ideas espoused by Kang, whom he had not yet met but whose *Forged Classics* he had read. When he learned from Liang the entirety of Kang's teachings based on the New Text exegesis, he discovered that he and Kang shared a great deal

---

102 Liang Qichao 梁启超, 'Xixue shumu biao 西学书目表 [A Bibliographical Table of Books on Western Learning (1896)]', in *Yinbingshi heji jiwaiwen* 饮冰室合集集外文 [*Essays not Included in the Collected Works from the Ice-Drinker's Studio*], ed. Xia Xiaohong 夏晓红, vol. 2 (Beijing: Beijing daxue chubanshe, 2005), 1121–58.

103 Liang Qichao 梁启超, 'Xixue shumu biao xuli 西学书目表序例 [Preface to A Bibliographical Table of Books on Western Learning (1896)]', in *LQCQJ*, vol. 1, 1999, 82.

104 Liang Qichao 梁启超, 'Du xixueshu fa 读西学书法 [On How to Read Books on Western Learning (1896)]', in *Yinbingshi heji jiwaiwen* 饮冰室合集集外文 [*Essays not Included in the Collected Works from the Ice-Drinker's Studio*], ed. Xia Xiaohong 夏晓红, vol. 2 (Beijing: Beijing daxue chubanshe, 2005), 1170.

in common in their thinking.[105] Thereupon he declared himself 'a disciple in spirit' (*sishu dizi* 私淑弟子) of Kang.[106] His admiration for Kang was such that he spoke effusively of Kang as being 'an equal of Confucius' (*Kongzi zhiwai wuou* 孔子之外無偶) and he saw in Kang the 'rebirth of the Buddha' (*yifo chushi* 一佛出世).[107] At the same time, Tan had also developed a deep interest in Buddhism and Christianity through reading Buddhist and Christian books as well as cultivating a close association with Buddhists and Christians. Liang recalled that by the time they met, Tan had become a great admirer of Jesus' teachings on love.[108] Like Kang during his youth, Tan also made efforts to integrate various components of his learning, Chinese and Western, scientific and religious.

Tan did not unthinkingly accept Liang's classification of knowledge in its entirety, however; he developed it into a three-tiered hierarchy, with science and religion being placed at the two ends. 'There are various kinds of sciences,' he wrote, 'and the most practical should be the natural sciences; there are different arts of management, and the most practical should be knowledge systems to promote people's rights. The profound principles of both management and learning can be found in religion, which is more comprehensive.'[109] When acquiring knowledge, Tan continued, one should begin with mastery of the natural sciences, which constituted the 'foundation of basic learning', and then proceed to the art of management. When solidly grounded in these two, one could then explore the 'profound, subtle truths of religion', the highest level of knowledge.[110] Tan found this progression from the study of nature to the acquisition of spiritual virtues in complete agreement with the teachings of Buddhism and Confucianism. In the Buddhist context, knowledge involved the 'transmutation of sense-consciousness' before that of 'thought-consciousness'. Similarly, the Confucian Classics stressed the point that the acquisition of knowledge relied on an examination of the nature of things, and Confucius had said that studies

105 Tan Sitong 譚嗣同, 'Zhishi shipian 治事十篇 [Ten Essays on Managing Affairs]', in *TSTQJ*, 1981, 445.

106 Liang Qichao 梁启超, 'Tan Sitong zhuan 谭嗣同传 [A Biography of Tan Sitong (1898)]', in *LQCQJ*, vol. 2, 1999, 231.

107 Tan Sitong 譚嗣同, 'Shang Ouyang Zhonghu 上歐陽中鵠 [Letters to Ouyang Zhonghu]', in *TSTQJ*, 1981, 475.

108 Liang, 'Tan Sitong zhuan', 232. For a discussion on the missionary influence on the reformers, see Zhi-yun Chen, 'Liang Ch'i-Ch'ao's Missionary Education: A Case Study of Missionary Influence on the Reformers', in *Papers on China*, vol. 16 (Cambridge, MA.: East Asian Research Center, Harvard University, 1962), 66–125.

109 Tan, 'Shang Ouyang', 462.

110 Tan, 464.

of what was below could help one get through to what was up above (*xiaxue shangda*).[111]

Thus, Tan attached the greatest importance to religion, on which depended, in his view, the success of the other two types of knowledge. 'When religion is not practised,' he wrote, 'management will deteriorate, which in turn will lead to the extinction of science.'[112] Therefore, it was no use discussing management and science without discussing religion, for otherwise 'the skills in management and accomplishments in science would be turned into lethal weapons.'[113] He implied that religion provided man with spiritual guidance for the application of social theories and scientific discoveries. On the other hand, science was considered no less important, for it could be 'the most efficient way of achieving the immediate end of national defence.'[114] In Tan's three-tiered hierarchy of knowledge, science and religion were united, with the former leading to the subtle truths in the latter, and the latter serving as a spiritual guide for the application of the former.

Tan's prioritization of religion in the system of knowledge reflected his religious zeal, which was shared by Liang and their friend Xia Zengyou 夏曾佑 (1863–1924), a lay Buddhist scholar who sympathized with the reformers. The three of them would read Buddhist texts and the *Book of Revelation* in the Bible together, and express their reverence for Confucius, the Buddha, and Jesus, whom they referred to as the Three Sages (*Sansheng* 三聖), the 'only three' who had attained divinity.[115] 'We were carried away by various religions,' Liang recalled about twenty years later, 'whose founders we regarded as being exalted above the common run of human beings. We revered and blindly worshipped them to such a degree that we decided no poems should be written without employing allusions from the Scriptures [...] of Buddhism, Confucianism and Christianity.'[116] As a result, their poems 'of new learning' were often full of terms, images or metaphors from the literature of different religions. Two lines by Tan, recorded by Liang, may be cited as an illustration:

> Three denials were uttered before a rooster began to crow,
> Don't fight against the Dragon and the Frogs over territory.[117]

111 Tan Sitong 譚嗣同, 'Renxue 仁學 [The Science of Benevolence]', in *TSTQJ*, 1981, 331.
112 Tan, 'Shang Ouyang', 464.
113 Tan, 455.
114 Tan, 455.
115 Tan, 'Renxue', 289.
116 Liang Qichao 梁启超, 'Shihua 诗话 [Literary Criticisms]', in *LQCQJ*, vol. 18, 1999, 5326.
117 Liang, 5326.

According to Liang, the dragon was interpreted as an allusion to Confucius and the frogs, his followers.[118] Although it is hard to see how they could connect the unflattering imagery of the frogs with the followers of Confucius, these lines are quoted merely to reveal Tan's passion for religion and the reverence and respect he and his youthful intellectual companions had for the founders of the three religions.

4.2 *Harmonizing the Three Religions*

The youthful idealists also cherished the ideal of unifying the three religions into one, an idea that Kang had proposed. Kang's contemporary critics accused of him of harbouring a desire to be the 'Pope' of the proposed unified religion.[119] While Liang considered Confucianism, or to be exact, Kang's version of Confucianism, deserving of being regarded as a superior universal religion for the world in the future,[120] Tan argued that the Buddha could unify Confucius and Jesus, all of whom were alike in their benevolence, but differed in the reasons for being benevolent.[121] The spiritual concept that Tan identified as the fundamental universal principle underlying Buddhism, Confucianism and Christianity was 'interconnection'; this principle of 'interconnectedness' was, according to Tan, the fundamental meaning of 'benevolence', through which the Way or Dao unites all as one. The sense of interconnection between self and others, Tan explained, which was derived from the Buddhist idea that there was neither a form of others nor of self, was shared by Confucianism and Christianity.[122] Here Tan was perhaps alluding to the Golden Rule: Do unto others as you would have them do unto you; or as Confucius put it, Do not do unto others what you would not have done unto you. With this spirit, Tan advocated the exchange or 'connection' between different peoples in the realm of religion to complement the already existing exchange in science, government and trade.[123]

In addition, Tan considered two things as being essential to all religions, that is, the virtue of compassion and belief in the soul. Drawing on Williamson's *Ancient Religions*, he emphasized that all religions, be they Western or Eastern, of the past or present, followed the two universal principles. No religion could, he asserted, be established without preaching these two doctrines, and for a

118 Liang, 5326.

119 Liang Dingfen 梁鼎芬, 'Kang Youwei shishi 康有为事实 [Facts about Kang Youwei]', in *Cong wuxu dao xinhai* 从戊戌到辛亥 [*From 1898 to 1911*], ed. Tang Zhijun 汤志钧 (Nanjing: Jiangsu guji chubanshe, 1990), 65.

120 Ding and Zhao, *Nianpu changbian*, 34–35.

121 Tan, 'Renxue', 289, 354.

122 Tan, 289.

123 Tan, 297.

religion to spread to the most barbaric and perverse regions, it had to speak of the soul in the most absurd manner if necessary.[124] In Tan's view, once the immortality of the soul and the existence of an afterlife were recognized, even the most benighted persons would know that there were infinite joys to delight in or bitterness to suffer after death; they therefore would not conceive of covetous or pessimistic thoughts only because of transient suffering or happiness in their present life. If they understood that besides the soul, the substance of the body would also continue to exist in various forms even when slain, Tan continued, they would be prepared to die for righteous causes and would have no fear in their minds.[125] This ideal can be regarded as an indication of Tan's stature as a highly idealistic yet principled thinker and social reformer. As Tan penned these lines, it is as if he was describing in advance the principle he was standing up for in his mind when he chose not to flee on the eve of his self-sought arrest and consequent death two years later.[126]

In discussing the soul and the afterlife, Tan stressed his belief in transmigration of souls, then a popular concept that was also a view held by Kang. It implied that every person in this present life had a former life and would have a future life. Even if one could usually not recall one's previous life, Tan reasoned, its recollection and knowledge would eventually come to him, probably on the 'Final Judgment Day' talked about by Jesus. Tan was aware that the concept of previous life and transmigration was not accepted in Christianity. He explained that because Westerners had no idea of the soul's 'previous karma', they conceived of a contrivance that all men were born sinful.[127] Tan considered the idea of original sin absurd: what had the ancestors' sin to do with their descendants? Even in man's law, he argued, punishments were not extended to the wife and children of an offender. How could Heaven's benevolence be outstripped by man's?[128] Tan was confident that Buddhist doctrines of transmigration, together with those of causation and retribution, could stand the test of time and could not be dismissed.[129]

4.3 *A Monistic Philosophy*

Eventually, Tan reduced his three-tiered system into a monistic philosophy of *ren* 仁 as expressed in his masterpiece *Renxue* 仁學 or *The Science of*

124 Tan, 309.
125 Tan, 309.
126 See also Liang, 'Tan Sitong zhuan', 233.
127 Tan, 'Renxue', 311.
128 Tan, 311.
129 Tan, 310.

Benevolence, which was published posthumously by Liang.[130] The composition of *Renxue* was, in Liang's view, a result of the author's reflections on the essence of Confucianism and Buddhism, his gleanings from the writings of philosophers, and his elaboration on the principles and goals set forth by Kang.[131] Tan considered that the principal property of *ren* was 'interconnectedness', which was expressed as 'oneness'.[132] From this perspective, any idea of a conflict in knowledge on any aspect of the universe, be they spiritual or material, was unacceptable to Tan. The medium of interconnectedness through which *ren* was made manifest, according to Tan, was the *yitai* 以太 or 'ether'.[133] The ether, however, was so closely related to *ren* that the two concepts often seemed interchangeable in Tan's philosophy. The nature of Tan's conception of the ether, that is, whether it is material or immaterial, has been a cause of major controversy among scholars. Considering the proper understanding of the ether to be a prerequisite for the study of *ren,* Tan devoted the first chapter of his *Renxue* to its discussion.

> Throughout the realms of physical phenomena, empty space and sentient beings, there is something extremely great and supremely subtle which adheres to everything, penetrates everywhere and connects everything, so that everything is permeated by it. It has no visible forms, no sound, no smell or taste. It has no name but we call it the ether (*yitai*).[134]

Tan then went on to unite in ether the central teachings of Confucius, Mozi, the Buddha and Jesus and the discoveries of science. He argued that in respect of its virtue, Confucius variously referred to it as benevolence (*ren*), as the origin (*yuan*) or as the nature (*xing*). Mozi referred to it as undifferentiated love (*jianai* 兼愛). The Buddha referred to it as the Buddha-Nature, and as compassion. Jesus referred to it as soul and as loving others as oneself and regarding

130 Tan, 'Renxue'. *Ren* refers to the virtue of goodness, humaneness or benevolence. Different scholars have rendered *Renxue* into various English combinations, including 'A Study of Benevolence', 'An Exposition of Benevolence' and 'The Science of Love'. These diverse renditions demonstrate the richness of the title and the possibility of different interpretations. Given that in *Renxue*, Tan identified the essence of *xue* as science, I prefer to use my own translation, 'The Science of Benevolence'. For an complete English translation of the work, see Chan, *An Exposition*.

131 Liang, 'Tan Sitong zhuan', 232.

132 Tan, 'Renxue', 291–93.

133 Tan, 291.

134 Tan, 293; translation taken from David Wright, 'Tan Sitong and the Ether Reconsidered', *Bulletin of the School of Oriental and African Studies* 57, no. 3 (1994) with minor adaptations.

one's enemies as friends. The scientists referred to it as the power of affinity (*aili* 愛力) or attraction (*xili* 吸力).[135]

Not only was the ether the very means by which creation came into being, Tan explained, but it was also the very sustainer of the universe and everything within it. He stressed that the ether was the substratum of all natural phenomena including ever-multiplying micro-organisms. Moreover, the ether was also the medium by which human thought was transmitted. He tried to illustrate the dynamics thereof through what he regarded as the mere manifestations of the interactions between ether, 'brain' and 'electricity'.[136] Clearly, Tan's ether carried material elements, but it was far from being purely material. Not only was the ether 'the origin of the elements', and the primal stuff that all things were made of, it could also be identified with the spiritual teachings of Sages, the power that maintained human relations, and the medium that spread thought. He went so far as to claim that the either could be considered as non-existent in a Buddhist sense.[137] Thus his ether is more of a spiritual concept, connecting the physical and spiritual realms.

What, one may wonder, were the sources of Tan's concept of the ether? It is, as is to be expected, a concept derived from a Western source. The ether is an ancient concept in the West. By the nineteenth century, it was widely used to refer to an assumed most elastic and subtle medium, which penetrated and permeated objects and spaces, through which electricity, light, magnetism, heat and even nervous impulses were believed to spread.[138] In confronting previous theories positing the existence of many types of ether, a general agreement emerged among physical scientists, such as Oliver Lodge and Joseph Larmor, on the existence of a single, pervasive ether uniting all natural phenomena.[139] Towards the end of the century the ether came to be generally recognized as an immaterial medium rather than consisting of minute particles with definite locations in space, and this concept of ether was able to survive in the early twentieth century when all other rival concepts were shattered by the relativity theory of Einstein (Kang would discuss this in his later years, as

135 Tan, 'Renxue', 293–94.

136 Tan, 294–95.

137 Tan, 306, 331.

138 G. N. Cantor and Hodge, M. J. S., eds., *Conceptions of Ether: Studies in the History of Ether Theories, 1740–1900* (Cambridge: Cambridge University Press, 1981), 1–2.

139 Cantor and Hodge, M. J. S., 258. See also Janet Oppenheim, *The Other World: Spiritualism and Psychical Research in England, 1850–1914* (Cambridge: Cambridge University Press, 1988), 371–90.

we will see in Chapter Five).[140] There were also attempts to explore the theological or spiritual significance of the ether in order to defend the coherence of science and religion, such as the attempts by a large number of eminent European scientists who joined the Society for Psychical Research founded in 1882.[141]

Tan's concept of the ether was shaped by his reading of the *Zhixin mianbingfa* 治心免病法, John Fryer's 1896 translation of Henry Wood's *Ideal Suggestion through Mental Photography* (1893).[142] Although Wood mentioned the ether only in passing in the original, Fryer elaborated on the concept in his translation by inserting a passage on the 'omnipresent' ether from David Well's *Principles and Applications of Chemistry*: 'Recently in Western countries it has been discovered that in the Myriad Things there is a fluid called ether….'[143] Fryer then emphasized the mental aspect of the ether and made it explicit that it was the ether that transmitted thought, just as sound was transmitted by the air.[144] In addition, Wood described love as a 'great universal spiritual law of attraction', which 'inheres in universal matter', and 'permeates universal Spirit'. By exercising the great force of thought focused on the attributes of God, one could heal not only oneself but also improve the situation of the world.[145] Fascinated by the translation, Tan commented that he had not been able to see a glimpse of the truths of Christianity until he read the work.[146] From Fryer's elaboration Tan drew a distinctive quality of his ether, namely, the power to convey thought between human minds, and he was convinced that mental power or thought was capable of transforming the world.[147] Tan expanded the

---

140 Edmund Whittaker, *A History of the Theories of Aether and Electricity* (Thomas Nelson and Sons Ltd, 1951), 303.

141 Balfour Stewart and Peter G. Tait, *The Unseen Universe: Or, Physical Speculations on a Future State* (Macmillan and Company, 1875), vii; Cantor and Hodge, M. J. S., *Conceptions of Ether*, 140, 147, 151; Richard Noakes, 'Ethers, Religion and Politics in Late-Victorian Physics: Beyond the Wynne Thesis', *History of Science* 43, no. 4 (2005): 415–55.

142 Henry Wood (1834–1908) was a pioneer and author of the New Thought movement. For more on him and his work, visit a website devoted to him: http://neweverymoment.com/henrywood/henrywood.htm.

143 Cited from David Wright, *Translating Science: The Transmission of Western Chemistry into Late Imperial China, 1840–1900* (Leiden: Brill, 2000), 381.

144 Wute Hengli 烏特亨利 [Henry Wood], *Zhixin mianbingfa* 治心免病法 [*A Method for Healing the Mind and Avoiding Sickness*], trans. Fu Lanya 傅蘭雅 [John Fryer] (Shanghai: Gezhi shushi, 1896), 1.13a-b; translated in Wright, 'Tan Sitong and the Ether Reconsidered', 30.

145 Henry Wood, *Ideal Suggestion through Mental Photography: A Restorative System for Home and Private Use, Preceded by a Study of the Laws of Mental Healing* (Boston: Lee and Shepard Publishers, 1893), 94, 112.

146 Tan, 'Shang Ouyang', 461.

147 Tan, 461.

ether into an omnipotent, omnipresent cosmic concept in his *Renxue*, and in so doing he was joining the ranks of Lodge and Larmor in the West.

Viewed from the perspective of Tan's cosmic ether, which carried both material and spiritual properties, knowledge can only be one, and science and religion cannot but be in harmony. If the ether is the origin of the universe and the cause of every phenomenon in the world, then the study of the natural world would inevitably lead to knowledge of the same single reality, the ether. If the ether is identified in its qualities with the virtues of benevolence, universal love, compassion and the concept of the soul, then the knowledge of the ether achieved through sciences would naturally lead to a surer grasp of the subtle truths in the great religions. Thus, both science and religion were harmonized when connected by a profound understanding of the ether. In Tan's philosophy, there was no true difference or separation in the universe, a universe which was defined by the single reality, the ether. He could always see the whole picture, for beneath the surface every object, every event or every phenomenon was interwoven and inextricably inter-connected with every other object, event or phenomenon. This somewhat mystical line of reasoning is consonant with the manner in which Kang conceived of the world, and as will be seen in the next chapter, Kang also made use of the concept of the ether in his reinterpretation of the all-important Confucian concept *ren* during his exile.

CHAPTER 3

The Great Unity

> For the sake of saving China, I have chosen to take risks and now I am faced with imminent death. This, it appears, is my destiny. However, can spirit ever die? I would like to entrust my elderly mother to the care of Junmian, a staunch and upright man. I believe he will spare no effort to look after my families and to promote our great Cause. Whoever among my other disciples arises and protects my families and our country, my spirit will surely bless him.
>
> — KANG YOUWEI[1]

∴

This is a translation of Kang's 'last note', penned on a British launch in the waters near Shanghai on 24 September 1898. He was travelling from Tianjin to Shanghai on a British steamer, utterly unaware that, as a consequence of the palace coup by the Dowager Empress, a reward of high price had been placed on his head, that a Chinese cruiser was speeding onward to intercept him, and that Chinese detectives and policemen were waiting on the wharf to arrest him.

Before the steamer pulled in, however, the quick-thinking British Consul-General, Byron Brenan, who feared that Kang's arrest under his jurisdiction might cause unwanted diplomatic complications, had ordered the immediate dispatch of a launch to meet Kang and transferred him to the British mail steamer Ballaarat – a most timely and fateful intervention, as it turned out.[2]

Upon learning about the death warrant issued by the Empress Dowager, charging him with the alleged crime of 'poisoning the Emperor', Kang sensed imminent danger of death and wrote on the spot his 'last note' addressed to one of his most loyal followers, Xu Qin 徐勤 (courtesy name: Junmian 君勉,

---

1 Kang, 'Wuxu lunzhou juebishu'.

2 A couple of sources in Chinese claim that the British missionary Timothy Richard played a role in Kang's rescue, but that is not substantiated by the relevant English sources, including Richard's own memoir. See Mao, *Woshi jianzhu*, 786; Timothy Richard, *Forty-Five Years in China: Reminiscences* (New York: Frederick A. Stokes Company, 1916), 265, translated into Chinese in 2005.

 | DOI: 10.1163/9789004468221_005

1873–1945), whom he had sent to Japan to head the Academy of Great Unity (*Datong xuexiao* 大同學校) established among overseas Chinese students in Yokohama in the previous spring.[3] Kang had a narrow escape, however.

Escorted by a British gunboat, he arrived safely in Hong Kong late at night on 29 September.[4] During his sojourn in Hong Kong, the *China Mail* made Kang's cause known internationally for the first time through an interview which was widely reported in other newspapers, including the English *North-China Herald* and the Chinese *Shen Pao* in Shanghai. By that time, it had become clear that the Emperor, who had previously been reported dead, was actually under house arrest. In that interview, Kang claimed that before leaving the capital he had received secret edicts from the Emperor, who, sensing that a palace coup was imminent, had instructed Kang to devise means for rescuing him.[5] Kang's strategy was to obtain international support, especially from Japan and Britain, to help reinstall the Emperor. At the invitation of Japanese Prime Minister Okuma Shigenobu, Kang journeyed to Japan, where he had the pleasure of meeting some of his disciples, including Liang Qichao. During their reunion they found time to write about their recent experiences. Kang updated his autobiography, *Woshi* 我史 (My History), of events in his life since 1895.[6] Liang, for his part, wrote the far-reaching *Wuxu zhengbian ji* 戊戌政變記 (An Account of the Coup of 1898), which was published in his newly-established political periodical in Yokohama, the *Qingyibao* 清議報 (China Discussions).

Both accounts were, of course, focused on the political events, and the total veracity of their narratives has long been debated. These accounts revealed that the narrow escape they each experienced had only served to strengthen

3 Jung-pang Lo, ed., 'Sequel to the Chronological Autobiography of K'ang Yu-Wei', in *K'ang Yu-Wei: A Biography and a Symposium*, trans. Jung-pang Lo (Tucson: University of Arizona Press, 1967), 178, 253–54.

4 Kang, 'Woshi', 102–3; see also 'Brenan to Sir C. MacDonald', in *Correspondence Respecting the Affairs of China* (London: Harrison and Sons, 1899), 307–8.

5 'Kang Yu-Wei's Story', *The North-China Herald*, 17 October 1898. The two edicts provided by Kang were translated into English and published along with the interview, but their authenticity would become a topic of heated controversy for a long time, both during Kang's lifetime and after his death. Scholars have now ascertained that the first edict was an adaptation by Kang of an edict addressed to one of his fellow reformers; the full truth of the second edict, however, is still a mystery. For an updated and comprehensive discussion, see Mao, *Woshi jianzhu*, 735–42.

6 Kang's autobiography, which had been read by only a handful of his disciples, was not published during his lifetime; shortly after his death in 1927, it began to circulate in manuscript copies before it was finally printed in the 1950s. For a detailed and up-to-date discussion, see Mao Haijian 茅海建, 'Kang Youwei zixie nianpu shougaoben yuedu baogao "康有为自写年谱手稿本"阅读报告 [A Report on Reading Kang's Manuscript of His Autobiography]', *Jindaishi yanjiu* 近代史研究 [*Journal of Modern Chinese History*], no. 4 (2007).

their faith in Heaven. Reviewing his ten-day escape journey to Hong Kong, Kang reckoned that he could have been caught and killed ten or eleven times, had he left the capital one day later, had he stayed a few hours longer in Tianjin, had he not changed plans and boarded a British steamer, had the cruiser giving chase not turned back halfway due to insufficiency of fuel, had the English authorities not been alarmed or had they not come to his rescue, etc. The fact that he had emerged unscathed by virtue of a string of remarkable coincidences, decisions and eventualities, was, he believed, the will of Heaven. These extraordinary occurrences amounted to a clear vindication of his belief that he had been entrusted a Heaven-sent mission. The whole episode strengthened his determination to prosecute his reform cause with renewed vigour despite the setbacks he had experienced.[7] This belief in heavenly intervention was shared by his disciples who were fellow survivors in exile. Drawing on Kang's account, Liang stressed that his teacher had been protected and preserved by Heaven, with 'gods and spirits' alongside to protect him during his narrow escape.[8] 'That both Mr. Kang and I have survived the extraordinary dangers,' Liang confided in a letter to his wife, who was left in China, 'seems to demonstrate the protection from spirits.... I am now even more convinced that Heaven has treated me with bounty, preparing me for future success.'[9] When he penned these lines, he may have been alluding to the traditional Confucian tenet that he who possessed great qualities would surely receive heavenly blessings: 'Heaven, in the production of things, is sure to be bountiful to them, according to their qualities.'[10]

With the downfall of the friendly Okuma administration, Kang failed to enlist the Japanese aid in rescuing his Emperor. As the pressure on the Japanese government from the Qing Court to expel him mounted, Kang decided to leave for Great Britain via Canada. Borrowing terms from the *Garland Sutra*, he wrote a farewell poem to his disciples, expressing his confidence in the imminent appearance of a 'garland China'.[11] While he was in London in June 1899, a motion was made, chiefly through the instrumental role of his friend Lord Charles Beresford, in the House of Commons for British intervention in China, but it was defeated.[12] Disappointed, he returned to Canada where there was a

---

7 Kang, 'Woshi', 106.

8 Liang Qichao 梁啟超, 'Wuxu zhengbian ji 戊戌政變記 [An Account of the Coup D'état of 1898]', in *Xuxiu Sikuquanshu* 續修四庫全書 [*A Sequel to the Complete Library in Four Sections*], vol. 446 (Shanghai: Shanghai guji chubanshe, 1995), 250 51.

9 Ding and Zhao, *Nianpu changbian*, 168.

10 *Zhongyong* 中庸 (*The Mean*) 17:3, in Legge, Chinese Classics I, 399.

11 Kang, 'Kang Nanhai shiji', 194.

12 Lo, 'Sequel to K'ang's Autobiography', 180, 255–56.

large population of overseas Chinese. On the strength of the supposed edicts, he established the *Baohuanghui* 保皇會 (Association to Save the Emperor) in Victoria in July 1899. The next few years saw a rapid expansion of the association, with branches established in Chinese communities across Canada and in the United States, Latin America and Southeast Asia. The expansion was assisted by the establishment of newspapers, schools and even property development and banking businesses. Travelling around and assisted by his disciples, Kang advocated the adoption of a constitutional monarchy for China. Inevitably, his party fell into dispute with the other party of their fellow outlaws, the *Tongmenghui* 同盟会 (Revolutionary Alliance) led by Sun Yat-sen 孫中山 (1866–1925), who aimed to establish a republic and called for the overthrow of the Manchu regime. Kang's group had the upper hand in the competition for overseas Chinese allegiance, before gradually losing ground to the latter.[13]

Since its establishment, the *Baohuanghui* had functioned as Kang's power base and, with donations from overseas Chinese, provided him with the necessary means to sustain his political activities. Taking advantage of the Boxer Uprising of 1900, Kang mobilized the entire resources of the association and sponsored an armed movement in central China to overthrow the Empress Dowager and to restore the imprisoned Emperor. The proposed plot, however, was discovered, and consequently thirty of his followers were caught and executed. This caused great sorrow to Kang, and made him resolve, at least for the immediate future, to abandon the use of violence as a means of achieving his reform program. He withdrew from political activities and entered a two-year period of semi-seclusion, taking up temporary residence in Penang, Malaya and subsequently at Darjeeling in northern India. He devoted much of his time to writing commentaries on the Confucian Classics, thereby completing his reinterpretation of Confucianism, a project which he had begun a decade earlier.[14]

1 Confucian Reinterpretation Completed

The first work Kang chose to annotate was the *Zhongyong* or *Doctrine of the Mean* (hereinafter referred to as the *Mean*), attributed to Zisi 子思 (circa

13 Lo, 181; for more details on Kang's overseas activities, see Robert Leo Worden, 'A Chinese Reformer in Exile: The North American Phase of the Travels of K'ang Yu-Wei, 1899–1909' (Ph.D., Columbia, Georgetown University, 1972); see also Zhongping Chen, 'Kang Youwei's Activities in Canada and the Reformist Movement Among the Global Chinese Diaspora, 1899–1909', *Twentieth-Century China* 39, no. 1 (2014): 3–23.

14 Lo, 'Sequel to K'ang's Autobiography', 185–88.

483–402 BCE), the grandson of Confucius. Originally one of the treatises collected in the *Rites,* it was singled out and annotated by Zhu Xi before being included in the Four Books. Accordingly, the rank of its author was exalted in the Confucian tradition, with his tablet being placed in the Temples of Confucius as one of the four associates to keep his grandfather company.[15] Zisi was given the title of the Transmitter of (the Teachings of) the Sage (*Shusheng* 述聖), for he was believed to have transmitted the Doctrine of Confucius to Mencius, the Second Sage (*Yasheng* 亞聖).[16]

After he finished the *Mean Annotated* (*Zhongyong zhu* 中庸注), Kang completed his own commentary on the *Annals*. In this work, which he regarded as a sequel to his two earlier works, the *Forged Classics* and *Confucius as a Reformer,* Kang maintained that in emending the original *Annals,* Confucius had 'encoded' in the text profound principles and subtle meanings, some of which were explained in the two 'truthful' commentaries of the *Gongyang* and the *Guliang* before being tampered with by the 'false' *Zuo Commentary* allegedly forged by Liu Xin. Galvanized by a sacred sense of mission, Kang attempted to 'decode' as many of those hidden meanings and principles as he could by tracing, with the assistance of its two truthful commentaries, what the Sage had deleted, adapted and added during the process of his emendations. In doing so, Kang claimed to have rediscovered the Doctrine of the Universal Peace and its principle of the Great Unity, which he considered were not only requisite for the advancement of China but also vital in order for the whole world to achieve a higher stage of civilization in conformity with doctrine of the Three Ages.[17]

While residing at the hill resort of Darjeeling in India, Kang continued his reinterpretation of the Confucian Classics by completing three more works, the *Analects Annotated* (*Lunyu zhu* 論語注), the *Great Learning Annotated* (*Daxue zhu* 大學注), and the *Esoteric Meanings of the Mencius* (*Mengzi wei* 孟子微). In all these works, he claimed that he had sought to bring out the true meanings and ideals of the Way of Confucius. He pointed out, for example, that the texts of the *Analects* contained hints of the doctrine of the Great Unity, an doctrine that was authentically inherited by Mencius. In annotating the *Mencius,* Kang took the unusual measure of reorganizing the sections and

15 It was recorded that Zisi 'directly reached to Heavenly virtue' (*zhida tianting* 直達天德) and possessed a mysterious power of discernment. For more discussion on this topic, see Legge, *Chinese Classics I*, 45.

16 Kang, 'Zhongyong zhu', 369.

17 Kang Youwei 康有为, 'Chunqiu bixiao dayi weiyan kao 春秋笔削大义微言考 [A Study of the Profound Principles and Hidden Meanings of the Emended Annals (1901)]', in *KYWQJ*, vol. 6, 2007, 4.

passages under eighteen thematic topics including nature and destiny, mind and body, benevolence and righteousness. With these accomplishments, he had finally completed his annotations of the Four Books. Around that time, however, the Four Books, along with other Confucian classics, were soon to lose their prestige among the intelligentsia in China by reason of the fact that the imperial service examinations, which had been abolished during the 1898 Reform but reinstalled afterwards, were again abolished in 1905 – this time for good.[18]

Kang claimed that he had annotated those Confucian texts (except the *Mencius*) in the early 1890s when lecturing in Guangzhou, but he noted that all the manuscripts had been lost or destroyed following the Coup of 1898.[19] His re-written annotations, however, were not published until after his return to China in 1913, by which time the manuscript of the *Great Learning Annotated* had been lost again.[20] Although many of his views expressed in those annotated works had already been made known in his earlier lectures and works, he was trying to elaborate on them in his annotations as well as to provide them with a scholarly, textual foundation. As the scope of those texts was broad, Kang's comments accordingly covered a variety of topics, including the 'divinity' of Confucius, Heaven, human nature, spirits and the soul, and involved the appropriation of Western scientific terms in his efforts at Confucian exegesis.

1.1 *Confucius as a Divine Teacher*

Earlier, as we have seen, Kang had attempted to deify Confucius by following indiscriminately the apocryphal account that Confucius, who was born of the 'Black God', was 'a being of divine intelligence' and 'a religious leader for the whole world'. In his latest commentaries, however, he only accorded a brief mention of the apocryphal claim of Confucius' miraculous birth, and further glossed it by stressing that the account was intended to mean that Confucius had been born of Heaven and had received the Mandate of Heaven.[21]

---

18 Kang was aware of the tendency, for he had commented in 1904 that the new school textbooks would disregard the Confucian Classics, which meant that the Chinese would renounce the founder of their religion (*jiaozhu*) and become a people without religion. See Kang, 'Yidali youji', 375.

19 Some scholars have rejected Kang's claim outright, while others have observed that Kang's claim was not groundless. My understanding is that Kang was very likely to have produced manuscripts of those works in the form of lecture notes.

20 Only portions of his commentary on the *Mencius* had been printed during his exile in the *China Discussion* in Japan by Liang Qichao in late 1902.

21 Kang, 'Zhongyong zhu', 386; Kang Youwei 康有为, 'Mengzi wei 孟子微 [The Esoteric Meanings of the Mencius (1901)]', in *KYWQJ*, vol. 5, 2007, 444.

Kang's firm belief in the 'divinity' of Confucius, however, remained unchanged. He considered Confucius a *shenren* 神人 or 'divine man', a term that was used by Mencius in his categorization of men into several ranks according to their virtue, with *shenren* and *shengren* 聖人 (sage) listed as the highest two. In Legge's translation: 'He whose completed goodness is brightly displayed is what is called a great man. When this great man exercises a transforming influence, he is what is called a sage. When the sage is beyond our knowledge, he is what is called a spirit-man.'[22] Legge translated *shenren* as 'a spirit-man' rather than 'a divine man', because he considered the latter rendition unacceptable on the grounds that it would apply to man a term appropriate to the actions of God.[23] In a sense Legge's statement was in accordance with the orthodox interpretation of the term by Zhu Xi, who held that by *shenren* was meant a sage – an unfathomable sage rather than an entity of a higher rank.[24]

Kang, however, categorically disagreed. In his commentary on the *Mencius*, he claimed that by *shenren* Mencius had meant a divine man whose station was above that of a sage, and had intended to reserve the term for Confucius alone. Unfortunately, Kang argued, Mencius was not properly understood by his posterity, and as a result, Confucius had been inaccurately called a sage for two millennia.[25] However, Kang failed to explain why Mencius himself called Confucius a sage in numerous places in his writings, and placed Confucius on a par with the previous sages including King Wen of the Zhou dynasty, King Tang of the Shang dynasty, and Chinese legendary sage kings Yao and Shun, with approximately 500 years between the appearances of two consecutive sages.[26]

To support his claim that Confucius had been regarded as a divine man, Kang turned to Mencius' illustrious contemporary, Zhuangzi, who had also classified men into several ranks, including, from higher to lower, the heavenly man (*tianren* 天人), the divine man (*shenren*), the perfect man (*wanren* 完人) and the sagely man (*shengren*).[27] According to Kang's interpretation, Zhuangzi ascribed to Confucius the status of the divine man (the spirit-man in Legge's translation) and designated him as the 'Holy King with Divine Intelligence'

---

22 *Mengzi* 孟子 (*The Mencius*) 7(2):25, in James Legge, trans., *The Chinese Classics II: The Works of Mencius*, 2nd ed. (Oxford: The Clarendon Press, 1895), 490.

23 Legge, 492.

24 Zhu Xi 朱熹, *Sishu zhangju jizhu* 四書章句集注 [*Interlinear Analysis of and Collected Commentaries on the Four Books*] (Beijing: Zhonghua shuju, 1983), 370.

25 Kang, 'Mengzi wei', 384–484.

26 *Mengzi* 7(2):38, in Legge, *Chinese Classics II*, 501–2.

27 *Zhuangzi* 莊子 33:11:1, in James Legge, trans., *The Texts of Taoism II: The Writings of Kwang-Tze, XVIII-XXXIII, The Thai-Shang Tractate of Actions and Their Retributions* (Oxford: The Clarendon Press, 1891), 214–15.

(*shenming shengwang* 神明聖王), while in comparison, Zhuangzi only considered Laozi, whose body of thought he had inherited, to be a perfect man.[28] Again, that was Kang's taking from Zhuangzi, for the latter was not explicit about assigning to Confucius a divine status. Kang did not stop there. In his laudatory description of Confucius, Kang even appropriated certain Buddhist expressions as well as the attributes of the first hexagram (*qian* 乾) described in the *Changes*:

> Possessing the creative power and having Nature under his command, he possesses a complete understanding of the connection between the end and the beginning. Mounting the carriage drawn by the six dragons at the proper times, he drives through the sky. With a pre-knowledge of numerous ages, he is capable of transformation by bending or extending himself, with limitless incarnations. – Can these attributes be ascribed to anyone but Confucius?[29]

Clearly, Kang portrayed Confucius as a divine figure, one who was mighty and omniscient. In his commentary on the *Analects*, he described Confucius in a way that was reminiscent of the way Jesus was presented in the Christian tradition. 'Confucius had a heavenly body,' he claimed, 'whose celestial soul reaches everywhere and whose intelligence resides in everything. Occasionally, he would come to the earthly world by taking on a human temple and content himself with his life by following the right path.'[30] In this, Kang seemed to have been influenced by both the Buddhist concept of reincarnation and the Christian metaphor of the 'Word becoming flesh'.

Having thus sought to establish the station of Confucius as a divine teacher, Kang proceeded to elucidate Confucius' teachings in their totality by turning to the *Mean*. He believed that this canonical classic had revealed the Sage's doctrinal discourse (*jiaolun* 教論) more profoundly than any other Confucian text; however, its import had, unfortunately, not been properly and adequately comprehended.[31] Kang made a good point in highlighting the doctrinal centrality of the *Mean* in the Confucian tradition. Composed in fewer than 2,000 characters, the treatise contains a masterly discussion of what we would

28 Kang, 'Mengzi wei', 384–484.

29 Kang, 483.

30 Kang Youwei 康有为, 'Lunyu zhu 论语注 [The Analects Annotated (1902)]', in KYWQJ, vol. 6, 2007, 450.

31 Kang, 'Zhongyong zhu', 369.

describe as the spiritual or religious dimension of Confucianism, or as the contemporary New Confucianist Wei-ming Tu termed it, the 'Confucian transcendence'.[32] The *Mean* begins with three definitional statements on three significant notions in Confucian philosophy, *xing* 性 (nature or human nature), *dao* 道 (path or way) and *jiao* 教 (teaching or instruction):

> What Heaven imparts to man is called human nature. To follow human nature is called the Way. Cultivating the Way is called teaching [jiao].[33]

In his commentary, Kang explained that human beings were born of Heaven in the sense that their nature, the spiritual essence of human life, originated in Heaven rather than with their parents. This human nature was shared by all human beings, and those who were enlightened ahead of others and therefore were more advanced at cultivating the Way would instruct the rest of them.[34] For Kang, a proper *jiao* or Teaching (or a religion, according to his own understanding of the term) was 'rooted in Heaven but designed for humanity' (*yuanyutian er yiyuren* 原于天而宜于人), and the Teaching of Confucius had perfectly manifested these two qualities. It was rooted in Heaven because its very foundation as well as object, human nature, was imparted by Heaven. It was designed for humanity because it instructed people by means of, to use Kang' own term, *rendao* 人道, literally, the human way or the way of being human (I will discuss this term in detail later).[35] Kang summarized the 'great, comprehensive, heavenly' Way of Confucius as follows:

> It has its root in the decree of Heaven (*tianming*) and, unfolded, finds its realization in the way of being human (*rendao*). It is based on the virtue of utmost sincerity, and, expanded, achieves the instruction and transformation of the people. It reveals the subtleties of ghosts and spirits (*guishen*) and all creatures, and makes manifest the evolution of the

32 Sze-kar Wan, 'The Viability of Confucian Transcendence: Grappling with Tu Weiming's Interpretation of the Zhongyong', *Dao: A Journal of Comparative Philosophy* 7, no. 4 (2008): 407–28.

33 Here I have adopted Wei-ming Tu's translation. Wei-ming Tu, *Centrality and Commonality: An Essay on Confucian Religiousness* (New York: State University of New York Press, 1989), 5. Cf. Legge's rendition: What Heaven has conferred is called the Nature; an accordance with this nature is called the Path of duty; the regulation of this path is called Instruction. Legge, *Chinese Classics I*, 383.

34 Kang, 'Zhongyong zhu', 369.

35 Kang, 370.

> Three Ages .... In its basic terms, it dwells on human relations and conduct, and the practices of government; in its subtle forms, it even transcends the attributes of Heaven that have no sound or smell...[36]

Kang stressed that Confucius did not leave spiritual beings out in his subtle teachings. Confucius was quoted in the *Mean* as discussing the 'abundant display' of the powers of ghosts and spirits (*guishen*): 'We look for them, but do not see them; we listen to them, but do not hear them; yet they enter into all things, and there is nothing without them.'[37] To gloss this passage, Kang entered a detailed discussion on the two notions of *gui* 鬼 (ghosts, demons) and *shen* 神 (spirits, deities), drawing on classical quotations while also interpolating his own interpretations. As is well known, there were diverse and often confusing accounts of the two notions in Chinese traditional thought and folk religion.[38] One version held that a person had two types of soul, the *hun* 魂 (the intelligent or spiritual soul) and the *po* 魄 (the corporeal or animal soul). After the physical death, the *hun* would go up to heaven and was called *shen* (spirits), while the *po* would return to the earth and was called *gui* (ghosts).[39] This account was close to the Confucianist idea expressed in certain Confucian Texts such as the *Rites,* with which Kang aligned himself.

Kang then went into a discussion of how various religions approach the popular belief in ghosts and spirits. 'The Buddha talked exclusively about ghosts,' he commented, 'and Jesus focused his discussions on spirits.'[40] Some might say, in this context, that Kang had grossly over-simplified the teachings of the Christ and the Buddha. However, according to Kang, Confucius 'taught about both ghosts and spirits and recognized their virtue' and regarded the 'combination of both ghosts and spirits' to be the most effective way for moral instruction of the people.[41] Here Kang was referring to a discussion in the *Rites* on how, for the teaching of the common people, rules were constituted making use of their belief in ghosts and spirits, and how temples and shrines were built for ghosts and spirits, with sacrifice ceremonies designed, in order to

36 Kang, 369.

37 *Zhongyong* 16, in Legge, *Chinese Classics I*, 397–98.

38 Stevan Harrell, 'The Concept of Soul in Chinese Folk Religion', *The Journal of Asian Studies* 38, no. 3 (May 1979): 519.

39 Farzeen Baldrian-Hussein, 'Hun and Po', in *The Encyclopedia of Taoism*, ed. Fabrizio Pregadio (Routledge, 2008).

40 Kang, 'Zhongyong zhu', 376.

41 Kang, 376.

strengthen the people in their observance of those rules.[42] This argument was made to support his repeatedly asserted view that the doctrine of Confucius was more comprehensive and more easily within the capabilities of ordinary human attainment than the Buddhist and Christian doctrines of, for instance, celibacy. Kang did not, however, lose sight of the fact that the later Neo-Confucianists were not fond of talking about spirits. He observed with regret that the discourses of bygone times on ghosts and spirits had sharply declined during the transition from classical Confucianism to Neo-Confucianism. He held the Neo-Confucianist masters accountable, noting that they had obtained their inspiration from the fourth-century Ruan Zhan's 阮瞻 treatise, *On the Non-existence of the Spirits* (*Wuguilun* 無鬼論). 'While Confucius stressed the widely-extended virtue of the spirits,' he wrote, 'the Cheng brothers and Zhu Xi attributed them to the energetic operations of heaven and earth, and Zhang Zai 張載 (1020–1077) consigned them to the two breaths of nature.'[43] Acknowledging that the more intelligent people became, the fewer spirits they would believe in, Kang nevertheless asserted that the existence of the spirits was irrefutable.[44]

Not content with repeating the classical view, Kang ventured into a discourse of his own conceptualisation on the nature of the human soul, that is, *hun* or the spiritual soul. He asserted that unlike the *po* that came from one's parents, the human soul came from Heaven, to which it would also return. In this connection, he explicitly equated the human soul with the Heaven-endowed human nature. He stressed that in essence the 'brilliant, pure, indivisible' soul was none other than the 'illustrious virtue' (*mingde* 明德), the 'virtuous nature' (*dexing* 德性) and the 'intelligent spirit' (*shenming* 神明) that were mentioned in various Confucian Classics; and he related the term to the Buddhist concept of 'hidden nature to be manifested' (*jianxing* 見性).[45] Kang explained that this human (spiritual) soul and the animal soul, which both resided in a human being, were locked in a constant struggle. The former was inclined towards manifesting such virtues as benevolence and righteousness,

---

42 Kang, 376. Cf. Huang Kan 黄侃, ed., 'Liji 禮記 [The Book of Rites]', in *Baiwen shisanjing 白文十三經* [*The Thirteen Classics: An Unannotated Edition*] (Shanghai: Shanghai guji chubanshe, 1983), 170; James Legge, trans., *The Texts of Confucianism IV: The Li Ki, XI-XLVI* (Oxford: The Clarendon Press, 1885), 221.

43 Kang, 'Zhongyong zhu', 376. Indeed, it was not until the time of the comparatively modern Song scholars that *gui* and *shen* were given a semi-physical, semi metaphysical signification. See a discussion in James Legge, trans., *The Texts of Confucianism II: The Yi King*, 2nd ed., The Sacred Books of the East, XVI (Oxford: Clarendon Press, 1899), 34.

44 Kang, 'Zhongyong zhu', 376.

45 Kang, 'Mengzi wei', 423, 438.

while the latter tended to be vulgar and greedy by nature. If a man's spiritual soul could subdue his animal soul, then he was a superior man (*junzi* 君子). If the former was obscured and suppressed by the latter, then he was a mean man (*xiaoren* 小人).[46] Eventually, the spiritual soul could get ensnared in worldly desires, losing its Heaven-endowed nature, and could even remain so forever.[47] Therefore, Kang reasoned, that it was incumbent upon a man to protect his Heaven-endowed nature from the temptations and corruptions of his animal soul. He should further nourish it in such a way as to expand its powers and make it become 'brilliant, illustrious, strong and mighty' (*guang ming gang da* 光明剛大), even capable of assisting the transforming and nourishing powers of Heaven and Earth. That, Kang commented, was what Mencius meant by the statement, 'Knowing his nature, he knows Heaven.... To nourish one's nature, is the way to serve Heaven.' Kang added that this was precisely the way to serve *Shangdi* (God).[48] Kang's view in this regard is comparable to that of the later 'New Confucianists' (*Xinrujia* 新儒家), who claimed that the religious character of Confucianism was expressed through 'ultimate self-transformation' or the infinite potential of each human being for self-transcendence.[49]

The spiritual soul of a man was immortal, Kang stressed. It would not change or perish with the death and decomposition of the body. In particular, the well-nourished soul, meaning, that of a superior man, was refined and strengthened by benevolence and intelligence accumulated in this life and would therefore undergo a myriad of changes without being even slightly affected. For such a soul, this life was no more than a transient phase, just like a reflection of the sun in a mirror.[50] To Kang, the immortal soul, after being disembodied, would wander everywhere before becoming incarnate again in this world – that was why it was described as 'wandering and returning soul'. This accounted for the doctrine of retribution, a deep-rooted notion in Chinese tradition, which Kang upheld. The goodness done or the evil perpetrated by one in his previous life,' Kang wrote, 'will be proportionately repaid in reward or punishment in this life.' Therefore, a person should be extremely careful about his deeds and actions in this life, for they would determine his well-being in his next incarnate life, even though ages of time may separate his two lives.[51] Again Kang identified this as yet another subtle teaching of Confucius, largely forgotten but now being made explicit by him.

---

46 Kang, 430.

47 Kang, 438.

48 Kang, 433.

49 Tu, *Centrality and Commonality*, 94.

50 Kang, 'Mengzi wei', 423.

51 Kang, 434.

In this entire endeavour by Kang to exalt the station of Confucius to that of a metaphysical divinity, and in his striving to restore the Teachings of Confucius to their pristine purity as he divined it, Kang seems to have been driven by what is known as a 'Messianic complex'. We might well characterise Kang's indefatigable endeavours as the expressed mania of a self-appointed Bodhisattva, a modern-day Sage driven by his desire to carry out his own Heaven-ordained mission, essentially of the same nature as that given to Confucius, mandating him, the late Qing reformist, to edify and rescue the people of China and the rest of humanity.

1.2 *Scientific Notions Appropriated*

What is physics, or what is a physicist? Kang reflected on the question by tapping into both ancient Chinese concepts and modern scientific notions. Physics, he declared, simply involved the investigation of 'Heaven-ordained laws' (*tianze* 天則). In his view, the concept of Heaven-ordained laws had been proposed in the *Odes*, which was quoted by Mencius. The verse in question means, literally, that 'Heaven produced human beings; there are things, and there are (their) laws'. Traditionally, however, the 'things' were interpreted narrowly as being what belonged to the human body and human relations. According to Zhu Xi, for example, the word 'things' referred to all societal relationships, such as that between father and son, as well as all man's bodily faculties like the hearing of the ears. The laws, therefore, referred to the function of every faculty and the duty of every relationship that every human being must seek to observe.[52] It was based on this understanding that Legge translated the verse as 'Heaven in producing mankind, gave them their various faculties and relations with their specific laws.'[53] Kang, however, deviated from his predecessors and interpreted the verse in such a way as to include everything in the world:

> In producing each and every thing, Heaven has endowed it with shape, colour, awareness, and nature. Thus each has its own measure, degree and boundary, with prescribed rules, which are called heavenly laws. This applies to everything in the world, be it water and clay filled with *qi*, trees and plants with growth, or birds and beasts with senses. A physicist, therefore, is simply someone that investigates the heavenly laws in things.[54]

52 Zhu, *Sishu jizhu*, 387.

53 *Mengzi* 6(1):8, in Legge, *Chinese Classics II*, 403. The verse was rendered as 'Heaven gave birth to the multitudes of people, so that they had faculties and laws.' in Fung, *Chinese Philosophy*, 1983, 1:34.

54 Kang, 'Mengzi wei', 426.

In this passage Kang put the 'things' in the world into three categories, the mineral, the plant and the animal (including humans). All of them, he affirmed, constituted the objects for study by physicists. The purpose of physicists, therefore, was to discover the Heaven-ordained laws so as to put them to purposeful human use. He gave two examples to illustrate his point, one from traditional Chinese medicine and one from Western science. The plant aconite (wolfsbane) possessed the attribute of being 'heaty', while rhubarb possessed the property of being 'cooling'; once their natural properties were discovered, they were capable of being used for medicinal purposes by Chinese physicians. Similarly, metal could conduct heat and electricity could travel long distances; following the laws of their nature, Westerners had made wires out of the metal and harnessed electricity to transfer sound via wires (i.e. the invention of the telephone).[55] It was not surprising that Kang's interpretation is reminiscent of the concept of 'natural laws' in the West. In the late nineteenth country, missionaries customarily employed it to support their argument for the existence of an infinitely intelligent God as the Creator. But as we will see later, Kang's concept of *Shangdi* (God) differed from that of an external 'Law-Giver' of the natural world.

After discussing plants and metals, Kang's discourse turned to the human realm. He equated the 'Heaven-ordained law' in human reality, which he also called 'Heaven-endowed nature', with the 'virtuous nature' spoken of in the *Mean* and the 'illustrious virtue' in the *Great Learning*. Citing the verse following the one quoted above, in the *Odes*, Kang argued that Man, the most noble of all creatures, had the unique natural capacity to 'love the admirable virtue' (*haoshi yide* 好是懿德). By 'virtue' was primarily meant benevolence (*ren*) and righteousness (*yi*), the former embodying the will of Heaven, and the latter the moral principle of Heaven. For man to love these virtues, he wrote, was like a magnet attracting the iron filings. In a way reminiscent of the approach adopted by Tan Sitong in his *Renxue* and, in all likelihood under Tan's influence, Kang then turned to the concept of the ether and his knowledge of electricity in an attempt to illustrate the underlying mechanism. It was the ether within man that attracted virtue, for virtue contained ether as well. 'That is just like the attraction between two things that carry positive and negative electricity [electrical charges] respectively,' he commented, 'for if they don't contain electricity themselves, they cannot attract each other.'[56]

55 Kang, 426.
56 Kang, 426.

Kang seemed to have regarded the traditional concept of *qi* as a pervasive ether that lent spiritual form to both electricity and lightning.[57] Western knowledge of electricity was introduced into China in the mid-nineteenth century. Electricity and electrotherapy were the subjects of MacGowan's *Bowu tongshu* 博物通書 (Philosophical Almanac, 1851) and Hobson's more influential *Bowu xinbian* 博物新編 (Treatise of Natural Philosophy, 1851), both of which regarded electricity as a key to understanding the human nervous system. For Kang, however, electricity was not merely a subtle material force, but also a mysterious spiritual force capable of generating power. In his appropriation of the scientific notion, he tried to connect the ancient concept of 'spiritual energy' (*shenqi* 神氣) with the modern concept of electricity. He quoted Confucius as saying, in the *Rites*, that Earth contained a spiritual or mysterious energy that produced the wind and thunder-clap (*fengting* 風霆), by which the seeds of forms were carried abroad, and the various things showed the appearance of life. Kang interpreted this 'spiritual energy' as 'electric spirit', arguing that etymologically both the characters for 'spirit' and 'electricity' shared the radical '*shen* 申', which meant 'to extend'. This electric power permeated man and every other creature and connected each and all. 'The invisible human consciousness,' he wrote, 'is like electricity that connects all, and that's why an idea born of one's sincerity may reach others, irrespective of the interval of hundreds of ages or the distance of millions of miles.'[58] 'The concept of benevolence,' he further observed, 'whose pictographic representation literally means two human beings in each other's company, conveys the meaning of mutual attraction; that is love, which is in reality electricity.'[59] Kang used 'ether' in a similar manner. 'The commiserating mind (*buren* 不忍),' he commented, 'is benevolence, is electricity, and is ether.'[60] The notion of a commiserating mind was emphasized by Mencius, who held that 'all men have a mind which cannot bear to see the sufferings of others'. This virtue, according to Mencius, was essential to Man and was the principle of benevolence. When a commiserating mind was widely practised, the emergence of a commiserating government to rule the kingdom would be a matter of course.[61]

---

57 Elman, *Science in China*, 399.

58 Kang, 'Zhongyong zhu', 376, see also 371.

59 Kang, 379.

60 Kang, 'Mengzi wei', 414. The discussion of the 'ether' and 'electricity' was not among the contents of the *Mengzi wei* when the manuscript was first published in 1902; Kang may have added these topics during his later revision of the work.

61 *Mengzi* 2(1):6, in Legge, *Chinese Classics II*, 201–2.

To interpret *ren* or 'benevolence' with 'electricity' and 'ether' was a radical departure from the conventional frame of reference in the Confucian tradition. This has been considered by some scholars to be an 'invention', or a new paradigm, of Kang, on the grounds that although Tan Sitong had used the concept of the ether in his written works before Kang, the former had actually borrowed the idea from the latter, who had been spreading that concept orally in his lectures during the 1890s.[62] This claim is speculative, at best, for no written records can be found to lend support to it. The lecture notes kept by Kang's disciples in the 1890s, for example, did not mention Kang's use of the 'ether'. Unless new evidence emerges to the contrary, it is more plausible that Kang had borrowed the idea from Tan, whose *Renxue* had been widely circulated by the time of Kang's composition of those commentaries. It does not matter, however, whether Kang or Tan had used the term 'ether' first. What is significant is that they were each open to integrating Western scientific notions with Chinese concepts in their reinterpretation of the Confucian tradition, an effort that demonstrated their holistic views on the material and spiritual aspects of the world.

Another term that Kang used in his reinterpretation of the Confucian Classics was 'heat' (*re* 热). Deviating from his earlier view that the world originated from the primal warm *qi* through the interactions of *yin* and *yang*, Kang now claimed that 'heat' was the beginning of everything. 'All things were born of and through heat,' he wrote in his commentary on the *Annals*, 'and in fact the earth was born of the sun.'[63] As heat was from the sun, he reasoned that that was why the *Changes* extolled 'the brilliant sun' (*daming* 大明) as the source of the beginning (of the earthly existence).[64] That was also why, Kang continued, Confucius established the ritual in the *Rites* that the Son of Heaven was supposed to regard the sun as the chief object of veneration (*zhuri* 主日) when offering annual thanksgiving and sacrifice to Heaven on the occurrence of the vernal equinox.[65] Beyond that, worshipping the sun was, in Kang's view, an

62 Wang Fansen 王汎森, *Zhiniu de diyin: yixie lishi sikao fangshi de fansi 执拗的低音 —— 一些历史思考方式的反思* [*An Insistent Low Voice: Some Reflections on the Modes of Historical Thinking*] (Beijing: Sanlian shudian, 2014), 85.

63 Kang, 'Chunqiu kao', 77.

64 Apart from 'the sun', the word '*daming* 大明' is capable of another interpretation, that is, 'to clearly understand', which Legge followed in his translation. See *Zhouyi zhengyi 周易正義* [*The Verified Meaning of the Book of Changes*] (Beijing: Peking University Press, 1999), 9; Legge, *Texts of Confucianism II*, 213 (appendix 1:1).

65 See Huang, 'Liji', 94. Legge's reconsidered translation of the verse under question is: '... the sun was considered (on the occasion) as the residence of (the spirit) of Heaven.' Legge, *Texts of Confucianism IV*, 427.

ancient institution that many countries observed in antiquity. A case in point was Zoroastrianism, whose followers in Persia still continued the practice of worshipping the sun.[66] To Kang, there was a fundamental agreement between the natural phenomenon of the sun being the source of earthly life and the religious practices of worshipping the sun in China and other countries.

Kang also made use of his scientific knowledge in his account of the principle of 'retribution' in relation to the recurring nature of the immortal human soul. 'The rays that reached the earth today,' Kang wrote, 'had actually been issued forth from the sun twelve years before.[67] In like manner, the cause of a certain retribution one has received today might have occurred hundreds of years before.' He added that it was not unusual for the cause to have an effect after so many years, and he offered an analogy in the physical world: 'Just like the travel time of the lightning, which takes quite a while to reach the earth despite its marvellous swiftness.'[68]

Kang's attribution of all the Five Classics to Confucius would sometimes bring into question the authoritativeness or infallibility of the Sage when statements in the classics were found to be in stark contradiction to the established findings of modern science. In such cases, Kang would appeal to what we may describe as the technique of explanatory accommodation. In the *Annals,* for example, there is a recorded instance of an eclipse of the sun, which occasioned the ritual of 'saving the sun' by 'drumming and sacrificing to the God of the earth'. The reasoning offered in the *Gongyang Commentary* was that the sun was eclipsed and offended by the moon, the spirit of the earth, so by performing the ritual, it was intended that the moon would be frightened and persuaded to withdraw.[69] This was in direct contradiction to scientific facts. As Kang acknowledged, 'the eclipse of the sun can easily be predicted and its mechanism is very easy to understand.' It was not, however, that the Sage did not understand it; it was just that he was offering an explanation that would accommodate the low intelligence of the people of ancient times by attributing the eclipse of the sun to the workings of spirits. After all, Kang concluded, to regulate the conduct of the people by referring to the power of the spirits and the need to please or appease them was the ancient Sage's way of moral

66 Kang, 'Chunqiu kao', 77.

67 This is an incorrect statement. It actually takes only 8 minutes for the sun's rays to reach the earth.

68 Kang, 'Mengzi wei', 434.

69 Gongyang Shou 公羊壽, He Xiu 何休, and Xu Yan 徐彥, *Chunqiu Gongyangzhuan zhushu 春秋公羊傳注疏 [The Gongyang Commentary on the Annals with Annotations and Commentaries]* (Beijing: Beijing daxue chubanshe, 2000), 199–200.

instruction during the Age of Disorder, a time of widespread ignorance and simple-mindedness.[70]

1.3 *In the Future World of* Datong

Not long after he completed the reinterpretation of the Confucian Classics, Kang Youwei also finished the writing of the *Datong shu* 大同書 or *Book of the Great Unity,* whose original manuscript he claimed had been written nearly two decades earlier and whose ideas he had shared with some of his students in the 1890s.[71] *Datong* 大同 or Great Unity, a term which has received multiple renditions in English, refers to the ideal state of society depicted in the *Rites*, a concept that was traditionally understood to have denoted the conditions of a 'lost paradise' which had prevailed under the legendary sage-kings who had reigned long before the birth of Confucius.[72] Kang, however, attempted to align the concept with the theory of the Three Ages, and recast it as Confucius's prophetic anticipation of a future ideal commonwealth that was to be realized in the Age of Universal Peace.

The central theme that ran through Kang's *Datong shu* was his foundational conviction that the goal of human life was to attain happiness through the removal of suffering. To seek pleasure and avoid suffering, Kang asserted, was the only correct, natural Way for all who were endowed with life, and even those who underwent sufferings out of their own volitional choice were likewise only seeking pleasure in the long run. The standard by which to judge the merits or demerits of various teachings and institutions, therefore, was the degree to which they were able to assist people in achieving that goal.[73] Many religions had fallen short of the standard, in his view, for some of their practices and doctrine ran contrary to the principle of seeking pleasure and avoiding pain, such as the extreme austerities advocated in the discontinued religion of Mozi or practised by adherents of various sects of Hinduism and Protestantism, celibacy in Catholicism and Buddhism, and the domination

70 Kang, 'Chunqiu kao', 85.

71 Kang, 'Kang Nanhai shiji', 136. Except the first two parts which were printed in 1913, Kang did not have the work printed during his lifetime, for he considered the time was not right. The entire book was published in 1935, eight years after his death.

72 For a discussion on the concept and its English translations, see Laurence G. Thompson, *Ta T'ung Shu: The One-World Philosophy of K'ang Yu-Wei* (London: George Allen & Unwin Ltd, 1958), 27–30; see also Bart Dessein, 'Yearning for the Lost Paradise: The "Great Unity" (Datong) and Its Philosophical Interpretations', *Azijske Studije* 5, no. 1 (2017): 83–102.

73 Kang Youwei 康有为, *Datong shu* 大同書 [*The Book of the Great Unity*] (Shanghai: Zhonghua shuju, 1935), 9.

FIGURE 3.1
Datong 大同 in Kang's Calligraphy, reproduced by the author from an Oxford Bodleian copy of Kang's *Datong shu* (Shanghai, 1935).

of women by men in Judaism and Islam.[74] It was Confucius, the Sagely King with divine intelligence, who, in the Age of Disorder, set up the progressive doctrine of the Three Ages in order to guide humanity to develop and advance towards the Age of Approaching Peace and finally enter into the Age of Universal Peace. Now, it was his duty, Kang was internally convinced, to carry forward the mission of Confucius by elaborating on the Way of the Great Unity for the Age of Universal Peace.[75]

74 Kang, 10.

75 Kang, 11. This may remind a reader versed in Western philosophy of Epicureanism or John Stuart Mill's Utilitarianism. Like Epicurus, Kang defined absence of all pain and anxiety as pleasure, but unlike Epicurus, he gave considerable attention to the satisfaction of bodily desires, and to the improvement of the happiness of others as well as oneself. In the latter aspect Kang was perhaps more like the social reformer Mill, whose primary goal was to seek the happiness of all. For a concise discussion on Epicureanism and Mill's theories, see Edward Craig, *Philosophy: A Very Short Introduction* (Oxford: Oxford University Press, 2002), 47–48.

In this ideal world, as conceptualized by Kang, happiness was maximized, and suffering was minimized. Kang accepted the Buddhist idea that life was basically characterized by various kinds of suffering, but he rejected the Buddhist solution, that is, the extinction of desires. As discussed in previous sections, Kang always held that what he conceived of as perfectly natural human desires should be acknowledged and allowed to be gratified in a seemly manner according to the teachings of *rendao*. In his analysis, the cause of suffering was from various 'boundaries' and inequalities, such as those between nations, classes, races, sexes, families, occupations, and governments. So, the way to avoid suffering and obtain happiness was to abolish these boundaries and establish equality. While his proposals to dissolve families and to deinstitutionalize marriage were considered too radical and even ridiculous by many of his readers, his scheme to abolish the barriers of nation states and form a united world government supported by a universal parliament seemed to be practically pertinent to the needs and exigencies of his times; his ideas anticipated and predated the call for the creation of the League of Nations (more on this in Chapter Five). The world of the Great Unity, therefore, would be one of peace, equality and high moral attainments. No place would be allowed, Kang stressed, for the cruel Darwinian principle of 'survival of the fittest', which only held true in the biological realm of the jungle, definitely not in the society of the Great Unity, where cooperation and collaboration were exalted above brutal competition.[76]

Kang's future world of the Great Unity could not be realized without a high degree of scientific and technological development either, for he believed that the increase of human happiness and decrease of suffering would largely depend on material inventions. The true purpose of material inventions, he argued, was to add to the sum of human happiness. Life in the world of the Great Unity should be very pleasant: much of unpleasant toil and tedious labour would be carried out by machines; travelling would be made easy by technologically advanced vehicles, including 'flying rooms and ships' propelled by electricity or a new fuel; public housing (there would be no private homes any more) would be equipped with mechanical devices or robots; 'smart beds' would be made so that romantic music would play automatically when people were making love on them.[77] Kang also predicted that sophisticated, advanced devices would be made to monitor the deeds of people, so that they would not dare to cheat or commit wrongdoings in private, just as if they were being

76 Kang, *Datong shu*, 357; for a concise discussion of the one-world philosophy of the book, see Thompson, *Ta T'ung Shu*, 37–55.

77 Kang, *Datong shu*, 441–47.

observed by the invisible spirits.[78] In the world of the Great Unity, honours and rewards would be offered to encourage people to pursue knowledge and strive to make new inventions, new theories and new techniques. 'In this way human life on earth will advance daily and there is no need to fear retrogression.'[79]

Applying the protean theory of the Three Ages to the analysis of religion, Kang predicted that the development of religion would undergo three stages as well, polytheism, monotheism and atheism.[80] He further divided the process of the unification of the nations into one world of the Great Unity into three successive stages, with a table showing the differences in the degrees of progression between and among them. In the first stage, when the first foundations of the Great Unity were laid, the adoption of a religion would be left to the decision of each state; people would revere Heaven, but they would revere the spirits still more. During the second stage, when the Great Unity was gradually coming into being, states would deliberate on the refinement of religious doctrines, and a new religion would be created through the process of combining the merits of the teachings of all the Sages; Heaven was to be worshipped, while reverence for various spirits would decline and disappear. During the third stage, when the Great Unity was achieved, no particular religion was to be upheld, but the most meritorious precepts of all religions would be selected and upheld. Even Heaven would not then be worshipped, and instead, the spirit of every man (i.e. the human soul) would be revered, as all the people would be recognized as being denizens of the heavens, all endowed with absolute equality.[81]

Kang further predicted that most of the existing religions would become extinct in the world of the Great Unity. One reason was that everyone would, as mentioned above, be equal, and no founder of religion was to be revered. It would be forbidden for a religious leader to gather a multitude to follow him, for the idolizing of an individual would destroy the principle of equality and cause people to regress to the Age of Disorder. Moreover, the increase of human knowledge would make it difficult to idolize a religious leader or indeed any leader.[82] The other reason was, Kang argued, that by the time of the arrival of the Age of Universal Peace, Confucianism, Christianity and Islam would have served their purposes and would no longer be needed in the new intellectual

78 Kang, 448.

79 Kang, 409, 414.

80 See the discussion of Huan-Chang Chen, *The Economic Principles of Confucius and His School* (New York: Longmans, Green & Co., 1911), 16–17.

81 Kang, *Datong shu*, 164; Thompson, *Ta T'ung Shu*, 126–27.

82 Kang, *Datong shu*, 427–28.

and social conditions then prevailing. Confucius' doctrine of the Three Ages would have been completely fulfilled and Confucianism would itself become unnecessary. In the case of Christianity, its four essential doctrines – reverence for Heaven (God), love for our fellow men, repentance of sin, and judgment after death – would become redundant. For people would naturally love one another and would naturally be without sin; their understanding of the natural workings of evolution would take away their reverence for Heaven (God), and their realization of the logical impossibility of limitless numbers of souls waiting in the empty space for the Last Judgment would make them refuse to believe in or behave according to a fear of a Doomsday. Likewise, in Kang's view, the excessive emphasis placed within Islam on the worship of Heaven (Allah) as well as such practices as the subjugation of women would have lost their appeal. 'The shore having been reached,' Kang explained by means of a Chinese idiom, 'the raft may be discarded.'[83]

There would remain certain Taoist and Buddhist practices, for in the perfect and joyful society of the Great Unity, men would have desire for none other than longevity and the life beyond. The study of the art of physical immortality (*xianxue* 仙學) could still be regarded as an extension of the law of the Great Unity, while the study of Buddhahood (*foxue* 佛學), was more profound and went beyond the world of the Great Unity. That would not be the end of it, for after the studies of immortality and Buddhahood would come 'the study of celestial peregrination' (*tianyou zihxue* 天游之學), on which Kang said he had written another book, but he stopped short of specifying it. The doctrine of celestial peregrination would become a major subject of Kang's talks and writings in his later years, which I will discuss in the final chapter.[84]

2 Liang's 'Change of Heart'

During his retreat at Darjeeling, Kang received disturbing news about the rise of different opinions among the ranks of his party, and, in particular, about the views of Liang Qichao in Japan. For one thing, Liang had lost faith in reforming the Qing court by urging it towards a constitutional monarchy, and had begun to approve of the call for a revolution to overthrow Manchu rule and to establish a republic, very much in line with the thought of rival revolutionaries led by Sun Yat-sen. Greatly alarmed, Kang wrote long letters to Liang and his other followers, explaining why a reformed constitutional government would serve

83 Kang, 452; Thompson, *Ta T'ung Shu*, 275.
84 Kang, *Datong shu*, 451–53; Thompson, *Ta T'ung Shu*, 275–76.

the best interests of China by maintaining its unity and peace. Drawing on the Confucian visionary theory of the Three Ages, he argued that China, which he considered was passing through the Age of Approaching Peace, could not directly leap in one bound onto the stage of a republican government, which was the appropriate institution for the Age of Universal Peace. As a result of Kang's efforts, Liang was convinced and returned to his previous position of promoting constitutional monarchy as an intermediate measure.

Liang's mind, however, remained unchanged in another area of emerging difference between him and Kang: the characterization of Confucianism as a religion. Liang's concept of religion had undergone a dramatic change around 1902 (when he turned 30), as a result of the influence of Western liberalism and the Japanese intellectual deliberations on religion. During the first few years of his exile in Japan, Liang, who had quickly learned to read in Japanese, immersed himself in Japanese translations of Western books. The authors he read included such major names in Western philosophy as Bacon, Descartes, Hobbes, Locke, Hume, Spinoza, Leibniz, Wolf, Comte and so on. Attracted to the idea of liberalism, he considered freedom of spirit or thought a most desirable value, the exercise of which required freedom from enslavement to the old learning of China.[85] At the same time, he kept abreast of the discourse on the nature and function of religions among the Japanese academics. He came to see the difference between the Chinese concept of *jiao* and the European concept of religion, expressed in Japanese by a neologism, *zongjiao*.[86]

2.1 *Confucianism Not a Religion*

The manifesto of this change in Liang's conception of religion was stated in his 1902 essay entitled 'To Preserve the Faith is Not the Way to Exalt Confucius'. Its import was that Confucianism was not a religion in the European sense, and that there was, in any case, no purpose to be served in making it into one. Liang remarked in the beginning that he was fighting against his old self, for what he was about to set forth ran counter to what he had proclaimed a few years

85 Liang Qichao 梁启超, 'Jinshi wenming chuzu erdajia zhi xueshuo 近世文明初祖二大家之学说 [On the Thought of Bacon and Descartes, the Two Pioneers of Modern Civilization (1902)]', in *LQCQJ*, vol. 4, 1999, 1030.

86 For the discussions about whether Buddhism and Confucianism were religions or philosophies in late nineteenth century Japan, see G. Clinton Godart, '"Philosophy" or "Religion"? The Confrontation with Foreign Categories in Late Nineteenth Century Japan', *Journal of the History of Ideas* 69, no. 1 (2008): 71–91. For a well-informed study on the Japanese intellectual impact on Liang, see, Zheng Kuangmin 郑匡民, *Liang Qichao qimeng sixiang de dongxue beijing 梁启超启蒙思想的东学背景* [*The Japanese Intellectual Background of the Enlightenment Ideas of Liang Qichao*] (Shanghai: Shanghai shudian, 2003).

before.[87] He defined religion in the European sense, adopting the Japanese neologism *zongjiao* instead of the traditional Chinese term *jiao*:

> By religion (*zongijao*) the Europeans mean superstitious, irrational belief (*mixin*). Its domain lies beyond the body, with the soul as its basis, with practices of worship as its rituals, with leaving this world as its aim, with the Kingdom or nirvāna as its desire, with reward or punishment in the next world as its way. These features are shared by all religions, be they big or small, refined or rudimentary.[88]

Having thus defined religion, Liang stated that religion was no longer desirable in this day, as it hindered freedom of thought and consequently, the progress of human society. Instead of the theory of the Three Ages, Liang now employed Comte's theory of Three Stages in his analysis of the nature and future of religion. 'To profess a religion,' he wrote, 'the most crucial requirements are to have absolute faith in its doctrine and to fight against heresies. The former tolerates no doubt and therefore suffocates the freedom of thought. The latter makes believers attach themselves to their own doctrine to the exclusion of other thoughts.' Therefore, despite its great contribution in the first stage of social progression, religion was no longer instrumental to human progress and its disadvantages might outweigh its advantages at the present, second stage. Liang was convinced that being a product of the first stage, religion would decline in the second stage, a phenomenon that had already been proved by the vicissitudes of Christianity in the West.[89]

According to Liang's new definition of religion, Christianity, Buddhism, Islam and even Daoism were considered religions, but not Confucianism. Liang argued that the teachings of Confucius were primarily concerned with statesmanship, ethics and morality, without such elements as irrational beliefs, practices of worship, and intolerance of doubts and differences of opinion. 'In short,' Liang wrote, 'Confucius was a philosopher, statesman, educator, but not a founder of a religion.' In addition, Liang continued, Confucius did not talk about the spirits and the life after death, and therefore was very different from the founders of religions in the West.[90] Here Liang obviously deviated from the views that Kang had espoused as well as his own earlier statements in the 1890s

---

87 Liang Qichao 梁启超, 'Baojiao feisuoyi zunkong lun 保教非所以尊孔论 [Preserving the Faith Is Not the Way to Exalt Confucius (1902)]', in *LQCQJ*, vol. 3, 1999, 765.

88 Liang, 766.

89 Liang, 766.

90 Liang, 766.

that Confucius had teachings on the supernatural world and that Confucianism had historically been regarded as a religion in history.

To take Confucianism out of the category of religion was now a must for Liang, because his concept of the usefulness of religion as an instrument for social advancement had changed from positive to negative; yet his basic allegiance to Confucianism remained firm. Excluding Confucianism from the category of 'religion' was, in Liang's view, indeed, better for Confucianism, for in his opinion Confucianism was more comprehensive and more open-minded than all other religions properly regarded as such, and because of its tolerance and openness, it could (as a body of principles for social governance and moral development) be further perfected by learning from the merits of the great religious and philosophical systems. There was no need, therefore, to attempt to preserve Confucianism by means of recasting it into a religion like Buddhism and Christianity.[91] Again, this was in stark contradiction to the view espoused and championed by Kang, a position which Liang himself had once wholeheartedly embraced and supported.

The notion that religion as a system of superstitious, irrational belief which was no longer conducive to meeting the urgent requirements and exigencies of the present age, however, represents only one aspect of Liang's revised conception of religion. He could not simply turn a blind eye to the important role religion played in society, be it in history or during his own lifetime. Perhaps to remedy the one-sided critique of religion in his previous essay, he discussed its various functions in a more comprehensive way in a subsequent essay on the advantages and disadvantages of religion and philosophy. 'Everything has two sides,' he started, 'if we only see its disadvantages and losses and therefore dismiss it in its entirety, we will fail to benefit from its advantages and gains.'[92] While he maintained the view that religion, being inclined towards superstitious, blind belief, was an obstacle to truth, he emphasized that people with religious zeal were more skilled at managing the world than philosophers, although the latter excelled the former in terms of their skill in exploring the principles in depth. In the annals of history, he continued, most of the extraordinary people, the heroes and heroines who had made great contribution to their nations, were chiefly motivated by their religious belief. Here Liang gave a long list of such people, including Oliver Cromwell, Joan of Arc, Amerigo Vespucci, George Washington, Abraham Lincoln, Cavour and William Gladstone

91 Liang, 769.

92 Liang Qichao 梁启超, 'Lun zongjiaojia yu zhexuejia zhi duanchang 论宗教家与哲学家之短长得失 [On the Advantages and Disadvantages, and the Gains and Losses of Religious Leaders and Philosophers (1902)]', in *LQCQJ*, vol. 3, 1999, 762.

in the West who were committed to their Christian faith, Yokoi Shōnan and Oshio Heihachiro in Japan, who both drew heavily on Zen Buddhism. In particular, he singled out his teacher Kang and his deceased friend Tan Sitong, whose Buddhist belief, in his view, had been instrumental in fuelling their endeavours to bring about the awakening of their fellow countrymen.[93]

Liang attempted to discuss religion in its relation to philosophy (*zhexue* 哲學, another Japanese neologism). He argued that there were two types of philosophy: materialist and idealist. The former could only produce learning, but the latter was capable of producing great men, as shown by the influence of the doctrines of Spinoza and Hegel on Napoleon, Bismarck and the Russian nihilists. Liang claimed that idealist philosophy was close to religion in its effect and indeed could be placed under the category of religion. He observed that religious thought was more conducive to the management of the world than materialistic philosophy for five reasons. First, religious thought was crucial to the unification of diverse minds. He was at pains to point out that although in the future when human civilization achieved the highest level and everyone was able to conduct himself well, there would be no need for religion, yet at present people did need religion to help maintain unity. Second, religious thought could instil a hope for the future amidst difficulties and suffering, as it taught happiness in the next world for the soul. Third, religious thought was capable of detaching people from their worldly cares and pursuits. Fourth, religious thought could prevent people from transgressions by its teachings which induced fear of punishment for wrongdoing. Lastly, religious thought encouraged people to be fearless in the face of great challenges, a fearlessness born of the belief that there was no death, since after the body died, the soul would exist for ever.[94]

At the same time, Liang argued that despite its benefits to a society, religion was not conducive to the advancement of scholarship and learning. In order for a society to develop healthily, it needed 'both people professing a religion and those who try to destroy it'. Interestingly, Liang called this a form of 'division of labour' and cited the laws of economics for his support, saying that the more labour was divided, the more advanced a society would become. By 'destroying religion' Liang meant to do away with the superstitious, irrational beliefs in religion rather than its moral teachings – morality belonged to the whole society rather than a particular religion; if morality was destroyed in the course of fighting against religion, that would cause great harm to the society. Here, Liang was calling for a twofold approach to religion: retain its moral

93 Liang, 762.

94 Liang, 762–64.

teachings and get rid of its blind beliefs. 'Religion and superstition often keep each other's accompany,' he commented. 'The moment superstition arises, truth becomes obscured, and as superstition continues, human intellect will not develop.' Therefore, he concluded that those in pursuit of scholarship had to fight against the enemy, superstition.[95] Liang's view as expressed in this essay has sometimes been considered by some scholars as the beginning of the discussion on the Conflict Thesis in the discourse on science and religion in the Chinese context.[96]

2.2 *Buddhism a Better Choice*

Liang considered it to be within one's right whether to believe in a religion or to fight against it. Which religion was the best if one wanted to choose one? Liang's answer was that it all depended on one's particular situation. Although religions varied in terms of their efficacy and degrees of perfection, what really mattered was the degree to which one was moved by the respective doctrine rather than its profundity and perfection. Whoever was confirmed in a religion would have the 'utmost devotion' (*zhicheng* 至誠), and this devotion was the very source of great courage and perseverance, and therefore of great achievements. 'Therefore,' Liang concluded, 'the great achievements of each and every extraordinary personage in the world are but a result of that utmost devotion. Only one who understands this truth can proceed to talk about religion.'[97]

Deep in his heart, however, Liang had his own definitive answer: Buddhism was the best choice, at least for the Chinese people. He elaborated on this idea in a follow-up essay entitled 'On the Relationship between Buddhism and Social Governance'. This essay stands out, in my view, among all of his writings as his most comprehensive, in-depth treatise on religion; it reveals Liang's basic views on Buddhism and Christianity, which he considered to be the archetypes of two categories of religion, and in many instances, it also reflects the views of Kang Youwei on the two religions. These views established the railroad tracks for Liang's future attitudes towards religions, from which he might occasionally deviate but would never be derailed.[98]

Liang began the essay by trying to address 'a big question': Could China develop without a religion or belief in a faith? He argued, in a tone different from that in his previous tow essays, that although religion would not be

95 Liang, 764.
96 Zhuo, 'Zhongguo zhishijie'.
97 Liang, 'Lun zongjiaojia', 764.
98 For Japanese influence on Liang's new appreciation of Buddhism, see Noriko, 'Liang Qichao, Late-Qing Buddhism'.

needed in a fully developed civilization, 'religion is a must' for today's world which was still far away from that perfect far-off future stage. Liang disagreed with the opinion that education could be a substitute for religion, an idea which was spreading among some of his associates in Japan. It might be an option in countries where universal education had been achieved and where good conduct had become second nature to the generality of the people. As China had not yet reached that stage of development, there was an 'absolute need' for the discussion of the question of faith and religion for China.[99]

If China must have a faith to sustain and support her development, which religion should the Chinese people put their faith in? In Liang's mind, Confucianism was no longer an option, for he now considered the Confucian tradition more of an ethical teaching than a religion, with its emphasis on practice instead of faith. This did not, however, in any way mean that Confucianism was not good. The teachings of Confucius might be much more effective in the more civilized future than in the present age which, in Liang's mind, still bordered on the barbarian. He also disregarded the option of Christianity, as suggested by those who were 'intoxicated with the Western wine' and thereby became misguided by 'the assumption that the Christian faith is the source of the power of the West'. He listed three reasons. First, historically Christianity had no emotional connections with the Chinese people and was not conducive to guiding the people; second, it was backed by and made use of by European powers and therefore might bring disaster to the nation; third, its doctrine was not so profound and refined as to be able to train the entire humanity. 'If we are to choose a new faith for our nation which had not possessed one,' he asked, 'why not turn to the loftiest and profoundest one?' By that he meant Buddhism.[100]

Liang then elaborated on six 'superior characteristics' of Buddhism. Firstly, Buddhism was an 'intellectual' faith rather than a superstitious one. While other religions stressed faith even without understanding, Buddhism aimed at true understanding through enlightenment and upheld the belief in the believers' attainment to Buddhahood. Second, Buddhism was committed to the well-being of all rather than the personal salvation of oneself. This, according to Liang, was most evidently testified by the mantra of the bodhisattva: 'I will not attain to the Buddhahood until everyone has attained it'.[101] Third, Buddhism was oriented towards this world rather than the other world. Liang refuted the

99 Liang Qichao 梁启超, 'Lun Fojiao yu qunzhi zhi guanxi 论佛教与群治之关系 [On the Relationship between Buddhism and Social Governance (1902)]', in *LQCQJ*, vol. 4, 1999, 906.

100 Liang, 906.

101 Liang, 907.

Neo-Confucian criticism that Buddhism was inclined to quietness and nothingness, arguing that the doctrine of the bodhisattva was to endeavour to turn this world into Heaven on earth by guiding and enlightening the wayward and lost souls.[102] Fourth, the Buddhist teachings on the soul were superior to those in other religions, in particular, Christianity. Buddhism taught that the soul would forever suffer or benefit from what it had done in this life. Knowing that there would be greater happiness or bitterness after death, people would so conduct themselves as to perform good deeds and avoid committing bad actions. Knowing that the soul would never perish, people may find it easier to die for righteousness without fear. This was, Liang declared, the 'applied Buddhism' advocated by Tan Sitong and exemplified by his unflinching courage.[103] Fifth, Buddhism upheld equality rather than differentiation. Other religions commanded the obedience of all to one, while Buddhism aimed at the equality of all. Good as they were, other religions were at most suitable to the Age of Disorder and the Age of Approaching Peace, but Buddhism suited all the three Ages. The last advantage of Buddhism was that it emphasized self-reliance rather than dependence. According to other religions, happiness and woe came from an outside agent, and that was why prayers and worship mattered. As for Buddhism: all-encompassing was the Buddhist doctrine of cause and effect, according to which each and every one was the bearer of the effects of previous causes as well as the creator of new ones. Therefore, we should always be cautious in considering what causes our own actions or deeds might bring about. Liang concluded that these six features were the very reasons why he believed in Buddhism himself. If Buddhism was that good and helpful, one may ask, how come India, its birthplace, failed to remain an independent nation? Liang's answer was very forceful: India's plight was not because of practising Buddhism, but because of failure to practise it.[104]

These are the major themes of Liang's three 1902 treatises on religion. The differences and even contradictions between the first and third are easily noticeable. In his first treatise, Liang was very critical of religions, including Buddhism, arguing that they would all be bound to decline in the future as they were superstitious in nature and not conducive to the progress of the human mind and the development of science, while stressing that Confucianism, which was not a religion but could have the advantages of a religion, would surely prosper on its own right by reason of its own merits as a moral system regulating interpersonal relationships. In his third treatise, however,

102 Liang, 907–8.

103 Liang, 108.

104 Liang, 908–10.

Liang affirmed that China did need a religion, and that religion should be Buddhism. In accounting for these apparent contradictions, Xia Xiaohong, a leading scholar on Liang, has commented that 'they do not mean Liang's views on religions have dramatically changed over the short span of a few months'. It is just that Liang, in his eagerness to convince people of an idea, would 'amass all proofs and arguments in his favour', without due considerations of possible contradictions to what he had written or would write.[105] There is certainly some truth in the comment. However, I would offer this observation: Liang's conception of religion had evolved over those months and had crystallised into a final comprehensive statement of his own credo in his third and final thesis. As Liang himself acknowledged, 'I do not mind criticizing myself of yesterday with myself of today'.[106] In other words, he had, over the years, formed a complex view of religion: he conceived of religious beliefs as being superstitious by their very nature but that religion as such was still needed in today's world which has not yet attained the perfect stage of human civilization; at the same time, he differentiated between religions, arguing that Confucianism was not a religion in the Western sense and that Buddhism was, in his considered opinion, a superior religion that was better suited to the particular requirements of the Chinese people. As mentioned earlier, this constituted his basic understanding of religion, and, on this point, his view remained more or less the same for the rest of his life.

While Kang was very swift and forceful in 'correcting' Liang's political inclination towards revolution, he appeared, for the time being, relatively quiet on his disciple's 'change of heart' about promoting Confucianism as a religion. As far as extant records indicate, Kang did not have direct discussion with Liang on whether Confucianism was a religion or not. The reasons for Kang's reticence are not entirely clear. Perhaps he thought that the proof he had set forth was complete and his arguments well-founded. Perhaps wisdom dictated silence so long as Liang remained 'politically correct'. Perhaps he did not want to enter into a debate with his own rank and file on an issue which he considered of great importance but lesser urgency at that time. Only after the passing of two years, among the records of his travels, did Kang refute Liang's view that Confucianism was not a religion, and explicate on the undesirability of the Japanese neologism *zongijao* as a fitting rendition of 'religion'. In contrast to Liang, Kang's belief in Confucianism as an archetype of religion only increased during his next round of world travels.

105 Xia Xiaohong 夏晓红, *Jueshi yu chuanshi: Liang Qichao de wenxue daolu 觉世与传世：梁启超的文学道路* [*Awakening the People and Passing Down to Posterity: The Literary Path of Liang Qichao*] (Shanghai: Shanghai renmin chubanshe, 1991), 134.

106 Liang, 'Qingdai xueshu', 3100; see also Liang, *Intellectual Trends*, 103.

3 Observations during World Travels

In April 1903, Kang learned about the news of the death of Grand Secretary Ronglu 榮祿, the favoured advisor to the Empress Dowager, whom Kang regarded as his arch opponent and nemesis. Feeling safer, Kang decided to end his retreat in northern India, and he left for a tour of Southeast Asia. He was happy to witness the further development of the branches of his *Baohuanghui* or Association to Save the Emperor. He convened a conference in Hong Kong for leaders of the Association, at which it was decided that renewed efforts would be invested in the diffusion of reformist knowledge among the Chinese people, both at home and abroad. Among other things, two newspapers, the *Shangbao* 商報 (Commercial News) and the *Shibao* 時報 (Eastern News), were to be started in Hong Kong and Shanghai respectively, and arrangements were discussed to reorganize the *Xinmin congbao* 新民叢報 (New People's Miscellany), which was started and run by Liang Qichao in Yokohama, and the book company, known in English as Quong Jee Sue Kouk Ltd, which was established in Shanghai but under the distant supervision of Liang as well.[107]

Having settled the affairs of the *Baohuanghui* and recovered his health, Kang embarked, from May 1904 onward, on an extended tour of Europe, visiting as many as eleven countries, namely, Italy, France, Hungary, Austria, Denmark, Norway, Sweden, Germany, Belgium, Holland and Great Britain. During his journey, he was accompanied by his interpreter Zhou Guoxian 周國賢 and joined by his second daughter Tongbi 同碧, then studying in the United States, and Lo Chong 羅昌, who was then an undergraduate student at Oxford and who would later marry Tongbi. Kang's sojourn in Europe ended in November, when he set sail from Liverpool for Canada. Shortly afterwards, he completed a book consisting of his accounts of and observations from his travels to the eleven European countries, which was published in Shanghai in 1905.[108]

During the next five years, while making his main residence in Vancouver, Canada, he made another four tours to Europe (August 1906 – March 1907, November 1907 – July 1908, March – April 1909 and July – August 1909) in addition to his frequent visits to the United States, Mexico and Brazil. He even purchased a small island near Saltsjobaden in Sweden as his resort home on the European continent. A daring entrepreneur as well, Kang, with the backing of the *Baohuanghui*, had property developments in Mexico, which were

107 Lo, 'Sequel to K'ang's Autobiography', 193–94.

108 Lo, 195–98.

successful until the Mexican revolution that started in 1910.[109] He visited new cities and revisited old ones, covering nearly all the countries in Europe. During some of these travels he was accompanied by his second concubine, Ho Chan-li 何旃理, a young American-born Chinese woman, who served as his English secretary after they got married in 1907. He wrote another series of accounts of his travels, most of which were not published until the 1910s.

Unlike his first visit to Europe in 1899, when his sole purpose was to secure political support to reinstall the Emperor, the aim of these later tours was, according to Kang, to examine the politics, culture and customs of Western countries so as to find the best remedies for the ills of China.[110] Among other things, his extensive travels had further convinced him that a constitutional monarchy was the most suitable form of government for China at its current stage of political development, that the Confucian religion enjoyed doctrinal superiority over other religions, and that China must follow the examples of European countries and seek material development through science and technology.

3.1 *Reflections on Religions*

As his own travel accounts show, Kang often sought to visit religious sites during his travels; he was keen to observe the local religious practices and customs, which in turn stimulated his reflections on religions in general and on Chinese religious practices in particular. This is most evidently manifested in the account of his sojourn in Italy. On 19 June 1904, he arrived in Rome and visited the impressive 'St. Peter's Temple', which he stressed was regarded by Europeans as the largest basilica in the world.[111] Apart from its size and grandeur, St. Peter's appeared unique in Kang's eyes for another reason: the decline of Catholicism and the rise of Protestantism were closely related to its construction, whose financing by means of the corruptive 'indulgences' had incurred the wrathful protestation of and precipitated the calls for reform issued by Martin Luther. On seeing the Chapel of the Confession, Kang remarked that the practice originated in Buddhism, and that confession was in reality the same as what Confucius described as inwardly accusing oneself for one's own perceived faults (*neizisong* 內自訟).[112]

---

109 Lo, 204. For a discussion on Kang's commercial activities in Mexico as well as Canada, see Worden, 'A Chinese Reformer in Exile', 188–225.

110 Kang Youwei 康有为, 'Ouzhou shiyiguo youji xu 欧洲十一国游记序 [Preface to Accounts of My Travels to Eleven European Countries (1904)]', in *KYWQJ*, vol. 7, 2007, 344.

111 Kang, 'Yidali youji', 360.

112 Kang, 362.

Unlike the 'inward confession' advocated by Confucius, Kang wrote, the Catholic practice of having priests hear the sins of penitents was disadvantageous, because it could lead to sexual abuse of women by the priests. The numerous cases of such sex abuses, Kang argued, contributed to the notoriety and consequent decline of Catholicism, very much in the same way that sexual misdeeds by Buddhist monks had debased and harmed Buddhism in Chinese history. Kang believed that the requirement of celibacy among the priesthood was the cause of the scandals in both Catholic and Buddhist settings, for he considered the human sexual urge a natural and legitimate need that should be properly expressed. In a way that brings to mind his earlier distinction between the *yin* and *yang* religions according to their attitudes towards physical desires, Kang opined that religions which suppressed human sexual desire would decline while those that acknowledged them would prosper. In this connection, he quoted three examples in the history of religion. Martin Luther had succeeded in establishing Protestant Christianity because 'he married a nun' himself and denounced celibacy in his teaching; Padmasambhava (蓮花生, literally, Lotus-Born) was able to spread Buddhism to Tibet because he allowed his followers to 'gratify their sexual desires by getting married'; the Japanese monk Shinran founded the flourishing Jōdo Shinshū sect because he 'took a wife and ate meat'.[113] Here Kang gave, in the words of Hsiao, 'a hedonistic interpretation' of the history of religion.[114] This interpretation was rooted in Kang's consequentialist understanding of the goal of human life as being to seek pleasure and happiness through the removal of all kinds of pain and suffering, as have already discussed in one of the previous sections on the central themes of his *Book of the Great Unity*.

Kang's visit to the Basilica of St. Paul outside the Walls was equally impactful and thought-provoking to him. Witnessing how Paul, an indirect disciple of Jesus, was so much revered and loved in the Christian tradition, he lamented the fact that, back to China, the temple of Mencius, a distant yet most illustrious disciple of Confucius, was not even properly maintained. Moreover, he stressed that Confucius himself had come under recent attacks from a small number of 'ignorant and presumptuous people' (*wangren* 妄人).[115] They would pick up certain new theories from Europe and use them to denounce Chinese morality and religion, mistakenly believing the former to be superior to the latter. In particular, Kang singled out and refuted one of their main criticisms – the claim that Confucianism was not a religion. Without specifying names, Kang

113 Kang, 362.

114 Hsiao, *K'ang Yu-Wei*, 30.

115 Kang, 'Yidali youji', 374.

was almost certainly alluding to the view voiced by Liang Qichao two years earlier. 'It is said that a religion (*zongjiao*) must speak of the spirits', he wrote, 'and therefore the teachings of the Buddha, Jesus and Muhammad are considered religions, whereas Confucianism is not, for Confucius did not speak of spirits.' That was a very ignorant and weird view, Kang argued, for it implied that China had been a country 'without religion' before the arrival of those three foreign religions. That, however, Kang said, was far from the truth, as 'the Chinese concept and character of *jiao* was an ancient one, which had been discoursed on in numerous places in ancient Chinese classics and their commentaries'.[116]

Here Kang was struggling between the Chinese concept of *jiao* and the Western term 'religion'. Unlike Liang who had shifted to the Western conception of religion, Kang sought to align the two concepts together. He argued that under these two generic terms could be grouped the various systems of moral teachings in the world, all of which shared the same goal of teaching people to live a moral life. 'The essence of any *jiao* (religion)', he stressed, 'lies in making men shun evil and do good.'[117]

On another occasion, during his 1904 visit to Cambridge University, Kang discussed the concept of religion with the Professor of Chinese Herbert A. Giles (1845–1935),[118] who was about to publish his small volume, *Religions of Ancient China*. According to Kang, Giles was very critical of James Legge's translations of Chinese classics, commenting that they had 'distorted' their original meaning, due to the translator's covert bias in favour of Christianity, and consequently, they had misguided European readers into looking down upon Chinese religion.[119] In his written account of that visit and interview, Kang attempted to define *jiao* and 'religion' in a broad sense. 'The Chinese word *jiao* means emulation (*xiao* 效),' he wrote. 'In other words, one who is enlightened ahead of others (*xianjuezhe* 先覺者) sets forth a doctrine (way), so that others can emulate the enlightened one by following the doctrine (way).' This was, Kang argued, exactly what was 'originally' meant by 'religion' (which

116 Kang, 374.

117 Kang, 374.

118 A British diplomat in China and reputable sinologist, Giles became only the second person to hold the chair of Chinese at Cambridge. Among other things, he was known for the widely used Wade–Giles Chinese romanization system which he had improved based on the work of his predecessor, Thomas Wade. See Herbert A. Giles and Charles Aylmer, 'The Memoirs of H. A. Giles', *East Asian History*, no. 13/14 (1997): 1–90.

119 Kang Youwei 康有为, 'Yingguo Jianbuliezhu daxue huawen zongjiaoxi Zhailushi huijianji 英国监布列住大学华文总教习斋路士会见记 [An Interview with Herbert Allen Giles, Professor of Chinese at Cambridge University of England (1904)]', in *KYWQJ*, vol. 8, 2007, 27.

he transliterated as *lilijin* and spelled out the English letters). It followed that there could be a variety of *jiao* or religions, and they did not have to be limited to those that were focused on doctrines pertaining to spirits. The fact that the concept of religion as being restricted to belief systems that subscribed to a belief in the existence of spirits was, in his view, a result of the limitation imposed by Christianity. Nevertheless, Kang argued in line with the definition in the *Mean*, a proper *jiao* was one that was based on an intimate grasp, by a Sage, of the Heaven-endowed human nature, and in that sense it could be said to have originated in Heaven.[120] In summary, Kang understood a proper religion (*jiao*) to be a set of teachings on cultivating the divinely-bestowed human nature expounded by one who had become enlightened ahead of others and which was passed down to posterity through a community of followers. With this analysis, he had made another attempt to harmonize *jiao* and 'religion', considering them to be different linguistic references to the same universal concept.

In this connection, Kang offered an uncompromising criticism of the Japanese neologism, *zongjiao*, which, as discussed in the previous section, had contemporaneously begun to be used to refer to 'religion' in China. Kang adopted an etymological approach in his argument. The Japanese, being fond of using two-character words, had coined the term *zongjiao* by combining *jiao*, meaning Buddhism, and *zong*, sects of Buddhism, but it was, he said, an awkward linguistic hybrid that did not make much sense. It could only be used to refer to doctrines that were focused on or made use of beliefs in the spirits, but it was not a suitable term for what he believed to be the true concept of 'religion' or *jiao*. Therefore, Kang urged Chinese scholars to disregard the use of *zongjiao* and to continue to employ *jiao*.[121] As I will show in later chapters, Kang would revisit and eventually modify his own position on this point.

3.2 Shendao *and* Rendao

The differences between religions, then, primarily consisted in the means by which they instructed the people. '*Jiao* (religions) vary in their ways of instruction. Some instruct people by means of *shendao* 神道 (the way of the spirits), others by means of *rendao* (the human way), still others by means of both of those ways.'[122] Here Kang resorted to the binary terms, *shendao* and *rendao*, to substantiate his analysis. *Shendao* is an ancient concept that can be traced back to the concept of *shendao shejiao* 神道設教 in the *Changes*: The Sages laid

120 Kang, 34.
121 Kang, 33–34.
122 Kang, 'Yidali youji', 374.

down their instruction for the people by means of the way of the spirits.[123] In its normal usage, *shendao* might mean either or both of these two things: 1) the teaching that spirits exist and they have power to reward virtue and punish evil; and 2) the ways in which spirits are supposed to act, especially in terms of their interaction with human beings.[124] It follows that a *shendao* religion was one that made extensive use the notions of hell and paradise, doctrines on the retributive powers of ghosts and spirits and various practices, so as to make people avoid evil and do good out of fear of punishment and desire for reward.

In contrast, by *rendao* or the human way, Kang meant a set of teachings that was capable of being practised by the whole community of human beings (ordinary human beings) instead of a mere portion of them (like priests, monks or hermits). These teachings nurtured proper human relations through the practice of moral principles and permitted adherents to satisfy human desires in a seemly and moderate manner.[125] These teachings for promoting moral conduct, interpersonal relationship and social harmony, Kang claimed, found their best expression in the Way of Confucius, which was designed for the very purpose of guiding and moderating people in meeting their natural needs on a daily basis, and that was why it was said in the *Mean*, Kang argued, that the Way could not be departed from even for a moment.[126] The Confucian teachings on the Five Relations (husband-wife, father–son, ruler–minister, elder-junior and friend–friend) and the Five Constant Virtues (benevolence, righteousness, ritual propriety, wisdom, and trustworthiness), for example, conformed to the Heaven-endowed human nature and were indispensable to human life. These made the Confucian Way superior to other doctrines that advocated practices, such as celibacy and eating no meat, which, in Kang's view, sought to suppress human nature, and therefore could only be practised by a few.[127]

Kang compared the Sages, the founders of religions, to be superb spiritual Physicians (*yiwang* 醫王), who would exercise their discretion to choose the most appropriate ways for their mode of instruction of the people according to their mental state and development. In ancient times, Kang explained, the people were ignorant, and they believed that everything beyond their clear

123 *Shendao* defies simple translation. Legge rendered it as 'the spirit-way', while Richard Wilhelm adopted 'the divine way'. Legge, *Texts of Confucianism II*, 230; Richard Wilhelm, *The I Ching or Book of Changes*, trans. Cary F. Baynes, 3rd ed. (London and Henley: Routledge & Kegan Paul, 1985), 486.

124 For a concise discussion on the term, see Robert F. Campany, *Strange Writing: Anomaly Accounts in Early Medieval China* (Albany: State University of New York Press, 1996), 149.

125 Kang, 'Zhongyong zhu', 373.

126 Kang, 370.

127 Kang, 369, 373.

perception was in the charge of spiritual beings. So the ancient Sages took advantage of what the people believed and thereby instilled in them a sense of fear. As a result, people would refrain from doing evil because they had something that they feared; similarly, they would turn toward goodness because they had something that they longed for. Hence most of the archaic religions set forth clear-cut doctrines on ghosts and spirits. Kang claimed that the Buddha, Jesus and Muhammad had inherited and retained those ancient doctrines to form such conceptions as the promise of paradise and the threat of hell in order to influence people in their moral choices, and as a result, their teachings had given rise to the growth and development of religious communities of followers and had been successfully promulgated far and wide.[128]

Similarly, many of the extravagant superstitious beliefs and practices in pre-Confucian China had been passed down to the Chinese people living during Confucius' time. Loathing the superstitious nature of the popular worship of various spirits and decrying its excessive influence on the people, Kang wrote, Confucius had undertaken to sweep it away by denouncing those various spirits which demanded worship, and had only preserved the sacrifices to Heaven, Earth, the gods of mountains and rivers, the god of the land and the god of grain. The reason why Confucius had spared and condoned these five sacrifices, Kang explained, was not so much because they were enduring truths, but because of their temporary utility as pragmatic concessions given after taking into account the needs of the princes, the ministries and the general populace of ancient times. As human intelligence progressed to a fuller degree, Kang predicted, those sacrifices (which were still being practised in Kang's lifetime) could and should, eventually, be abolished as well. With this thesis, Kang presented his own progressive outlook on religion, arguing that Confucianism constitute an advanced and superior form of religion:

> Previously, the way of the spirits (*shendao*) was employed to govern the primitive people, but when progress had been made [in human intelligence], the human way (*rendao*) was increasingly employed. This signified cultural advancement. It can be concluded, therefore, that as a founder of a religion Confucius represented an advanced stage in cultural progress.[129]

That was the reason, Kang explained, why Confucius rarely talked about spirits and advised paying due respect to them but kept them at a distance. This,

128 Kang, 'Yidali youji', 374.

129 Kang, 374; my translation, based on the translation in Hsiao, *K'ang Yu-Wei*, 114.

however, should not mislead one into doubting the Sage's faith in Heaven or *Shangdi* (God). Kang asserted that Confucius stressed reverence for the Supreme Deity in such a direct and emphatic manner that it even eclipsed the reverence evinced in the monotheist teachings of Christianity and Islam.[130] With this statement, Kang was trying to make the point that the Teaching of Confucius contained not only the truths that were enshrined in other religions, such as reverence for God and the immortality of the soul (part of *shendao*), but also the merits in its doctrine that were absent in Christianity and Islam, such as the use of the human way (*rendao*) in its moral instruction and the progressive doctrine of the Three Ages. It followed that Confucianism was a superior system of religious thought which was best suited to the needs of both the Chinese people and the people of other lands. Confucianism was, in Kang's view, not only good for the present time but also for the future of the world. This is a theme that Kang would repeat in a number of places among the various accounts of his thoughts during his travels, and as we shall see, in his later writings as well.

3.3 *On Material Reconstruction*

Around the time he completed the records of his travels, Kang wrote a book of a different kind, *Wuzhi jiuguo lun* 物質救國論 (On Saving China Through Material Reconstruction), in which he extolled the function and utility of science and technology. The work consisted of twenty essays, in which he promoted the need for development of commerce, industry and technology to enrich and modernize China. Unlike the other works written by him during the period of his exile, most of which did not get printed until the era of the Republic, this work, *Material Reconstruction,* was published in Shanghai in 1908. Its publication and circulation within China were possible because the book was focused on material development and contained little about political reforms and debates. Even before its publication, its main ideas had been made public through Kang's talks to the Chinese community in North America. On the evening of 16 March 1905, for example, he had given a talk to the members of the Los Angeles branch of the *Baohuanghui,* during which he spoke about the importance of science and technology to China.[131]

He made it clear in the preface to *Material Reconstruction* that during his travels he was awed by the level of material civilization in the countries of Europe, and he was increasingly convinced that China would be destroyed by the foreign powers unless she attained the same level of industrialization as

130 Kang, 'Yidali youji', 375.

131 Lo, 'Sequel to K'ang's Autobiography', 198.

had been achieved in the West. China's spiritual inheritance, in the form of its morality and philosophy, was in no way inferior to that of Western countries. The expedient and urgently needed key to national salvation, in his opinion, lay in 'material learning' (*wuzhi zhixue* 物质之学). By 'material learning' he meant a wide range of things, among which were, primarily, scientific technology and such practical, specialized branches of learning as chemistry, optics, electricity, mechanics, astronomy, geology, arithmetic, botany and biology.[132] Not only did Kang believe, as was clearly shown in his avid acquisition and advocacy of Western learning, that science could generate new knowledge and new understanding, thus contributing to personal development, but he was also convinced that science was essential for the survival and prosperity of a nation, as demonstrated by the examples of Japan, the United States and the European countries.

In Kang's opinion, the true basis of the wealth and strength of Western countries lay in material learning, something that the Chinese had, in his view, failed to grasp despite various earnest attempts in their quest for China's wealth, prosperity and power. From the 1860s onward, Kang wrote, the leaders of the Self-Strengthening Movement had mistakenly assumed that the strength of Western countries resulted from armaments and warships, and had then endeavoured to improve China's arms. After the military defeat of 1894 by Japan, the Chinese came to understand that Europe's and America's strength lay in an educated citizenry which in turn depended on a good school system. As a result, the whole country concentrated on establishing new schools. Since 1898, more and more Chinese had imagined that the foundation of wealth and strength of Europe and America lay in the Westerners' profound philosophy and in their ideas of freedom and revolution. Thus, they discarded China's traditional teaching and learning and embraced those Western concepts, such as freedom and revolution. All these efforts, according to Kang, missed the point, for Western countries had actually become strong because of the development of science and technology while the root of China's current malady and weakness lay solely in its failure to promote science and technology.[133] In today's world, he further added, countries prospered, declined or even collapsed depending upon the level of the development of science and technology they attained.[134] 'Indeed science is the first and foremost means for saving China,' he wrote at the end of the book. 'Even if other things are to be left unattended

---

132 Kang Youwei 康有为, 'Wuzhi jiuguo lun 物质救国论 [On Saving China Through Material Reconstruction (1904)]', in *KYWQJ*, vol. 8, 2007, 67 .

133 Kang, 63.

134 Kang, 63, preface.

to, science should never be neglected.'[135] This emphatic statement on science, in retrospect, brings to mind the similar views of the later New Culture intellectuals, who nevertheless would not give credit to Kang for his pioneering advocacy of science. For this reason, Kang has been acclaimed, and rightly so, by Hsiao as 'an early though unhonoured prophet of a new faith' in science.[136]

While advocating the rescue of China through material reconstruction and simultaneously engaging in business activities himself, Kang never neglected the political domain of his endeavours. He welcomed the Qing Court's 1906 announcement of preparations for constitutional rule. The imperial decision was stimulated by the fact of Russia's defeat by Japan in their 1905 war, which was regarded as a 'victory of constitutionalism over monarchism' (as expressed by the famous scholar-turned-industrialist, Zhang Qian 張謇), and was a response to increasing domestic demands to adopt constitutionalism.[137] Like other constitutionalists, however, Kang was not satisfied with the government's proposed gradual and time-consuming process of preparation; instead he petitioned for the immediate convening of a parliament.[138] Eventually, the constitutionalists, both at home and abroad, joined hands and formed a party, Friends of the Constitution (*Xianyouhui* 憲友會), which was legally registered in China in early 1911. Kang acclaimed this as a milestone in China's political history as well as the culmination of his efforts in this direction starting with the *Qiangxuehui* in 1895.[139] Earlier, in his renewed efforts to promote constitutional monarchy, Kang was greatly saddened by the news of the death, in 1908, of Emperor Guangxu, whom he believed to have been murdered by the Empress Dowager, who, as history records, herself strangely died one day after the Emperor. As the Emperor had died without an heir, one of his nephews, the three-year-old Puyi 溥儀 (1906–1967), was chosen to succeed to the throne, with his father as the regent. Kang promptly pledged his allegiance to Puyi, and proceeded forthwith to request the imperial Court to abrogate the arrest warrant naming him and his followers.[140]

135 Kang, 95.

136 Hsiao, 'Excursion into Science', 397.

137 See John K. Fairbank and Kwang-Ching Liu, eds., *The Cambridge History of China*, vol. 11 (Cambridge: Cambridge University Press, 1980), 140, 388–89.

138 Kang Youwei 康有为, 'Zhonghua Xianzhenghui qiaomin gongshang qingyuanshu 中华宪政会侨民公上请愿书 [A Petition of the Chinese Constitutional Association and Overseas Chinese (1908)]', in *KYWQJ*, vol. 8, 2007, 411.

139 Kang Youwei 康有为, 'Minzhengbu zhun Diguotongyidang zhuce lun 民政部准帝国统一党注册论 [On the Registration of the Imperial Unification Party by the Ministry of the Interior (1911)]', in *KYWQJ*, vol. 9, 2007, 192.

140 Lo, 'Sequel to K'ang's Autobiography', 212–20.

The reign of the child Emperor, however, was short-lived. Increasing anger and despair were provoked by the appointment of a cabinet dominated by members of the Royal Family. To the constitutionalists, the cabinet appointment revealed the Court's half-hearted commitment to real constitutionalism. As for the revolutionists, they took it as a justification for intensifying their call to overthrow the Manchu rulers. These sentiments served as a catalyst for the revolutionary movement and, consequenetly, the rapid fall of the Qing dynasty.[141]

141 For the fall of the Qing dynasty and the Republican revolutionary movement, see Michael Gasster, 'The Republican Revolutionary Movement', in *The Cambridge History of China*, ed. John K. Fairbank and Kwang-Ching Liu, vol. 11 (Cambridge: Cambridge University Press, 1980), 463–534.

CHAPTER 4

A State Religion

Kang Youwei was taken by surprise when news of a revolt at Wuchang (which triggered what came to be known as the Revolution of 1911) reached him in Suma, Japan, where he had recently gone to be with Liang Qichao. Within a period of less than two months after the Wuchang revolt, a majority of Chinese provinces had declared their independence of Manchu rule. To appease the revolutionary movement, the frightened Qing court quickly reshuffled the 'royal cabinet' by appointing the Chinese general Yuan Shikai 袁世凯 (1859–1916) as its prime minister and hurriedly announced the 'Nineteen Major Articles'. According to the articles, a Constitution was to be drafted by the National Assembly, the power of the Emperor would be radically curtailed, a parliament would be elected, and all political exiles were pardoned.[1] Kang and Liang welcomed these changes, which in their view had already turned the polity into a constitutional monarchy, and they therefore urged the revolutionaries to strike a peace deal with the government, so as to avoid the dangers of political chaos and national disunity.[2] The government's policy changes, however, were 'too little, too late' for the revolutionaries, who proceeded to declare the establishment of the Republic of China in Nanjing on 1 January 1912, with Sun Yat-sen as Provisional President. In an effort to avert a civil war, a compromise was struck between Yuan and the revolutionaries, whereby the presidency was handed to Yuan after he had pledged to support the republic and undertaken to secure the abdication of the Qing monarch.[3]

1 A 'Titular Monarchical Republic'

Kang, however, was deeply disappointed at the adoption of a presidential system by the new Republic. It was not that he was against the idea of a republic. On the contrary, he stressed that a popular government was what Confucius

1 Chuzo Ichiko, 'Political and Institutional Reform, 1901–11', in *The Cambridge History of China*, ed. John K. Fairbank and Kwang-Ching Liu, vol. 11 (Cambridge: Cambridge University Press, 1980), 398.

2 Kang Youwei 康有为, 'Yu Li Yuanhong, Huang Xing, Tang Hualong shu 与黎元洪、黄兴、汤化龙书 [A Letter to Li Yuanhong, Huang Xing and Tang Hualong (1911)]', in *KYWQJ*, vol. 9, 2007, 210–17.

3 For more on the birth of the republic, see Ichiko, 'Political and Institutional Reform', 515–34.

 DOI:10.1163/9789004468221_006

had envisioned for a future China and what he himself had expounded in his *Book of the Great Unity*.[4] The problem was, Kang argued, China was not yet ready for a presidential form of republican government, for he reasoned that in line with the doctrine of the Three Ages, the system of government should advance in stages as the society evolved, and constitutional monarchy should come in as an intermediate stage between despotism and democracy. Even if China was to become a republic, there were several other forms of republican rule to choose from. To copy the American version would lead to power struggles for the presidency and result in crippling disunity. In his view, the United States was successful with a presidential democracy because of its unique qualities, especially the fact that its founding members were all Puritans, who 'desired only to save the people' without 'thirsting for positions of power'.[5] He urged his countrymen to learn from the lessons of those countries in which the American model had been tried and had failed. The French adoption, for example, had resulted in a devastating revolution followed by decades of chaos and turmoil, primarily due to, in his view, the absence in France of a Puritan 'moral basis' to check contentious bids for the lucrative presidency. Kang predicted that an immediate Chinese presidential Republic would lead to similar dire consequences.[6]

The best choice for post-revolution China was to become, in Kang's words, a 'titular monarchical republic' (*xujun gonghe* 虚君共和). This was the term he had invented to describe the system of government practised in such countries as Canada and Australia (Kang would elsewhere sometimes use the same term to refer to the British parliamentary system, which he had earlier on called a 'constitutional monarchy'). A titular monarch, Kang argued, would function as a point of loyalty and allegiance for all the peoples of China, thereby warding off potential contentious bids by individuals for the position of President, thus preserving order during the period of transition. Disputes would then be fought out 'verbally and in ink' in Parliament rather than with 'guns and blood on the battle fields'.[7] Kang explained that republican rule represented the third stage of monarchy, succeeding the stages of a despotic monarchy and a constitutional monarchy, and he explained that what he was proposing was in essence a democratic polity with a titular monarch.[8] 'It is a blessing from

4 Kang Youwei 康有为, 'Gonghe zhengti lun 共和政体论 [On the Polity of a Republic (1911)]', in *KYWQJ*, vol. 9, 2007, 241.

5 Kang Youwei 康有为, 'Jiuwang lun 救亡论 [A Treatise on Saving China (1911)]', in *KYWQJ*, vol. 9, 2007, 233; Kang, 'Gonghe zhengti', 242.

6 Kang, 'Gonghe zhengti', 243.

7 Kang, 241, 245, 247.

8 Kang, 247.

Heaven', Kang wrote, 'for post-revolution China to adopt this wondrous system, one that possesses all the benefits of a republic but not the chaos and conflict it often entails.'[9] To occupy the position of titular monarch, Kang personally considered the abdicated Manchu emperor, Puyi, as a natural choice.[10] Aware of the widespread anti-Manchu sentiments, however, Kang had proposed an alternative candidate, Duke Yansheng, who, as a descendent of Confucius, was indisputably qualified for the position. Kang compared this proposed arrangement with the Papacy in Italy, although he noted that the latter was not a hereditary office and the Pope wielded much more power.[11] Kang did not spell out one nicety: having a descendant of Confucius serve as titular monarch would also support his other campaign to make Confucianism a state religion (to be discussed later).

Kang's proposal was enthusiastically supported by Liang, who elaborated on the proposed system – for which he even offered a name in English, 'Public Kingdom' – in his famous treatise on the reconstruction of post-revolution China.[12] Shortly after the revolution started, Kang and Liang circulated their proposal in China through telegrams, letters and emissaries. It was, unsurprisingly, turned down as unpractical and inopportune both by Yuan Shikai, who was seeking to make himself President, and by revolutionary leaders including Zhang Binglin 章炳麟 and Huang Xing 黄興.[13] Liang soon accepted the political realities and let go of the titular monarchy concept. At the invitation of Yuan, Liang returned to China and accepted a position in the government. He played a key role in the merging of several constitutionalist parties into the Democratic Party (*Minzhudang* 民主黨) to contest the first national elections for the formation of the National Assembly. His party, however, was defeated by the Nationalist Party (*Guomindang* 國民黨) which had evolved from the revolutionary *Tongmenghui*, the bitter rival of the *Baohuanghui*.[14]

Unlike Liang, Kang remained adamantly sceptical about the infant presidential Republic, whose poor performance only served to convince him of the potential advantages of his proposed titular monarchical republic. He had not

9 Kang, 248.

10 Kang, 'Yu Li Yuanhong', 211–12; Kang, 'Gonghe zhengti', 249.

11 Kang, 'Yu Li Yuanhong', 211; Kang, 'Jiuwanglun', 237; Kang, 'Gonghe zhengti', 248.

12 Liang Qichao 梁启超, 'Xin Zhongguo jianshe wenti 新中国建设问题 [On the Reconstruction of New China (1911)]', in *LQCQJ*, vol. 8, 1999, 2437.

13 Ding and Zhao, *Nianpu changbian*, 565–78.

14 Alarmed by the electoral victory of the Nationalist Party, Yuan's associates assassinated the party's charismatic leader Song Jiaoren 宋教仁 in early 1913, which triggered a short-lived armed resistance against Yuan led by Sun Yat-sen. After putting down the revolt, Yuan dissolved the National Assembly by expelling its Nationalist members in early 1914.

merely invented the term to appease and appeal to the revolutionaries; he genuinely believed in it. Not only was the system well-tested in a number of countries, but the installation of a titular monarch also carried, in his mind, an elegant spiritual connotation consonant with his championing of Confucianism as a religious system of social governance. To institute a 'sacred and inviolable' yet politically powerless 'monarch' was in a sense like setting up a clay or wooden idol in a temple for people to bow to with all due reverence. The former (the monarch) would avert a contest over the position of Sovereign, in the same way as the latter (the idol) would serve to resolve disputes among the worshippers. Kang emphasized that such a system conformed to the ancient principle of 'teaching people by means of the way of the spirits' (*shendao shejiao*). 'Spirits are between existence and non-existence,' he wrote, 'but they cannot be disposed of.' That was why, he explained, the ancient Sages had used the retributive power of the spirits and deities to edify the common folk and cultivate their morality.[15]

Kang's opponents, however, did not take the trouble to notice the subtle distinctions he tried hard to clarify between a titular monarchical republic and a despotic monarchy. Abhorring the Manchu regime, they branded Kang as an obstinate monarchist, hence an enemy of the new Republic. In his letter to his Minster of Education Cai Yuanpei 蔡元培 (1868–1940), for example, the then Provisional President Sun Yat-sen did not consider Kang a suitable candidate to be invited to join the new government, because of 'his persistent opposition to the Republican ideals'.[16] Among the revolutionaries, there remained a strong hostility towards Kang, and there had even been occasional assassinations of his 'monarchist followers' in China, which eventually required an edict by Sun to forbid those acts of violence.[17] Given those conditions, it is under-

15 Kang, 'Yu Li Yuanhong', 211; Kang, 'Jiuwanglun', 237; Kang Youwei 康有为, 'Ni Zhonghua minguo xianfa cao'an 拟中华民国宪法草案 [A Proposed Draft Constitution for the Republic of China (1913)]', in *KYWQJ*, vol. 10, 2007, 47. I am aware that Kang's views in this regard bear a degree of similarity to the notion of political theology in the West, but as this book is focused on the Chinese context, and given the constraints of space and time, I have to leave the topic for future research. For further reading, please refer to Carl Schmitt, *Political Theology: Four Chapters on the Concept of Sovereignty* (University of Chicago Press, 2005); David Wong, 'Confucian Political Philosophy', in *The Oxford Handbook of the History of Political Philosophy*, ed. George Klosko (Oxford: Oxford University Press, 2011), 771–88.

16 Sun Zhongshan 孙中山, *Sun Zhongshan quanji 孙中山全集* [*Complete Works of Sun Zhongshan*], vol. 2 (Beijing: Zhonghua shuju, 1982), 19.

17 'Guangdong Chen dudu ji gesheng dudu jian 廣東陳都督及各省都督鑒 [Cable to Governor Chen of Guangdong and Governors of Other Provinces]', *Linshi zhengfu gongbao 臨時政府公報* [*Provisional Government Gazette*], *No. 1*, 29 January 1912.

standable that Kang did not immediately go back to China until late 1913, when he returned to Guangdong to attend his mother's funeral, thereby ending his fifteen-year exile.[18]

2 The Confucian Movement

The successful governance of a country, in Kang's view, depended on two complementary factors: the agency of the spirits in the invisible realm as well as the art of government in the visible world of human affairs, with the former providing a moral foundation for the latter. To nourish and maintain the morality of the populace, ancient Chinese Sages had set up temples for the worship of deities and spirits, and, in like manner, the founders of Buddhism, Christianity and Islam had instructed people in the name of God or a supreme being.[19] In addition, the importance of morality would increase in a society as its system of government evolved. As Montesquieu had pointed out, Kang wrote, an autocratic government lived and ruled by might, a constitutional state by honour, and a republic by virtue.[20] Kang further quoted James Bryce (1838–1922), whom he called 'my friend', as saying that the American system 'would not have succeeded but for the presence of a mass of moral and material influences stronger than any political devices'.[21] Like Bryce, Kang was a firm believer in the religious origin of morality and in the moral basis of society.

Kang considered 'religion' (*jiao* 教) and the state (*zheng* 政) as the 'two wheels' of the chariot of a country.[22] Post-revolutionary China was in great danger, Kang said, because her wheel of politics was undergoing upheavals while the wheel of religion was disintegrating. As stated earlier, Kang had already 'appointed' Liang as his right-hand man to commit himself to the arena of

18 He also had the remains of his brother, executed in Beijing in 1898, brought back and buried beside their mother. In addition, he took the occasion to rebuild his ancestral cemetery which had been destroyed by imperial order as a direct consequence of the failure of the Reform of 1898.

19 Kang, 'Yu Li Yuanhong', 211; Kang, 'Jiuwanglun', 237; Kang, 'Ni xianfa cao'an', 47.

20 Kang Youwei 康有为, 'Zhonghua jiuguo lun 中华救国论 [On Saving China (1912)]', in *KYWQJ*, vol. 9, 2007, 325.

21 Kang, 325; cf. James B. Bryce, *The American Commonwealth* (New York: Macmillan, 1915), 357. The civic virtues required included a sense of reverence and love of and a commitment to the rule of law. Kang's acquaintance with James Bryce may have occurred when the latter served as British ambassador to the United States from 1907 to 1913.

22 Kang, 'Zhonghua jiuguo lun', 316. Similarly, Bryce commented that 'Morality with religion for its sanction has hitherto been the basis of social polity.' Bryce, *The American Commonwealth*, 725–26.

politics. For repairing the other wheel, Kang the self-appointed sage reformer had formulated a clear vision and a remedy for the ills besetting China: to recast Confucianism, set up Confucian churches on a national scale, and reinstate Confucianism as a state religion. He had trumpeted the call during the reform movement of the late 1890s, completed the reinterpretation of the Confucian Canon in the early years of his exile, and designed a blueprint for the proposed religious revival of China. Now he needed a handpicked lieutenant to execute his religious agenda. Kang assigned this role to his young disciple Chen Huanzhang. Chen would go on and become the chief evangelist for Kang's views both in the English-speaking world and in China. It is therefore appropriate, at this juncture, to outline Chen's presentation of Kang's Confucianism.[23]

2.1 *Chen's Presentation*

Chen was born in Guangdong, the province which had also produced Kang and Liang. After securing a good foundation in classical Chinese, he briefly attended Kang's academy in the late 1890s, where he consolidated his knowledge of Confucianism. It is noteworthy that the impulse to worship Confucius had come to Chen's mind in his childhood, long before he heard of Kang's advocacy of Confucius as a founder of a religion.[24] In 1904 Chen obtained the highest metropolitan decree (*jinshi*) before he journeyed to the United States to study economics at Columbia University. In 1911, he obtained a Ph.D. in economics, thereby becoming the first Chinese scholar to have secured 'doctoral' degrees in both China and the West.

Chen's dissertation, *The Economic Principles of Confucius and His School*, offered a detailed account of classical Chinese economic thought in light of Western economic principles. Since its publication in 1911, Chen's work has remained an important reference and has been reprinted repeatedly.[25] Its underlying goal was decidedly of a religious nature: to promote, in his own words, the 'religion founded by Confucius'.[26] That message got through to readers and reviewers. 'No one can read these pages', wrote Henry S. Seager, Profes-

---

23 Gongyang, He, and Xu, *Gongyangzhuan zhushu*.

24 Chen Huanzhang 陈焕章, 'Jiamu Li taifuren...shilue 家母李太夫人...事略 [A Biographical Sketch of My Mother, Mrs. Chen, née Li (1929)]', in *Chen Huanzhang wenlu 陈焕章文录* [*Collected Writings of Chen Huanzhang*] (Changsha: Yuelu shushe, 2015), 419.

25 Chen, *Economic Principles*. It received positive reviews upon its publication, from such prominent scholars as John M. Keynes, 'Review', *The Economic Journal* 22, no. 88 (1912): 584–88. It was also listed as key reference in Max Weber's classic *Confucianism and Taoism*. Since 1970, it has been reprinted or reissued more than 10 times by British and American publishers.

26 Chen Huanzhang 陳煥章, *Kongjiao lun 孔教論* [*On the Confucian Religion*] (Shanghai: Shangwu Press, 1912), 65; Chen, *Economic Principles*, 4.

sor of Political Economy at Columbia University, 'without becoming convinced that Confucianism is a great economic, as well as a great moral and religious, system.'[27] Chen devoted his opening chapters to a lengthy introduction to Confucius and the Confucian school, an introduction which was very much in line with the interpretation propounded by Kang, to whom Chen acknowledged his 'great indebtedness'.[28] Chen's account was the first full summary in the English language of Kang's Confucianism, presented with scholarly moderation and great clarity, occasionally offering his own adaptations to some of Kang's views.

Quite unconventionally, Chen categorized Confucian concepts into two general classes, one based on the 'law of variety', concepts changeable with the needs of the times, the other on the 'law of unity', precepts unchangeable and rooted in human nature.[29] The most significant of the first type was the doctrine of the Three Ages (which he rendered as the 'Principle of the Three Stages'). As for the unchanging law of unity, Chen argued that it had several designations. 'From the religious point of view,' he wrote, 'the unity is called God; from the philosophical point of view, it is called *Yuan*; from the ethical point of view, it is called love.'[30] *Yuan* 元, rendered as 'the great and originating power' by Legge, was represented by *Qian* 乾, the symbol of Heaven in the *Changes*.[31] Chen considered Legge's translation 'obscure', for it failed to make clear that both concepts, *Qian* and *Yuan*, meant 'God'. In addition, he asserted that the all-important Confucian virtue, *ren* (benevolence), referred to nothing else other than 'love'.[32] 'These three names [*Yuan, Qian*, and *ren*] are based on the same principle, because it is a unity,' he went on to explain. 'The unity of the universe is nothing but love. The cement of the universe is called God or *Yuan*, and that of society is called love.'[33] Here Chen had developed upon Kang's views. While Kang sought to establish the unitary nature of the concept of *Shangdi* or 'God', and sometimes struggled over the relationship between 'God/Heaven' and '*Yuan*', Chen went further by unequivocally equating God with the concepts of *Qian* and *Yuan* in classical Chinese thought. He considered *Yuan* to be the philosophical name for God, explaining that Confucius preferred the term *Yuan* to the more personal 'God' because the Sage was writing from the philosophical rather than the religious point of view.[34]

27 Chen, *Economic Principles*, ix.
28 Chen, xii.
29 Chen, 15.
30 Chen, 20–21.
31 *Yi*, Appendix 1:1; Legge, *Texts of Confucianism II*, 213.
32 Chen, *Economic Principles*, 21, note 1.
33 Chen, 20–21.
34 Chen, 58.

Chen continued to expand on the concept of *Yuan* in a theological way, relating it to such themes as creation, the origin of man and the fatherhood of God. Quoting from the *Changes* and the classical Gongyang scholar He Xiu, he stressed that *Yuan* was the 'infinite breath' that brought everything into being, and was the ruling power governing the whole universe. 'This is,' Chen declared, 'the theory of creation in Confucianism.'[35] If *Yuan* was the origin of everything, it followed that humanity, like the heavens and earth, also originated in *Yuan*. Building on the comments by another Gongyang scholar Dong Zhongshu, Chen speculated that man might originally have come from *Yuan* either at the same time with or even before the creation of the heavens.[36] It seems that by 'origin of man' Chen meant a spiritual genesis: the spirit of man was already with *Yuan* before the creation of heaven and earth. Chen considered this Confucian doctrine of the origin of man as a principle of 'the highest theological stage', for it 'made every man free from supernatural power' and dependent only upon his own conscience. In this light, he argued that one may call *Yuan* Father instead of God, and regard all other men and even all the heavens as his brothers.[37] To illustrate this concept, Chen quoted a famous passage from the *Guliang Commentary*, for which he offered an English translation: 'The female alone cannot give birth; the male alone cannot give birth; and God [Heaven] alone cannot give birth. The three must unite together, then there is a birth. Therefore, we may call anyone either the son of his mother, or the son of God [Heaven].'[38] In Chen's view, this might well be regarded as the Confucian doctrine of Trinity, because it required the union of father, mother, and God to give birth to a human being (compare Kang's discussion in Chapter Five). This concept of the Confucian Trinity, it may be observed, is certainly not the same as the Nicene Creed of the Holy Trinity. However, the key point in Chen's reasoning is: the fatherhood of God naturally included the brotherhood of men, and the two constituted the most important principles among Confucius' sociological and moral teachings.[39]

'It is the habit of the Chinese', Chen wrote, 'to seek from [Confucius'] writings sanction or guidance in the determination of important questions.'[40] The

35 Chen, 59. Cf. *Yi*, Appendix 1:1. Legge, *Texts of Confucianism II*, 213.

36 Chen, *Economic Principles*, 59.

37 Chen, 59. This is reminiscent of the statement in the *Mean* that human nature originated in Heaven.

38 Chen, 60. For the original quote, see 'Chunqiu Guliangzhuan 春秋穀梁傳 [The Guliang Commentary on the Annals]', in *Baiwen shisanjing* 白文十三經 [*The Thirteen Classics: An Unannotated Edition*] (Shanghai: Shanghai guji chubanshe, 1983), 18, 3rd year of the reign of Duke Zhuang.

39 Chen, *Economic Principles*, 58.

40 Chen, 37.

Canonical principles of Confucianism were of value not only for scholarly study, Chen argued, but also for their application to practical affairs, and his work on the economic principles of the Confucian school was presented as a testimony to the above claim. The work was a 'scientific' testimony as well: not only was it presented as a study that employed the methods of Western science, but its subject was also the 'science of economics', for which he found a 'scientific term' from the *Changes* – *licai* 理財, a concept likened to the scientific administration of wealth.[41] Hence, his definition of economics as the science of 'administering wealth according to the principle of justice, for the sake of men who live collectively'. Closely related to economics, Chen stressed, were the moral and political sciences, for the ability to formulate rules of economics as to what was right and wrong could only come through a body of moral teachings.[42] From this we can see that Chen used the word 'science' in a broad sense which covered both social and natural sciences. 'All the social sciences relate man,' he wrote. The principles of such sciences as sociology, politics, ethics, religion and justice could all, in his opinion, be found in the 'Confucian Bible'.[43]

With the introduction to his *Economic Principles,* Chen had summarised the cardinal principles espoused by Kang, and presented in a systematic way his own depiction of Confucius as a founder of an advanced religion. In order to do justice to Confucius and promulgate the Sage's teachings, however, words and theories must be realized in the material world. Kang's strategic agenda for saving China called for formal organizations and institutions akin to Christian churches, and in this respect Chen played an instrumental role.

2.2 *The Confucian Religion Association*

Chen had embarked on early efforts to institutionalize his own religious agenda. Emulating Kang's *Shengxuehui,* established in 1897 in Guilin, Chen had set up in his hometown a society of a similar nature, though on a smaller scale, the *Changjiaohui* 昌教會 (Society for Promoting the Religion [of Confucius]) in 1899.[44] Its membership was open to all the villagers, men and women, young and old. He also built a village school; he explained that the *Changjiaohui* served as the root of the school, and the school was the fruit of

41 Chen, 47.

42 Chen, 49.

43 Chen, 49.

44 Although a student of Kang, Chen did not seem to have been affected by the consequence of Kang's failed reform, probably because he was too young and too new a student of his to have gotten involved in Kang's political activities in Beijing in 1898.

the society. 'There is no greater joy', he wrote in a song for the school, 'than to emulate the Sage. Our school aims to promote the Sage's religion and to renew the people.'[45] Later, Chen repeated his practice in New York, where he also set up a *Changjiaohui* in order to promote Confucianism among overseas Chinamen and individual foreigners who may want to investigate Confucianism.[46] He once commented that if translated into English, *Changjiaohui* would be something like 'Confucian Church'. In his introduction to the society, he refuted 'misconceptions' about Confucianism, arguing that China had fallen behind in modern times not because of Confucianism but because of failure to practice true Confucianism, and that Confucius did talk about spirits as well as human beings and his religion was indeed the most perfect on the earth, superior to Buddhism and Christianity.[47] Together with the society in his hometown, the *Changjiaohui* in New York was designated a branch of the national Confucian Religion Association when the latter was established in 1912.

In early 1912, Chen returned to post-revolution China, determined to devote himself to his religious cause. The dramatic changes in the new Republic made Chen and Kang feel the urgent need to revive Confucianism as a religion. Two issues were uppermost in their minds. One was the disturbing news that the Ministry of Education had removed the traditional mandatory requirement that in all state schools the Confucian Classics be studied and rituals of reverence to Confucius (including the ritual practice of *kowtow*) be performed.[48] The other equally disturbing trend was the spread of the idea that Confucianism was not a religion, a denial first voiced by Liang Qichao ten years earlier. This idea was gaining currency in China as the Western conception of 'religion', together with the borrowed Japanese neologism *zongjiao*, became more widely accepted among scholars and the media. The Ministry of Civil Affairs, for example, did not consider Confucianism as a religion in the directives it

---

45 Chen Huanzhang 陈焕章, 'Gaoyao Yanzhou Yingchuan xuetang xiaoge 高要砚洲颖川学堂校歌 [Songs for the School of Yingchuan Village, Xianzhou, Gaoyao (1929)]', in *Chen Huanzhang wenlu 陈焕章文录* [*Collected Writings of Chen Huanzhang*] (Changsha: Yuelu shushe, 2015), 453.

46 Chen Huanzhang 陈焕章, 'Zhi yinbing xuezhang shu 致饮冰学长书 [Letter to Liang Qichao]', in *Chen Huanzhang wenlu 陈焕章文录* [*Collected Writings of Chen Huanzhang*] (Changsha: Yuelu shushe, 2015), 2.

47 Chen Huanzhang 陈焕章, 'Meizhou huaqiao chuangli changjiaohui xuyan 美洲华侨创立昌教会叙言 [Introduction to the Changjiaohui Established Among the Chinese in America]', in *Chen Huanzhang wenlu 陈焕章文录* [*Collected Writings of Chen Huanzhang*] (Changsha: Yuelu shushe, 2015), 363–66.

48 'Putong jiaoyu zanxing banfa 普通教育暫行辦法 [Provisional Regulations for General Education]', *Linshi zhengfu gongbao 臨時政府公報* [*Provisional Government Gazette*], *No. 4*, 1 February 1912.

issued. Those moves, which were welcomed by the more Western-minded or pro-revolutionist scholars, disturbed many Confucianists, who feared that the Confucian tradition, the Faith of China, might be discontinued. Thereupon, a variety of societies emerged throughout the country in order to 'venerate Confucius' and 'defend Confucianism', often backed by provincial officials and regional generals.[49]

Although he remained in Japan, Kang kept abreast of the latest developments in China. To counter those deleterious trends, in a letter to Chen dated 30 July 1912, Kang proposed the 're-establishment' of a *Kongjiaohui*, which literally means the 'Confucian Religion Association' or 'Confucian Church'.[50] His vision was for this national society to establish its branches first in big cities and then in each locality, with Confucian instructors or pastors giving regular sermons to a congregation of men and women, old and young. Kang promised Chen all his support, moral and financial, as well as the requisite human resources. He suggested that Chen simply commence operations after enlisting the support of a couple of venerable Confucian scholars, such as Kang's old friend Shen Zengzhi.[51]

Kang's proposal must have been music to Chen's ears. Chen had actually long cherished the same idea as the one Kang now proposed, and before the arrival of Kang's letter, he had already obtained the support of Shen and of Kang's old foe Liang Dingfen.[52] Chen also publicized the idea of a Confucian Association through his talks. Eventually, the Confucian Religion Association (hereafter referred to as the CRA) was officially established in Shanghai on 7 October 1912 (corresponding to the birthday of Confucius by lunar calendar reckoning). Chen was elected one of its five secretaries, while its presidency was intentionally left vacant (pending the return of Kang). Its mission statement proclaimed a twofold aim: to 'promulgate the Confucian religion' and to 'contribute to the betterment of the society'.[53] It soon secured official recog-

49 These societies included, for example, the *Zongshenghui* 宗聖會 or Worshipping the Sage Association established in north-western Shanxi and the *Kongdaohui* 孔道會 or Confucian Way Association set up in eastern Shandong.

50 Kang Youwei 康有为, 'Yu Chen Huanzhang shu 与陈焕章书 [Letter to Chen Huanzhang (1912)]', in *KYWQJ*, vol. 9, 2007, 337. By 're-establishment' Kang probably meant a continuation to the *Shnegxuehui* he had set up in Guilin, Guangxi in 1897.

51 Kang, 337.

52 Chen Huanzhang 陈焕章, 'Shen Yi'an xiansheng zhuan 沈乙盦先生传 [A Short Biography of Mr. Shen Yi'an]', in *Chen Huanzhang wenlu* 陈焕章文录 [*Collected Writings of Chen Huanzhang*] (Changsha: Yuelu shushe, 2015), 393. Thus Shen, a Metropolitan degree holder who had held a high official position in late Qing, became the most prestigious figure among the initiators of the Association.

53 Chen, *Kongjiao lun*, 98.

nition from the government, which paved the way for its rapid expansion. Its membership was growing fast; it even attracted some foreign members, such as Reginald Johnston (1874–1938), then Commissioner of the British-leased territory of Weihaiwei in northern China. Johnston wrote that he was 'thoroughly in favour of' the objects of the CRA and that he believed it would be a 'terrible misfortune' for the Chinese people to cease holding Confucius in reverence. Meanwhile, he rightly predicted that the Association might expect harsh criticism and opposition from Christian missionaries and their converts.[54]

As a national organization, the CRA declared in its constitution that branches and sub-branches would be set up across China and at overseas Chinatowns. This vision was soon turned into reality in an impressive way. According to the statistics published in its monthly, the *Kongjiaohui zazhi* 孔教會雜誌 edited by Chen, the CRA had had over 140 branches by early 1914. Its Beijing branch was particularly impressive. Set up in early 1913, it was supported by over two hundred Confucian scholars and politicians, including Yan Fu, Liang Qichao and Liao Ping. Given this new development, it had been decided at the first Confucian Congress held in September 1913 that the CRA's headquarters were to move to Beijing. The next year, however, Kang, who had assumed the position of its president, decided to move the headquarters to Qufu, the birthplace of Confucius. He explained that the religion of Confucius was not merely for a particular country, but for the whole world – beyond China, Confucius was also widely revered in Japan, Korea and Vietnam. Therefore, it made more sense for the headquarters of the CRA to be permanently instituted at the Holy Land of Qufu. The more important reason, which remained unspoken, was the worsening political situation in the capital, where President Yuan was preparing to make himself Emperor, a move that Kang and the CRA opposed.[55]

A pressing concern of the CRA was to defend the religious nature of Confucianism, as the opposite view had been gaining ground. In his second introduction to the Association, Kang revisited some of those arguments expressed in the earlier account of his travels: refuting the 'erroneous view' that Confucius should not be regarded as a founder of a religion, continuing to loathe the term

54 Zhuang Shidun 莊世頓 [R. Johnston], 'Zhuang Shidun jun yuanshu 莊世頓君原書 [Letters from Mr. R. Johnston]', *Kongjiaohui zazhi* 孔教會雜誌 [*The Confucian Association Monthly*], no. 3 (1913). Johnston would later serve as tutor and advisor to Puyi, who was allowed to continue living in the Forbidden City after his abdication.

55 For a detailed discussion, see Zhang Songzhi 张颂之, 'Kongjiaohui shimo huikao 孔教会始末汇考 [A Review of the History of the Confucian Religion Association]', *Wenshizhe* 文史哲 [*Journal of Chinese Humanities*], no. 1 (2008): 55–72.

zongjiao, the Japanese 'mistranslation' of 'religion',[56] and maintaining the view that *jiao* and 'religion', if properly understood, were equivalents. He stressed that religions differed not so much in *what* they taught than *how* they taught it. As the people developed to a more civilized stage, the founder of their religion, such as Confucius, would come away from superstitious concepts (*shendao*) and employ primarily the 'human way' (*rendao*) by focusing his teachings on nurturing proper human relations and establishing higher forms of social order through the practice of moral principles. At the same time, Kang claimed that even according to the narrow sense as expressed by the term *zongjiao*, Confucianism was a religion, for Confucius 'revered Heaven and worshipped God' and discoursed on spirits.[57]

Similar arguments were promulgated by Chen in his *Kongjiao lun* 孔教論 (A Treatise on the Confucian Religion), a collection of his two talks and his introduction to the CRA. The talks were given at the American missionary Gilbert Reid's International Institute, and he spoke on the topics 'Confucianism, a Religion' and 'China under the Republic Still in Need of Confucianism'. Highly influential and widely circulated, the book was regarded by some scholars as the 'Manifesto of the Confucian Religion'.[58] Unlike Kang, Chen accepted the neologism of *zongjiao* as an adequate translation of the English word 'religion'. He admitted that most English definitions of 'religion' in general denoted beliefs in the spirits (*shendao*). With this understanding, he set out to defend the view that Confucianism was a *zongjiao* or religion in its very essence. Emphasizing that Confucianists should ask themselves what was meant by religion/*jiao* instead of turning to Westerners or followers of other religions, he went straight to the Confucian definition of *jiao* in the beginning of the *Mean*, which we have discussed in Chapter Three in reference to Tu's considered translation in English: 'What Heaven imparts to man is called human nature. To follow human nature is called the Way. Cultivating the Way is called teaching [jiao].'[59]

Chen asserted that the word 'Heaven (*Tian*)' referred to in the *Mean* meant God (*Shangdi*), and 'cultivating the Way', therefore, was to set right 'words and thoughts' according to what God had imparted, human nature. This represented Man's striving for the divine union 'between man and Heaven' by

56 However, Kang was not always successful in resisting the occasional need to use the term. He would use *zongjiao* interchangeably with jiao in some of his later writings, such as his 1917 treatise on republicanism. See Kang Youwei 康有为, 'Gonghe pingyi 共和平议 [A Dispassionate Treatise on Republicanism (1917)]', in *KYWQJ*, vol. 11, 2007, 33.

57 Kang, 'Kongjiaohui xu (er)', 345–46.

58 Chen, 'Confucianism Encounters Religion', 124.

59 Huang, 'Liji', 196; Tu, *Centrality and Commonality*, 5.

following the will of Heaven. 'Is this not religion?' Chen asked emphatically.[60] Like Kang, Chen was in fact trying to redefine the contemporary concept of religion or *jiao* by drawing on the ancient Confucian Classics or what he would prefer to call the Confucian Bible. He explained that the Chinese word *jiao* contained three senses, namely, religion (*zongjiao*), education (*jiaoyu* 教育) and edification (*jiaohua* 教化). 'Confucianism has all three connotations,' he wrote, 'but it is essentially a religion.' That's why he insisted on using the term 'Confucian Religion' rather than the 'Confucian Way', for the former demanded, and was capable of producing, 'faith' on the part of followers, a condition that in his view was a prerequisite for a true believer's practice of any religion.[61]

Chen then argued that Confucius was a founder of what must rightfully be regarded as a religion, a station that had been accepted by Confucius himself and acknowledged by his disciples and followers, by his contemporaries and the people of later generations, and even by foreigners. In portraying Confucius as a divine figure, a divinely-inspired revealer of religious truth, Chen had to explain away the contradictory remarks by Confucius acknowledging the limits of his own knowledge. Chen said that this was simply a sign of the superlative humility evinced by Confucius. The reason that Confucius had not exalted himself to the same station as Jesus and Muhammad was because the society he lived in was much more civilized than the societies of human beings living during the times of Jesus and Muhammad, which rendered the excessive use of *shendao* or the way of the spirits less necessary.[62] Having asserted the point that Confucius was a founder of a religion, Chen offered arguments intended to show that Confucianism had been properly regarded as a religion in terms of its designated name, lineage, rituals, codes of conduct, temples, Holy Land as well as the Confucian doctrines on the immortality of the soul, the existence of spirits and deities subordinate to God and the principle of divine retribution.[63] Throughout the discussion, Chen did not attempt to distinguish between the Chinese *Shangdi* and the Christian 'God'; instead, he followed Kang in equating the two terms and by articulating a universal concept of Divinity.

2.3 *The State Religion Campaign*

Going beyond simply defending (and, indeed, propagating) the religious nature of Confucianism, the CRA endeavoured to have Confucianism adopted as the state religion of the Republic. In the Provisional Constitution of 1912,

---

60 Chen, *Kongjiao lun*, 2.

61 Chen, 93.

62 Chen, 6–7.

63 Chen, 16–26.

'freedom of religion' was granted to the people, but no 'state religion' had been specified.[64] To many Confucianists, however, that article was positively and definitely detrimental to the interests of Confucianism, because it was often used as an excuse for diminishing the influence of Confucius in the educational system and in contemporary society. A requisite way to remedy the situation, they believed, was to make Confucianism a *guojiao* 國教 or state religion; this they proposed to achieve by having this status recognized and protected in the 'Permanent Constitution' that was about to be drafted.

When the CRA was established, it did not explicitly state in its constitution its intention to advocate for Confucianism to be constitutionally installed as a state religion. However, that intention was implied in Kang's introduction to the Association, in which he noted with regret that the new republican government had failed to make Confucianism the state religion, a status that Confucianism had enjoyed *de facto* for thousands of years.[65] In a talk given at the CRA, Chen likewise restated his earlier view that religious freedom had been practised in traditional China when Confucianism was the *de facto* state religion. He stressed that specifying a state religion and protecting religious freedom were compatible objectives, like the 'two wings of a bird' or 'two wheels of a chariot',[66] the latter of the which was probably a phrase Chen borrowed from Kang. Chen's call was echoed by many other Confucianists. One Confucian scholar, for example, argued that of the four major religions professed by people in China, Confucianism was much better qualified to be nominated as the state religion than Buddhism, Islam and Christianity.[67]

Kang deliberated further on this issue in his *Proposed Draft Constitution,* which was published by instalments in his newly-launched personal magazine, *Buren* 不忍 (meaning 'a commiserating mind'). He stated that freedom of religion should be constitutionally protected, but added that 'Confucianism should be made the state religion' except in Tibet and Mongolia where

64 For an English translation of the 1912 Provisional Constitution, see B. L. Putnam Weale, *The Fight for the Republic in China* (New York: Dodd, Mead and Company, 1917), 401–7.

65 Kang Youwei 康有为, 'Kongjiaohui xu 孔教会序 [Introduction to the Confucian Religion Association (1912)]', in *KYWQJ*, vol. 9, 2007, 341.

66 Chen Huanzhang 陳煥章, 'Mingding yuanyou zhi guojiao bingbu aiyu xinjiao ziyou zhi xinmingci 明定原有之國教並不礙於信教自由之新名詞 [Specifying the Previous State Religion in Clear Terms not Contradicting the Neologism for Religious Freedom]', *Kongjiaohui zazhi* 孔教會雜誌 [*The Confucian Association Monthly*], no. 1 (1913): 1–12.

67 Chen Baozhen 陳寶徵, 'Kongjiao yi dingwei guojiao lun 孔教宜定為國教論 [On Confucianism Being Qualified for Being the State Religion]', *Kongjiaohui zazhi* 孔教會雜誌 [*The Confucian Association Monthly*], no. 1 (1913): 27–30.

'Buddhism can concurrently be regarded as a state religion'.[68] Kang shared Chen's view on the compatibility of adopting a state religion and guaranteeing the principle of religious freedom. He explained that religious freedom was a hard-won principle in Europe, acquired through much bloodshed and many wars – the Crusades against the 'foreign' religion (Islam) followed by the Thirty Years' War (1618–1648) between the Catholics and the Protestants. By contrast, religious freedom had been practised for over two millennia in China, where Daoism, Buddhism, Islam and Christianity were all allowed to spread alongside Confucianism, the *de facto* state religion in times past. There had never been religious wars fought on Chinese soil, a situation that was due to Confucius, who taught the principle of 'transmitting his religion with tolerance'.[69] Unfortunately, Kang lamented, Confucianism was now faced with the danger of being neglected and even disregarded by the Chinese people, so, he argued, it was high time that Confucianism should be enshrined as the state religion. Not only would a state religion not impinge on religious freedom, but it would also have the effect of strengthening a country by serving as a unifying force. Chinese history had proven Confucianism's tolerant and therefore unifying nature. Furthermore, many European countries had specified a state religion in their constitutions while upholding the principle of religious freedom. The same constitutional arrangement could be adopted in China, Kang stressed.[70]

Kang contended that Confucianism was a better candidate for serving as a state religion than any of the other major religions that were practised in China. Beautiful as they were, the Buddhist and Christian religions were not very helpful to the government of a state as they aimed at individual cultivation and redemption through 'revering God and nourishing the soul', while the religion of Confucius was not only more all-embracing in scope but it also paid more attention to the concerns of this world.[71] Some scholars have argued that here Kang had emphasized the lack of 'a concern with the transcendental' as 'a fundamental advantage of Confucianism'.[72] However, Kang would not have agreed that Confucianism was lacking a transcendental aspect. He maintained that Confucianism embraced both 'the human way' (*rendao*) and 'the way of the spirits' (*shendao*), but with a distinctive focus on the former, and that it

68 Kang, 'Ni xianfa cao'an', 81.

69 Kang, 82.

70 Kang cited the practices of many Europeans countries where either Catholicism or Protestantism was exalted above other religions. For example, he emphasized that at Oxford and Cambridge students were required to attend morning service in honour of Jesus Christ, as required by the Church of England.

71 Kang, 'Ni xianfa cao'an', 82.

72 Chen, 'Confucianism Encounters Religion', 147–48.

was that focus, not the presence or absence of a transcendental concern, that distinguished Confucianism from other religions. Kang concluded his discussion by specifying what was implied by making Confucianism a state religion: to continue the traditional rituals of worshipping Confucius by officials; to restore the requirement that the Confucian Classics be studied at all state schools; and for the state to sponsor and protect a Confucian Church.[73] When he made the last proposal, Kang may have had the newly-established CRA in mind.

The calls made by Chen and Kang were duly followed up by the CRA. On 15 August 1913, the Association sent a petition to the newly-instituted bicameral National Assembly,[74] requesting that Confucianism be adopted as the state religion in the proposed Constitution. The petition, which was drafted by Chen and later published in the Association's *Monthly*, set forth several arguments for their position. First, Confucianism had, historically, always served as the basis of morality for the Chinese people. 'The establishment of a state', Chen wrote, 'is based on morality, and morality is regulated by none other than religion.'[75] Chinese morality was rooted in Confucianism, a moral tradition that was inherited from the ancient sage kings. Confucius' teachings on morality were enshrined in the Confucian Bible. Therefore, to sustain Chinese morality required that Confucianism be upheld as a state religion. Second, Confucianism had been the state religion of China for the past two millennia. This was the view that Chen had expressed in his earlier writings, such as in his 1911 *Economic Principles* and his *Kongjiao lun*.[76] To make Confucianism a state religion of the republic, then, was to continue a time-honoured tradition, a tradition that met the expectations of the general populace. The third argument, the principal focus of Chen's petition, was that a state religion would not impinge on religious freedom:

> A state religion and the freedom of religion can go hand in hand and complement each other. If religious freedom is granted but no state religion is instituted, that would give too much freedom to the people and the consequent centrifugal forces of societal change would cause the

73 Kang, 'Ni xianfa cao'an', 82.

74 The Assembly was called together on 8 April 1913 by Provisional President Yuan with a twofold aim: to elect him President of the Republic of China and to make a Constitution to replace the provisional one.

75 Chen Huanzhang 陳煥章, 'Kongjiaohui qingyuanshu 孔教會請願書 [A Petition of the Confucian Religion Association]', *Kongjiaohui zazhi* 孔教會雜誌 [*The Confucian Association Monthly*], no. 6 (1913): 13.

76 See Chen, *Economic Principles*, 42–47.

> nation to go off balance. On the other hand, if a state religion is instituted but religious freedom is not allowed, that would mean too much interference in the spiritual life of the people and the pressure so exerted would destabilize the conditions of peace and tranquillity in a country.[77]

Chen pointed to the common practice in many modern countries simultaneously to specify a state religion and constitutionally protect freedom of religion. To support this, he quoted relevant constitutional provisions from as many as eleven countries. He added that although countries such as the United States, Great Britain, France and Russia did not specify a state religion in their constitutions, they practised either Protestantism, Catholicism or Eastern Orthodox Christianity as their *de facto* state religion. Now that the new Republic had made a provisional Constitution, in which freedom of religion was granted but a state religion was not specified, that situation had caused confusion among the people, who were thereby led into thinking that the principle of religious freedom was meant to get rid of Confucianism. To arrest that negative inference and dispel popular misapprehension, Chen argued, the new Constitution should, 'in explicit terms', adopt Confucianism as the state religion in addition to constitutionally upholding freedom of religion as a modern principle of democracy, so that Chinese morality would be consolidated, and the new Republic could be established upon a solid, lasting foundation.[78]

The petition was co-signed by Xia Zengyou 夏曾佑, Wang Shitong 王式通 as well as Yan Fu and Liang Qichao. Xia, who had been a long-time friend and ally of Kang and Liang, was serving as head of the social education department at the Ministry of Education. Wang was a legal scholar who, like Xia, had also served in President Yuan's administration. It was interesting, however, to see the name of Yan Fu among them. For one thing, he had earlier distanced himself from Kang's call to 'preserve Confucianism' in the late 1890s. For another, he did not consider Confucianism as a 'religion' *per se*, as he explicitly stated as early as in 1899.[79] That was also the view that Liang had voiced in the early 1900s in Japan, questioning the asserted religious character of Confucianism. Indeed, Liang had argued that to specify a state religion would interfere with the desirable principle of religious liberty.[80] Does the fact that both Yan and Liang had now agreed to join Chen in proposing that Confucianism be made a

---

77 Chen, 'Qingyuanshu', 1913, 14.

78 Chen, 21.

79 Yan Fu 严复, *Yan Fu ji 严复集* [*Collected Works of Yan Fu*], ed. Wang Shi 王栻, vol. 4 (Zhonghua shuju, 1986), 850.

80 Liang, 'Baojiao feisuoyi zunkong'.

state religion mean that they had changed their position and now considered Confucianism as a religion? The answer is, arguably, no. Rather, their support for the petition may be viewed as an effort to resist the contemporary tendency to remove Confucius from schools and reduce the influence of Confucianism in society. Although they did not consider Confucianism a religion, both men held Confucius in high esteem and regarded his teachings as being vital for sustaining Chinese morality and maintaining social order in China.

The CRA's petition immediately received popular support across the nation. Its branches and sub-branches promptly dispatched circular telegrams to support the petition and they sent delegates to the capital to present their petitions to the government and the National Assembly. Its overseas branches joined in the campaign by sending telegrams to President Yuan.[81] The petition was also seconded by a large number of prominent individuals, such as Vice President Li Yuanhong 黎元洪 (1864–1928), who was in turn seconded or supported by nearly twenty military and civil governors from a dozen provinces.[82] In addition, several members of the National Assembly also publicly voiced their support for the proposal.[83] For many of the supporters, it should be observed, their emphasis was not to seek to assert the religious nature of Confucianism, but to acknowledge the need to buttress the role of Confucianism in sustaining Chinese morality and social unity.

At the same time, voices of opposition were mounting, especially from those with a Christian background. Realizing that the movement to install Confucianism as a state religion, if successful, would give an advantage to Confucianism and possibly hold in check the growth of Christianity in China, many Christians immediately initiated a 'counter-campaign'. They made 'every possible effort' by means of public meetings and newspaper articles to protest against the proposition. To coordinate their efforts, a committee of seven leaders was set up. Among other things, they also sent a petition to the National Assembly asking for the rejection of the state religion proposal.[84]

On 27 September, corresponding to the birthday of Confucius in the Chinese lunar calendar, a proposal based on the petition of the CRA was brought up on the agenda for deliberation by the 'Constitution-drafting Committee',

81 For a collection of these telegrams, see ZDLDG 中国第二历史档案馆 [China Second Historical Archives], ed., *Zhonghua minguoshi dang'an ziliao huibian* [*ZMDZH*] *3: wenhua* 中华民国史档案资料汇编 *3:* 文化 [*Compilation of Archival Materials Pertaining to the History of the Republic of China 3: Culture*] (Nanjing: Jiangsu guji chubanshe, 1991), 51–52.

82 Chen, 'Confucianism Encounters Religion', 161.

83 See reports in *Kongjiaohui zazhi* 孔教會雜誌 [*The Confucian Association Monthly*], vol 1, no. 7 (Aug 1913); vol. 1, no. 9, (Oct 1013).

84 'A State Religion for China?', *The North-China Herald*, 30 August 1913.

FIGURE 4.1 Hall of Prayer for Good Harvests (*Qiniandian*), Temple of Heaven, Beijing. Estimated date of the photo: 1900–1910.
IMAGE COURTESY OF DAVID HUTTON-POTTS AND HISTORICAL PHOTOGRAPHS OF CHINA, UNIVERSITY OF BRISTOL (https://www.hpcbristol.net/visual/p001-040).

consisting of 60 members elected from both the Senate and House of Representatives. The committee chose as the venue for their conferences the Hall of Prayer for Good Harvests (*Qiniandian* 祈年殿) at the former imperial Temple of Heaven (*Tiantan*). In the magnificent, solemn circular hall, which had been considered as the most sacred spot in the Ming and Qing dynasties, and in whose centre was placed the tablet of *Huangtian Shangdi* (August Heaven the Supreme Ruler), the Chairman of the committee, Tang Yi 湯漪, was inspired to urge his fellow members to draft an enduring Constitution, 'as solid as' the circular triple-gabled building, for the new Republic.[85]

The state religion proposal was heatedly discussed, but no agreement was reached. The debate continued in a meeting held on 13 October, when a vote was eventually taken. According to the minutes, six members spoke for the proposal, while seven stated their opposition to it. As a result, the proposal failed

85 Wu Zongci 吳宗慈, *Zhonghua minguo xianfashi* 中華民國憲法史 [*A History of the Constitution of the Republic of China*] (Shanghai & Beijing, 1924), 24.

to pass, having received as few as eight votes. As an alternative, the majority of the committee voted for an amendment to be added to Article 19, stating that 'in the education of citizens, the Way [*dao*] of Confucius shall be adopted as the great principle for the regulation of persons'.[86] The Chinese wording, the *dao* of Confucius, deliberately differed from the *jiao* (Religion) of Confucius (*Kongjiao*), as proposed by Kang and Chen. Furthermore, the article avoided the mention of a 'state religion'; it actually limited the role of Confucianism to the realm of education. As expected, the news that the state religion proposal had been rejected by the Constitution-drafting Committee disappointed the numerous supporters of the proposal, who called for a reconsideration of the proposal through telegrams and letters, but all in vain. Before the Constitution drafted by the committee (known as the Temple of Heaven Draft) was submitted to the National Assembly for examination and approval, however, the assembly itself was abruptly dissolved in the beginning of 1914 by President Yuan in his bid to secure more presidential power.

Some hold the view that President Yuan supported the state religion movement and exploited it to advance his scheme to become Emperor. This view was not only circulated among his contemporaries (including some members of the Drafting Committee), but it has also been upheld by a number of later scholars.[87] Admittedly, from the early days of his presidency, Yuan had taken measures to 'venerate Confucius'. In his edict of 22 June 1913, for example, he declared that he would solicit opinions about resuming the abolished ceremonies to offer ritual sacrifice to Heaven and to worship Confucius. He extolled the Sage for his 'republicanism-oriented' doctrines of the Three Ages and the Great Unity in a manner reminiscent of statements made by Kang.[88] Several months later, the 'Political Council', which Yuan had set up in place of the National Assembly, resolved that the traditional rites and activities in Confucian temples be revived across the country. Then, in the Autumn of 1914,

86 In Chinese, 國民教育以孔子之道為修身大本. The English translation is taken from Wei-tung Pan, *The Chinese Constitution: A Study of Forty Years of Constitution-Making in China* (Washington, D.C.: The Institute of Chinese Culture, 1945), 159.

87 See, Wu, *Zhonghua xianfashi*, 38, 108. See also Chow's misleading statement that Yuan became an 'enthusiastic patron of the Confucian movement' in Tse-tsung Chow, 'The Anti-Confucian Movement in Early Republican China', in *The Confucian Persuasion*, ed. Arthur F. Wright (Stanford: Stanford University Press, 1960), 291.

88 *Zhengfu gongbao* 政府公報 [*Government Gazette*], No. 404 (23 June 1913). Yuan's edict was challenged by over thirty members of the National Assembly, who blamed Yuan for 'copying' Kang's ideas and for interfering with the principle of freedom of religion, insisting that the issue of 'worshipping Confucius' should be resolved by the Assembly instead. See also ZDLDG 中国第二历史档案馆 [China Second Historical Archives], *ZMDZH 3: Wenhua*, 3–5.

Yuan personally presided over a grandiose ceremony to worship Confucius at the Confucian Temple in Beijing, an act that officially resumed the traditional biannual sacrifices offered to the Sage, known as *dingji* 丁祭.[89] Because these measures were similar to those suggested by Kang and Chen, it was easy to assume that Yuan supported their state religion campaign. However, as Hsi-yuan Chen has convincingly argued, Yuan was careful to distinguish his 'veneration of Confucius' from promoting Confucianism as a religion, let alone a state religion.[90] That was understandable, for the former was able to help him obtain legitimacy and advance his monarchist scheme, while the latter could induce unwanted protests from those who opposed the state religion proposal and could conceivably lead to a potential power struggle between religion (the Confucian Church) and the machinery of the state (unless he, as President, were to be designated head of the proposed Confucian Religion).

It was also true that the CRA welcomed Yuan's reinstatement of the traditional ceremonies to worship Confucius, but they kept at a distance from him politically, and many of the prominent members of the CRA were against Yuan's monarchical movement and his increasingly evident ambition to declare himself Emperor. For Kang, the attempt by Yuan to restore a despotic monarch differed dramatically from Kang's proposal in favour of a titular monarchy. He and Liang were among the first to oppose the monarchical movement. Their follower and associate, General Cai E 蔡锷, initiated a revolt in South China, followed by a number of other provinces which declared their independence. Sun Yat-sen, who had fled to Japan, also made efforts to overthrow Yuan. Besieged by protests and rebellions, Yuan was forced to change his newly-installed monarchy back to a republic in early 1916.

After Yuan's death from uraemia a few months later, the National Assembly was reconvened in September, at which meeting the resumption of the examination of the Temple of Heaven Draft was placed at the top of its agenda. Chen Huanzhang again sent a petition to the Assembly, urging that Confucianism be made the state religion and that '*Kongdao* [the Way of Confucius]' in the amendment be changed to '*Kongjiao* [the Religion of Confucius]'.[91] Meanwhile, contrary proposals to abandon ceremonies for the worship of Heaven and Confucius were also put forward to the Assembly for deliberation. In

89 See See *Zhengfu gongbao* 政府公報 [*Government Gazette*], No. 499 (24 September 1913), 10; No. 563 (27 November 1913), 1; No. 636 (14 February 1914), 21–22; No. 860 (26 September 1914), 1–2.

90 Chen, 'Confucianism Encounters Religion', 165–72.

91 Chen Huanzhang 陳煥章, 'Kongjiaohui guojiao yijianshu 孔教會國教意見書 [The Views of the Confucian Religion Association on State Religion]', *Ching Shih Pao* 經世報 [*World Managing Monthly*] 2, no. 3 (1913): 9.

response, Kang sent letters to the Assemblymen as well as to the President and the Premier, urging them to show respect to the tradition and disregard those 'absurd' proposals.[92] As was the case during the Drafting Committee period, supporters and opponents of the state religion proposal at the plenary sessions of the reconvened Assembly debated three inter-related issues: Was Confucianism a religion? Was it qualified to be the state religion? Would adopting Confucianism as the state religion interfere with the principle of religious freedom? The underlying issue to be resolved as a pre-requisite to any meaningful discussion on the three major issues was: How was 'religion' or *zongjiao* to be clearly defined in the context of Chinese thought and practice? No consensus, however, was reached on any of these issues. During the first reading of the draft Constitution, the controversial proposal was again brought up for deliberation by the Assembly of more than eight hundred members. When it was put to a vote, the proposal failed to win the support of the majority. Nor was it passed during the second reading. Instead, a compromise amendment was passed, stating that Chinese citizens had the freedom to 'honour Confucius or profess any religion'.[93] This was tantamount to an announcement of the ultimate failure of the state religion campaign. Deeply disappointed, Chen immediately criticized the Assembly in an open letter for having 'slain the life of' Confucian Religion in a twofold way: not only did it fail to make it the state religion, but it also disqualified Confucianism as a religion.[94]

The Temple of Heaven Draft was not confirmed by heavenly blessings: before the second reading was completed, the National Assembly was again dissolved due to the power struggles among regional warlords. On 1 July 1917 General Zhang Xun 張勛 (1854–1923) launched a coup in Beijing to restore the abdicated Puyi to the throne. Kang collaborated in this effort, and he must have considered it a good opportunity to promote his proposal of a democratic polity with a titular monarch. In support of his actions he cited the Restoration of the English monarchy in 1660 following the Cromwellian Protectorate. Kang reasoned that in order to regain her stability and avoid further upheavals, China should follow the English model and restore Puyi as titular monarch.[95] To his

92 Kang, *Zhi guohui yiyuan shu* 致国会议员书, KYWQJ, vol. 10, 318–320.

93 Wu, *Zhonghua xianfashi*, 170; Pan, *The Chinese Constitution*, 192.

94 Chen Huanzhang 陳煥章, 'Guohui xianfa huiyi duiyu Kongjiao dageming jinggao guoren shu 國會憲法會議對於孔教大革命敬告國人書 [A Letter to Compatriots on How the Constitutional Session of the National Assembly Has Slain the Life of the Confucian Religion (1917)]', in *Kongjiao shinian dashi* 孔教十年大事 [*Events on Confucian Religion during a Decade*], ed. Ke Huang 柯璜 (Taiyuan: Zongshenghui, 1924), 34a–36b.

95 Kang Youwei 康有为, 'Zhongguo shanhou yi 中国善后议 [On the Reconstruction of Post-Revolution China][1916]', in KYWQJ, vol. 10, 2007, 274–75.

great disappointment, he soon found that his collaborator, General Zhang, was only paying lip service to constitutionalism, and the vast majority of the whole nation, including many of his early followers such as Liang, were against the Manchu restoration. In less than two weeks, Puyi the emperor was forced to abdicate again, and Kang had to seek refuge in the United States legation in Beijing. A warrant for his arrest was issued by the restored Republican government, and was not lifted until March the next year. At the national level, the attempted restoration ushered in a decade of warlordism, with regional leaders fighting to dominate the government and the presidency changing hands frequently.[96] Kang's participation in the ill-fated restoration struck a heavy blow both to his political credibility and to his moral standing, after which he seemed to have been able to exert little influence upon Chinese political and intellectual trends. As a result, he resigned his post as president of the CRA, after which Chen Huanzhang became its *de facto* leader.[97]

3 In the Name of Science

As can be seen from the above discussions, the campaign for Confucianism to be adopted as a state religion had also given rise to a countervailing force, an anti-Confucian movement. Among other things, many of its critics drew on the authority of science in their opposition to 'religion', a term that had by then become associated or even equated with 'superstition'. The revival of the state religion campaign and Kang's participation in the Manchu restoration of 1917 had attracted intense opposition from Chen Duxiu and his fellow New Culture intellectuals. With the increasing commitment of these thinkers to a radical form of scientism, their iconoclastic attacks on religion in general, and on Confucianism in particular, firmly established the Conflict Thesis in the mainstream of Chinese thought.

3.1 *Religion and Superstition*

By the early 1910s, the neologism *zongjiao*, which had been widely accepted as a legitimate translation of 'religion', was considered to refer predominantly to

96 John King Fairbank and Merle Goldman, *China: A New History, Second Enlarged Edition* (Harvard University Press, 2006), 257–87. It was not until 1927 that warlord-divided China was brought to a nominal unification by a united front of the reorganized Nationalist Party and the newly-established Communist Party, both inspired by Leninist Russia.

97 The Association underwent a revival in the early 1920s under Chen's leadership. For his contribution, Chen was elected vice-president of the newly initiated World Conference for International Peace Through Religion in the late 1920s.

a system of belief in spirits and deities. As Wang Jingfang 王敬芳, a member of the National Assembly, pointed out, the mention of religion would immediately remind a contemporary Chinese of the concept of *shen* 神 (spirits, deities).[98] That was why in the state religion campaign the attitude of Confucius toward the existence of spirits was heatedly debated. Even the same saying of Confucius on 'respecting spiritual beings but keeping aloof from them' (*Analects* 6:20) was interpreted differently to support opposing arguments. Defenders of the religious nature of Confucianism stressed Confucius' reverence for spirits or his acknowledgment of deities, while opponents of this view highlighted his exhortation to keep a distance from them.[99] In this connection, the opponents argued that in the context of Confucianism *jiao* was meant to refer to *jiaohua* (education and transformation) rather than *zongjiao*, while the supporters contended that *jiao* encompassed both, stressing that moral instruction of the people was the very purpose of religion. For Kang and Chen, to define religion merely in the sense of belief in the spirits was inadequate, but even if that definition was accepted, they maintained, Confucianism could properly be regarded as a religion because there were numerous places in the Confucian Canon where spirits and deities were discussed.[100]

Closely related to the association of 'religion' with the need, as a matter of definition, to subscribe to a belief in spirits, was the increasingly politicized notion of *mixin* or superstition as a negative force in the fervour of nationalistic and political effervescence surrounding the anti-imperialist sentiments. As stated in the previous chapter, *mixin* was another borrowing from Japanese terminology. Like *zongjiao,* it became popular in China in the beginning of the twentieth century largely due to its frequent use by Liang Qichao in his influential journalistic articles. Initially, the term was employed in the literal sense of 'irrational faith' in contrast to 'rational faith', and whether a person's faith convictions were rational or irrational could be settled by pointing to 'reason' as the standard for distinguishing between the two. As we have already seen, Liang held that without the sanction of reason religion would give rise to *mixin,* and yet he pointed out that religion was not the same as *mixin.* He stressed that superstitious beliefs in religion would obstruct the progress of

98 Wu, *Zhonghua xianfashi*, appx. 39.

99 See also Chen, 'Confucianism Encounters Religion', 76.

100 Chen Huanzhang 陳煥章, 'Kongjiaohui wei xianfa qicao weiyuanhui foujue guojiao jinggao quanguo tongbao shu 孔教會為憲法起草委員會否決國教警告全國同胞書 [Letter to Compatriots on the Rejection of the State Religion Proposal by the Constitution-Drafting Committee (1917)]', in *Kongjiao shinian dashi* 孔教十年大事 [*Events on Confucian Religion during a Decade*], ed. Ke Huang 柯璜 (Taiyuan: Zongshenghui, 1924), 83a–87b.

the human intellect and scholarly pursuits, but he did not regard *mixin* as an absolutely negative force, for he acknowledged the achievements of certain historical figures as a result of the transformative power of their persistent, irrational faith.[101]

As time passed by, however, the term *mixin* acquired increasingly negative connotations in the minds of Chinese intellectuals. A foreign religion was regarded, in the minds of patriotic Chinese, as promoting certain elements of 'blind faith' or unquestioning obedience to sometimes inexplicable dogma and occasionally outright superstition. To many of the educated elite, superstitious religious ideas were abhorrent, for they considered these religious ideas to be part of an imperial tool that had been employed to convert, rule and oppress the Chinese populace, and they now regarded superstition as a great hindrance to the progress of scholarship and science in China. Moreover, the term *mixin* came to also be used to designate various practices of popular beliefs, such as animism, reincarnation, divine retribution, and geomantic omens, all of which were often considered signs of ignorance. Accordingly, in early twentieth China, when a concerted campaign was mounted against whatever might qualify as a form of 'superstition', it had become a revolutionary tool to transform or to reform the political, intellectual and popular traditions of China. In this connection, superstition was often regarded as an intrinsic part of 'religion' and sometimes even as a surrogate name for religion, in the sense that both entailed 'irrational faith' in supernatural forces.[102] It should be noted that, paradoxically, there was at the same time another trend to separate religion from superstition, with science becoming the touchstone in differentiating them.[103]

Given this context of the movement to expunge superstition from the cultural life of the Chinese people during the period of revolutionary struggle, the advocacy of Confucianism as a religion, not to mention a state religion, raised the eyebrows and incurred the opposition of many intellectuals. As expected, a recurring view expressed in the Drafting Committee and during the debates at the Assembly was that to qualify as a 'religion', a religious body of teachings had to contain elements of superstition and such doctrines as 'divine retribution'. On this basis, some Assemblymen dismissed the claim that Confucianism was a religion, because it contained nothing superstitious. Even some of the those

101 Liang, 'Lun zongjiaojia', 764–66.

102 For more details, see Shen Jie 沈洁, 'Fanmixin huayu jiqi xiandai qiyuan 反迷信话语及其现代起源 [The Discourse against Superstition and Its Modern Origins in China]', *Shilin* 史林 [*Historical Review*], no. 2 (2006): 30–42.

103 Goossaert and Palmer, *Religious Question*, 50–51.

who wanted to exalt Confucianism hesitated to endorse the state religion proposal for fear of narrowing the scope of Confucianism and even of degrading Confucius, which was exactly what Liang had argued a decade earlier.[104]

In response, the supporters of the state religion campaign adopted two seemingly contradictory approaches. Some accepted that superstition was a part of religion, and argued that Confucius also condoned the use of divine retribution for 'stealthily ruling the people'.[105] Some rejected the view that superstition was a prerequisite attribute of a religion, stressing the fact that Confucius' focus on morality and human relations made it a more advanced form of religion. In his open letter collectively addressed to the President, the National Assembly, the State Council, governors, local assemblies, and the 'gentlemen' in the military, educational, journalist, agricultural, commercial and industrial circles, Liao Daochuan 廖道传, a Confucian poet and local magistrate from Guangdong, argued that 'religion' referred to Sages' teachings that cultivated the Way for the people, to which superstitious belief was not at all indispensable. He stressed that Confucian transcendence (seeking to understand Heaven and Earth) was to be achieved through living this life rather than in a next world.[106] In a similar vein, Chen Huanzhang added that 'superstition' was 'a term used to discredit religion', for what was taken as 'faith' by believers was often unjustly dismissed out of hand as 'superstition' by non-believers.[107]

In fact, Chen and Kang had both long held the view that Confucius, the founder of a new religion, was *against* the superstitions that were prevalent in China's old religion of polytheism.[108] Prior to Confucius's time, Chen argued, people had believed in a myriad of spirits, which fell into four classes, those of heaven, of earth, of the dead and of all material things, but above them all there was a Supreme God (*shangdi*). Under this ancient form of Chinese folk

---

104 See Wu, *Zhonghua xianfashi*, 106–7.

105 Wu, appx. 20.

106 Liao Daochuan 廖道传, 'Qing zun Kongjiao wei guojiao shang dazongtong dengshu 请尊孔教为国教上大总统等书 [A Petition to President (et al.) to Enshrine Confucianism as State Religion]', in *Zhonghua minguoshi dang'an ziliao huibian 3: wenhua 中华民国史档案资料汇编3: 文化* [*Compilation of Archival Materials Pertaining to the History of the Republic of China 3: Culture*], ed. ZDLDG 中国第二历史档案馆 [China Second Historical Archives] (Nanjing: Jiangsu guji chubanshe, 1991), 47. Although it was not known how widely the letter was circulated, it is recorded as having been received by the office of the President.

107 Chen, 'Kongjiaohui wei xianfa', 84a.

108 Although Kang did not use the neologism *mixin,* he meant the same things that were later covered by the term. Chen, on the other hand, used both 'superstition' and *mixin* in his English and Chinese writings.

religion, superstition prevailed in the whole empire.[109] The communication between the spirits and men were achieved through the intermediary priesthood of shamans, which was in fact a body of scholars. Their profession was further divided into six categories, astrology (*tianwen* 天文), the almanac (*lipu* 曆譜), the five elements (*wuxing* 五行), milfoil and tortoise (*shigui* 蓍龜), miscellaneous fortune-telling (*zazha* 雜占), and physical laws (*xingfa* 形法) which studied the features of geography, of human beings, and of things. Although the six professions were traditionally classified as magic (*shushu* 數術), Chen argued that 'they were really a mixture of magic and science', a feature that had been obscured and indeed forgotten in modern times.[110]

Chen claimed, as Kang had, that as a religious reformer, Confucius had swept away the old superstitions; that was why Confucius disliked talking about spiritual beings. Instead, Confucius stressed the individual cultivation of one's Heaven-endowed nature, aiming by this purposeful discipline to achieve the highest standard of harmony with Heaven and Earth. The religion of Confucius, Chen argued, was a Way of self-cultivation worthy of being held up as being a more advanced stage in religious practice and in the advancement of human history.[111] In other words, Chen claimed that Confucius replaced the old religion (polytheism) with his own set of teachings. This, he added, should not lead one into thinking that Confucius was irreligious, for the Sage had recognized the usefulness of the old religion and retained some of the old elements in his writings. Chen then cited quotations from the Confucian Texts that stressed 'the way of the spirits' (*shendao*), many of which Kang had cited and discussed.[112] By doing so, however, Chen left the reader with a somewhat schizoid image of Confucius, characterized by both his unwillingness to talk about spirits and yet his recognition of their usefulness as agents of retribution as part of a process contributing to the moral instruction of the people.

---

109 Referring to the same period, Feng Youlan has commented that the Chinese at that time were 'superstitious and ignorant', and they 'had religious thought but no philosophy', and the religion and spirits they believed in were similar to those of the Greeks. He has also argued that belief in a myriad of spirits had precedence over the concept of a supreme God. Fung, *Chinese Philosophy*, 1983, 1:24.

110 Chen, *Economic Principles*, 40. For another discussion on the five professions, see Fung, *Chinese Philosophy*, 1983, 1:26–30.

111 Chen, *Economic Principles*, 40–41. Chen considered this form of religion not only new to ancient China, but also new in today's Western world, which was 'only now' witnessing the appearance of such forms as the Ethical Culture Society. Nevertheless, he acknowledged that the Society contained essentials found in the teachings of Aristotle and the stoics.

112 Chen, 41; Kang, 'Kongjiaohui xu (er)', 346.

In this regard, Kang's arguments in his treatise on the *Annals* were clearer and more convincing. Applying his progressive theory of the Three Ages to his studies of religion, Kang asserted that Confucius had turned extravagant polytheism (the first stage) to measured polytheism (the second stage) by dramatically reducing the number of spirits that were deemed worthy of receiving worship and sacrifices, in an effort to gradually prepare the populace for the third stage, the eventual practice of monotheism. To be specific, Confucius preserved only five sacrifices: those offered to Heaven, Earth, mountains, rivers and ancestors, and he denounced any other type of worship, such as animal worship.[113] The problem was, Kang argued, that Confucius' teachings on the issue had been forgotten, and extravagant polytheism had been revived as manifested in various 'improper worships and sacrifices' (*yinsi* 淫祀), practices that he had earlier denounced in memorials to the Throne in the late 1890s. Thus Kang was, in his own way, a strong advocate of the need to eradicate superstition from Chinese cultural life and religious practice. Although he condemned many of the folk practices of the Chinese countryside, he did not consider as superstitions certain other beliefs, such as reincarnation and divine retribution, which were often denounced as superstition by critics and opponents who were his contemporaries.

When pressed for a distinction between religion and superstition, science was sometimes called on to play a role by Chinese intellectuals in the 1910s. That which was against science was superstition rather than religion. As will be seen in our later discussions, however, the problem was that there was no consensus, let alone agreement, on what 'science' was, or what its boundaries were. For Kang, science was an important source for intellectual and material development, but it had its limits. Many of his contemporaries of the younger generation, on the other hand, sharply disagreed; and they opted for scientism.

3.2 *Scientism versus Religion*

While most of the early advocates of science, like Kang and Liang, had not been trained in science, the 1910s witnessed the appearance of young Chinese scientists trained in the United States with financial support from the Boxer Indemnity Scholarship Programme.[114] Many of them, while still studying abroad, joined in the effort to promulgate science in China not only as a source

113 Kang, 'Chunqiu kao', 135.

114 The programme, known in Chinese as 庚子賠款獎學金, was established by the US government with the excess payment of the Boxer Indemnity. During the two decades following its inception in 1909, the scheme had sponsored around 1,300 Chinese students to study in the United States.

of material development, but also as a powerful tool to reform traditional cultural practices and do away with the traditional mental attitudes, which they thought were frequently superstitious, sometimes extremely so. In the practice and pronouncements of many of these proponents of science, the power of science was often exaggerated, and science was almost invariably placed in opposition to religion as well as superstition.

The most vocal promoter of modern science was a new monthly, the *Kexue* (Science). It was initiated in the beginning of 1915 by a group of Chinese students at Cornell University, who had organized the Science Society of China (*Zhongguo kexueshe*中國科學社) in the previous year, in their effort to emulate the American Association for the Advancement of Science. Its founding members included Ren Hongjun 任鸿隽 (H. C. Zen), Yang Quan 杨铨, Zhao Yuanren 赵元任 (Y. R. Chao) and Hu Shi 胡適, who would achieve prominent positions in politics, education and scientific research after they returned to their motherland. The society itself was moved to China in 1918 and became the largest association of scientists in China prior to the official *Academia Sinica* (1928). Ren, who had once served as secretary to Sun Yat-sen prior to his studies in the United States, would go on to head the department of specialized education at the Ministry of Education before taking charge of an organization that supervised a principal source of financial support for scientific activities in the Republic; Yang would serve as General Secretary of the Academia Sinica; Zhao, though trained in mathematics and physics, turned himself into an internationally known linguist; and Hu Shi, who switched to philosophy and became a disciple of John Dewey, became one of the leaders of the New Culture Movement.[115]

Although edited remotely in the United States, the monthly was printed in Shanghai and was distributed from there. It was stated clearly in its preamble that the journal aimed to spread the latest scientific knowledge as well as popular science so as to appeal to the general public. The journal would cover physical and natural sciences and their applications, and publish brief biographies of scientists, but it would exclude political and metaphysical discussions. It stressed the importance of 'rectifying names', saying that concerning Western scientific terms, they would make deliberate choices on the most appropriate translations from existing Chinese ones, or produce new terms if they had not

---

115 See Peter Buck, *American Science and Modern China, 1876–1936* (Cambridge: Cambridge University Press, 1980), 91–98; Zuoyue Wang, 'Saving China through Science: The Science Society of China, Scientific Nationalism, and Civil Society in Republican China', *Science and Civil Society*, Osiris, 17 (2002): 291–322.

been previously translated or if the old ones were found inadequate.[116] Naturally, one of the key terms to be first decided on was the Chinese translation of 'science'. As the journal's title indicates, the editors had decided on *kexue* rather than *gezhi*. In this way, it helped to put an end to the coexistence of the two as interchangeable terms for 'science'. It has been acknowledged by Chinese historians of science that since 1915 when the journal of *Kexue* was started, the term *kexue* had completely replaced *gezhi* as the standard translation for 'science'.[117]

Beyond a proper Chinese translation for 'science', the contributors to the *Kexue* also attempted to offer a proper definition of the term. A typical view was from Ren, the chief editor of the journal. 'Science is a generic term for systematic knowledge,' he wrote. 'In its broad sense, any system of rational knowledge that offers an account for a thing or phenomenon can be called science.' However, he emphasized that in its modern usage, the term was conventionally employed in a narrower sense, that is, it referred to 'systematic, interrelated knowledge of a certain phenomenon, whose general principles are induced from experiment-based observations and induction.' According to this definition, astronomy, physics, chemistry, physiology and psychology, for example, were considered as sciences, while history, fine arts, literature, philosophy and theology, for instance, were not. Ren stressed that the defining feature of modern science was its method, in particular the inductive method. This method was powerful and effective, because for one thing it was 'experimental', consisting of such steps as experiment, observation, hypothesis and verification, and for another, it was 'progressive', giving rise to general principles from which new knowledge and principles could be deduced. 'The essence of science', Ren summed up, 'lies in its method rather than its objects.' In his opinion, it was precisely the lack of the scientific method that had hindered the birth and rise of science in China.[118] In a similar vein, Hu Mingfu 胡明複, then a Ph.D. candidate in mathematics at Harvard, argued that the scientific method represented the very spirit of science and could make an even greater contribution to mankind if it was applied to such other domains of human life, such as thought, morality and culture.[119]

116 'Liyan 例言 [The Preamble]', *Kexue* 科學 [*Science*], no. 1 (1915): 1–2.

117 See Wu, *Shenme shi kexue*, 11. The monthly continued for 45 years until 1960, one year after the Science Society of China was dissolved under Communist rule. In 1985, the *Kexue* monthly was revived and has remained in publication until the present time.

118 Ren, 'Shuo Zhongguo wu kexue'.

119 Hu Mingfu 胡明復, 'Kexue fangfa lun yi 科學方法論一 [On the Method of Science: I]', *Kexue* 科學 [*Science*] 2, no. 1 (1916): 719–28.

The *Kexue* monthly grew quickly in its influence among Chinese intellectuals and played a very significant role in the popularization of science. Like Kang Youwei, the contributors of the *Kexue* extolled the practical benefits that science brought to mankind through technological achievements, and they therefore vigorously advocated saving China though science. They highlighted the fact that it was modern science that had given rise to the industrial revolution, which in turn contributed to Europe's rapid development in wealth and power. Therefore, it was of utmost importance for Chinese schools to pay great attention to science education, if China was to catch up with her Western counterparts.[120] Unlike Kang, however, they clearly distinguished science from its industrial applications. Echoing Huxley's rejection of application as the goal of science, they stressed that the natural purpose of science was not to put scientific discoveries to practical uses, but to 'explicate the truth of the world' through the exercise of man's innate abilities and to become the master of Nature, rather than its slave. They argued that this purpose of science resonated with the goal of education, which was, in the words of Matthew Arnold, 'to know ourselves and the world'. They stressed that the effect of science was not limited to the natural and physical domains, for science could also play a role in conjunction with the liberal arts in guiding human behaviour and social organization. Malthusianism and Darwinism were often-quoted examples in this regard. In education, therefore, more attention should be given to the method of science rather than the actual knowledge of science, and above all to the training of a scientific mind.[121] When Kang studied and advocated science three decades earlier, he had also emphasized the importance of science to one's intellectual development, but his focus was more on the principles discovered by science than on the method of science itself. In hindsight, the discussions on the definition, method, aim and spirit of science in the *Kexue* could be seen as a belated making up for what had been largely neglected and consequently found to be lacking in the earlier phases of introduction of Western sciences into the Chinese intellectual consciousness in the late nineteenth century.[122]

---

120 Ren Hongjun 任鴻雋, 'Kexue yu gongye 科學與工業 [Science and Industry]', *Kexue* 科學 [*Science*] 1, no. 10 (1915): 1089–99.

121 Ren Hongjun 任鴻雋, 'Kexue yu jiaoyu 科學與教育 [Science and Education]', *Kexue* 科學 [*Science*] 1, no. 12 (1915): 1343–52.

122 For a good account of the contribution of the *Science* and the Science Society of China to the development of science in China, see Zhang Jian 张剑, 'Cong "kexue jiuguo" dao "kexue jianguo" 从"科学救国"到"科学建国"[From "Saving the Nation" to "Building Up the Nation" through Science]', *Ziran bianzhengfa tongxun* 自然辩证法通讯 [*Journal of Dialectics of Nature*], no. 3 (2016). Those papers of general discussions on science had

Not only was science touted as the key to material development through technology and as the primary means of improving human health and longevity through medicine and food production, it was also regarded as a way to achieve intellectual emancipation and freedom through the scientific explanation of natural phenomena, and even of advancing moral development, via the explication of economic and sociological principles. Science, therefore, was not only the solution to China's material underdevelopment, but it was also the remedy for her spiritual backwardness. Without mentioning its name, Confucianism was criticized as outdated and no longer useful: 'Even China's morality and classical learning,' it was stated in the 'Foreword', 'long cherished as national essence for thousands of years, have become corrupted and lost efficacy in their entirety.' Confucianism was again held accountable by Ren for China's failure to give rise to modern science. China, so to speak, had lost both the material and spiritual means of maintaining her independence and existence. 'The solution, therefore, is science, and science alone!'[123] Here was an embryonic form of a statement that can be characterized as a strident commitment to scientism, according to which every aspect of the universe is knowable only by the methods of science and every problem can ultimately be solved by science.[124]

Although the *Kexue* attempted to limit its scope to the physical and natural sciences and it excluded metaphysical (including religious) discussions, it was difficult for the contributors to ignore religion in relation to the development of science. Not infrequently, religion was juxtaposed with superstition and placed as the conceptual opposite of science. 'In medieval Europe, religion and superstition obstructed the progress of human knowledge,' the author of the foreword to the first issue noted. 'We would have remained ignorant today, if it had not been for such scientists as Galileo, who loved truth and persevered in its pursuit, and who, with bloodshed, wrestled and fought with religion for freedom of thought.'[125] The case of Galileo was again highlighted

been collected and published in one volume in 1919, then expanded and reprinted a decade later. See also Zhongguo kexueshe 中國科學社, ed., *Kexue tonglun 科學通論* [*General Discussions on Kexue*] (Shanghai: Zhongguo kexueshe, 1934).

123 Ren Hongjun 任鴻雋, 'Fakanci 發刊詞 [Foreword]', *Kexue 科學* [*Science*] 1, no. 1 (1915): 3–7. While acknowledging the fact that advances in science had assisted the ongoing European war, its special issue on war and science argued that science had played a bigger role in preserving and achieving peace. Yang Quan 楊銓, 'Zhanzheng yu kexue 戰爭與科學 [War and Science]', *Kexue 科學* [*Science*] 1, no. 4 (1915): 353–58.

124 For more on the definition of scientism in the Chinese context, see Kwok, *Scientism in Chinese Thought*, 20–21.

125 Ren, 'Fakanci'.

in a following article in the same issue on the life and work of 'the pioneer of modern astronomy and experimental science'. It delineated how Galileo, at an advanced age, was tortured by the Catholic Church,[126] and was forced to promise on oath not to discuss the 'anti-Bible' view that the earth moved around the sun.[127] The Chinese readers were quick to discern the Conflict Thesis that was clearly conveyed therein. Upon reading its first issue, for example, Cai Yuanpei, the former Minister of Education, expressed his appreciation of its promotion of 'pure science' to the exclusion of religion. He applauded Galileo as a model for scientists, who in his opinion 'spearheaded the war between science and religion'. 'Science is founded on experiments,' he continued, 'and is fundamentally incompatible with religion, a product of fabrication.'[128]

It is understandable that the Conflict Thesis was propounded in a magazine edited by a group of Chinese students at Cornell. One should remember that the university's former president, Andrew Dickson White, was credited with being one of the chief promoters of that thesis through his influential book, *A History of the Warfare of Science with Theology in Christendom*, which since its first publication in 1896, was reprinted many times and remained a popular work for decades. 'As Andrew White pointed out,' it was stated in a 1916 *Kexue* article on the method and spirit of science, 'history shows that struggles and conflicts between science and religion have invariably resulted in the good of both.'[129] The article noted that the very spirit of science was to seek truth and dispose of what is shown to be false, and that the history of science in the West consisted of its 'constant fights' against old superstitions, customs, religions and morality. This warfare was welcome, because it contributed both to the progress of science and to the 'purification' of religions and morality.

The Conflict Thesis was further promoted by another new and even more influential magazine, the *Xinqingnian* 新青年 or the *New Youth* (*La Jeunesse*),

---

126 It should be noted, however, that Galileo was threatened with torture but not actually tortured, though the latter was widely believed.

127 Yang Quan 楊銓, 'Jialileiwo zhuan 加里雷倭傳 [A Biographical Sketch of Galileo]', *Kexue* 科學 [*Science*] 1, no. 1 (1915): 99–105. Unlike many of the later accounts of Galileo, however, the essay acknowledged that Galileo had professed Catholicism and 'was most sincere in his belief'. It further quoted Galileo as 'often' saying that 'science and Scriptures, if correctly understood, agree with each other'.

128 Cai Yuanpei 蔡元培, 'Fu Ren Hongjun han 复任鸿隽函 [A Letter in Reply to Ren Hongjun (1915)]', in *Cai Yuanpei quanji* 蔡元培全集 [*Complete Works of Cai Yuanpei*], vol. 10 (Zhejiang jiaoyu chubanshe, 1998), 249–50.

129 Hu, 'Kexue fangfa lun', 723; cf. White, *Warfare of Science with Theology*, viii. White's original wording reads, 'In all modern history, … all untrammelled scientific investigation, no matter how dangerous to religion some of its stages may have seemed for the time to be, has invariably resulted in the highest good both of religion and of science.'

established in late 1915 by Chen Duxiu. As a masterly publicist and propagator of ideas, Chen has been compared to Voltaire in his role as the 'grand impresario of modern thought' in China.[130] The magazine soon became the vehicle for intellectual change in China and spearheaded the New Culture Movement, an intellectual campaign for the rejection of Confucian values, which were held responsible for the weakness of the nation.[131] It was also a campaign for the embrace of egalitarianism and other Western values, which Chen articulated through the personae of 'Mr. Democracy' and 'Mr. Science'. He trumpeted the call to criticize and reject traditional culture and to build a new one based on the scientific outlook in his leading article 'My Solemn Plea to Youth' which initiated the first issue of the *New Youth*. He urged China's young minds to be scientific and not speculative, for science conformed to both objective phenomena and to human reason, while speculation, whose products included religion, 'not only overreached objective phenomena but also banished subjective reason'. The former characterized modern progress and was the force behind European ascendency; the latter, on the other hand, belonged to the past – the days of ignorance – but was still prevalent among the less enlightened people of the modern day. If the Chinese people were to progress from their unenlightened state, they had to have recourse to opportunities to practise science, which alone could correct all the fanciful notions and irrational beliefs in their traditional culture.[132] Chen was emphatic about the power of science:

> In today's ever-changing world, every new occurrence and every minute thing is subject to the scrutiny of scientific laws; the efficacy of science will subject all human thoughts and actions to reason and thereby get rid of superstitions and still the capricious winds of ignorance.[133]

By subjecting all human thoughts and actions to science and reason, Chen was explicitly advocating a strong scientism. Yet he had never been trained in science; for him, science was more of a weapon against superstition and a way to expose a lethargic mode of thought than a dynamic force in conquering nature.

---

130 Kwok, *Scientism in Chinese Thought*, 68.

131 Many of those anti-traditional, anti-Confucian themes had been advanced by the Anarchist groups of Chinese overseas students in Paris and Tokyo about a decade earlier. See Robert Scalapino and George T. Yu, *The Chinese Anarchist Movement* (Berkeley: Center for Chinese Studies, Institute of International Studies, University of California, 1961), 9.

132 Chen Duxiu 陳獨秀, 'Jinggao qingnian 敬告青年 [My Solemn Plea to Youth]', *Xinqingnian 新青年 [New Youth Journal]* 1, no. 1 (1915): 5 of 1st article.

133 Chen, 5 of 1st art; see also Kwok, *Scientism in Chinese Thought*, 64–65.

Equally forceful was Chen's attack on religion, which was taken as a symbol for various forms of unfounded belief: 'If the existence of the gods and spirits of heaven and earth cannot be proved accurately, all religions are nothing but deceitful idols: the god Amida is false; Jehovah is false; and the Supreme Lord of Heaven [head of the Taoist Pantheon] is false.' 'All kinds of God, Buddhas, Immortals, Spirits revered by various religions are useless,' he concluded.[134]

In disposing of the very basis of religious belief by reducing ultimate reality to the material atom and through drawing on an 'all-powerful science', Chen presented a clear conflict model of science and religion. 'In order to support Mr. Science [in China],' he categorically declared, 'we cannot but oppose [...] old religion [beliefs in gods and spirits].'[135] Many of his fellow leaders of the New Culture Movement joined him in promulgating science and denouncing religion in their eager efforts to get China onto the tracks of modern progress. And they did so with supreme confidence, or, one might say, even with a certain hubris. As they applied Auguste Comte's Law of Three Stages to their analysis of the Chinese society, they concluded that China still remained in the stage of religion and superstition, far away from the positive or scientific age. To get rid of religion and to promote science, they felt, were objectives that were fully justified and imperatively necessary.[136] As expected, the New Culture iconoclasts would inevitably come into conflict with Kang's insistence on reforming the Confucian religion, and, in particular, his campaign to establish Confucianism as a state religion.

3.3 *Looking for Substitutes*

Initially, in his hostile scrutiny of Confucian values, Chen Duxiu did not attack Kang Youwei. In fact, he had once been an admirer of both Kang and Liang and was an avid reader of their works during his formative years. He turned from being a reformer to becoming a revolutionary during the years he spent in Japan in the 1900s, but he did not join the anti-Manchu *Tongmenghui* led by Sun Yat-sen. He later became an advocate of modern (European) civilization, which he considered owed three significant debts to the French, namely,

134 Chen Duxiu 陳獨秀, 'Ouxiang pohuai lun 偶像破壞論 [On the Destruction of Idols]', *Xinqingnian 新青年* [*New Youth Journal*] 5, no. 2 (1918): 89; translated in Kwok, *Scientism in Chinese Thought*, 71.

135 Chen Duxiu 陳獨秀, 'Benzhi zuian zhi dabianshu 本誌罪案之答辯書 [A Reply to the Charges Against Our Magazine]', *Xinqingnian 新青年* [*New Youth Journal*] 6, no. 1 (1919): 10–11.

136 Chen Duxiu 陈独秀, 'Xuyi 序一 [Preface One]', in *Kexue yu renshengguan 科学与人生观* [*Science and the Philosophy of Life*], by Zhang Junmai 张君劢 et al. (Shenyang: Liaoning jiaoyu chubanshe, 1998), 3.

democracy as seen in Lafayette, evolutionary theory in Lamarck (the predecessor of Darwin), and socialism in Saint-Simon and Fourier.[137] Despite changing his mind on a number of issues, he remained appreciative of the early contribution of Kang and Liang to the 'enlightenment' of the Chinese people.

All of that changed in late 1916, when Chen read Kang's letter to the President and the Premier, which was published in a newspaper on 20 September. As was mentioned earlier in this chapter, Kang requested that the government restore the practice of worshipping Confucius on one's knees (which had recently been nullified by the Ministry of Interior Affairs); that it continue the ritual of offering sacrifices to Heaven; and that Confucianism be made the state religion by the Constitution.[138] Deeply incensed by Kang's statement that 'the Chinese would degrade themselves to the status of barbarians if they discontinued worshipping Confucius as founder of their religion', Chen immediately published a refutation of Kang's letter in the following month's issue of his *New Youth*. Apart from criticizing what he considered to be Kang's logical contradictions, Chen made several counter-claims. First, the Chinese were not a religious people and Confucianism was not a religion. 'Religion aims at salvation of the soul and is oriented to the other world,' he wrote, 'while Confucius did not talk about ghosts or death, so his doctrine concerning human nature and the way of Heaven is philosophy rather than religion.' Secondly, for the state to exalt Confucianism above other religions would interfere with the modern principles of religious freedom and of the separation between state and church. Third, given its 'inseparable association' with despotism, Confucianism was not only incompatible with republicanism but with modern life. Last but not least, religion was no longer valuable in the age of science. 'The majority of scientists have disregarded religions as fictions,' he claimed, 'and the knowledgeable Ernst Haeckel is their contemporary representative.'[139] In a sense, Chen Duxiu and Kang agreed on one point, that is, that orthodox Confucianism as it had been practised in imperial China was no longer suitable for modern China in a world of rival nations. They differed in their solutions, however. While Kang attempted to reform Confucianism to conform to the needs of the times, Chen wanted to get rid of it.

---

137 Chen Duxiu 陳獨秀, 'Falanxi renmin yu jinshi wenming 法蘭西人民與近世文明 [The French People and Modern Civilization]', *Xinqingnian* 新青年 [*New Youth Journal*] 1, no. 1 (1915): 2nd article.

138 Kang Youwei 康有为, 'Zhi Li Yuanhong Duan Qirui shu 致黎元洪段祺瑞书 [A Letter to the President and the Premier (1916)]', in *KYWQJ*, vol. 10, 2007, 316–17.

139 Chen Duxiu 陳獨秀, 'Bo Kang Youwei zhi Zongtong Zongli shu 駁康有為致信總統總理書 [A Refutation of Kang Youwei's Letter to the President and the Premier]', *Xinqingnian* 新青年 [*New Youth Journal*] 2, no. 2 (1916): 3rd article.

The points outlined above were further elaborated in Chen's subsequent articles. In his further discussion on the question of Confucianism as a religion, Chen proposed substituting science for religion.[140] 'It is my faith', he declared, 'that man will rely on science as the proper way to obtain true understanding and to regulate their actions in the future, and all religions are to be disposed of.' This conviction came from his understanding of the evolution of 'laws'. According to his view, there were two types of laws in the world, natural and man-made. Natural laws were universal, everlasting and inevitable; man-made laws were partial, transient, and fortuitous. The former could be revealed by science, while the latter were contained in religions, morality and civil laws. Man's greatest task and utmost goal, therefore, was to 'correct' all man-made laws and make them come into agreement with natural laws, laws that had been and would continue to be discovered by science. In this sense, he argued that the mysteries of the universe and of human life, which had so far been 'inaccurately' accounted for by religions, would one day be correctly explained by science. 'Therefore,' he concluded, 'I propose substituting science for religion as our true faith. It may take a long time to solve all the mysteries for us, but it will eventually.'[141]

Chen's view was echoed by Cai Yuanpei, who was then President of Peking University and who had invited Chen to head the university's division of arts and humanities. During a 1917 talk by Cai, whose notes were published in the *New Youth*, he proposed 'substituting aestheticism for religion'. He claimed that religion was a question of the 'bygone age', for 'the components of religion have now been accounted for by modern science' (that is why he considered Kang's Confucian movement an outdated, futile effort). In his view, religion was important in the past because it was able to provide people with knowledge (*zhishi* 知识), altruistic will (*yizhi* 意志) and emotional support (*qinggan* 情感). With social development and scientific advances, however, all the mysteries of the world, including the origins of different species, could be accounted for by science in a convincing way. These in turn could help people

---

140 In the nineteenth century there were several Western proposals in favour of substituting science for religion as the basis of society, the most notable being that of Auguste Comte. For a discussion on the similarity and relation between Chen's views and those of Comte's, see Gao Like 高力克, 'Yi kexue dai zongjiao: Chen Duxiu de kexuezhuyi zongjiaoguan 以科学代宗教: 陈独秀的科学主义宗教观 [Substituting Science for Religion: Chen Duxiu's Scientism-Oriented Concept of Religion]', *Shixue yuekan* 史学月刊 [*Journal of Historical Science*], no. 1 (2017): 89–97.

141 Chen Duxiu 陳獨秀, 'Zailun Kongjiao wenti 再論孔教問題 [My Second Discussion on the Question of Confucianism as a Religion] [1917]', *Xinqingnian* 新青年 [*New Youth Journal*] 2, no. 5 (1917): 1st article.

understand that moral principles were not God-given, but could be deduced from human experience and the findings of biology, psychology and sociology. Thus, religion had already been deprived by science of the functions of conferring knowledge and providing guidance through ethical principles. As for the remaining function of providing emotional support, Cai argued that it could be substituted for as well with an aesthetic education including fine arts and music. That way, modern people would be able to enjoy all the benefits originally offered by religion without having to suffer from its disadvantages.[142] 'In an age of science,' he claimed categorically, 'all knowledge and every moral issue can be secured or solved through science, and they have nothing to do with religion whatsoever.'[143]

As can be seen, there was a big difference between Kang and Chen in their conceptions of science as well as of religion. For Kang, science was of great importance for education and material development. That was a position he had always maintained, and he stressed how advances in sciences and their technical application could contribute to a country's material and cultural development.[144] But he never placed science above religion or exalted it to position where it could work to answer all questions and meet all the needs of human society to the total exclusion of religion. Kang considered science, religion and politics as important domains of human endeavour and emphasized their complementary roles in the development of a country. For Chen and his fellow New Culture intellectuals, however, science was something greater than the natural sciences and their materialization in the form of technology. Nor was it merely a weapon to fight against superstitions and religions. Chen developed a form of positivist philosophy, that is, he advocated the use of science to solve all issues in human life and society. As for those problems that could not yet be solved, they would be resolved by science in due course. In that sense, he advocated the substitution of religion by science and called science the true faith.

Rarely did Kang respond to his young critics directly. For one thing, he would not deign to argue with them. He did not regard them as being qualified for judging someone like himself, who had, in his own words, 'gained great insights and

142 Cai Yuanpei 蔡元培, 'Yi meiyu dai zongjiao shuo 以美育代宗教說 [On Substituting Aestheticism for Religion]', *Xinqingnian* 新青年 [*New Youth Journal*] 3, no. 6 (1917): 2nd article. For a English discussion on Cai's aestheticism, see Laurence G. Thompson and O. Briere, trans., *Fifty Years of Chinese Philosophy 1898–1950* (London: George Allen and Unwin, 1956), 31–32.

143 Cai Yuanpei 蔡元培, 'Tongxin 通信 [Correspondence]', *Xinqingnian* 新青年 [*New Youth Journal*] 3, no. 1 (1917).

144 Kang, 'Gonghe pingyi', 33.

rich knowledge from being immersed in reading Chinese books for decades and travelling to as many as thirty-one countries'. 'How overconfident are those', he wrote dismissively in a 1918 letter to a friend, 'who have neither read much [in the Chinese classics] nor travelled abroad and yet have presumed to take on the role of being my judge and examiner!'[145] Furthermore, he had already refuted, in his earlier defence of the religious nature of Confucianism, criticisms like those put forward by Chen Duxiu. In a written exchange with a Japanese friend, he commented that it was not worthwhile to further refute the claim that Confucianism was not a *zongjiao* or religion. He observed that it was ironic that European and American authors considered Confucianism a religion while the Chinese intellectuals insisted on its not being one.[146] There was, however, one criticism that Kang had not fully elaborated on yet, that is, the Conflict Thesis. This he would tackle in detail in his later years, during his lecture tours and his enunciation of 'celestial peregrination'.

---

145 Kang Youwei 康有为, 'Zhi Zhang Yishan shu 致章一山书 [A letter to Mr. Zhang Yishan] [1918]', in *KYWQJ*, vol. 11, 2007, 82.

146 Kang Youwei 康有为, 'Yu Riren moujun bitan 与日人某君笔谈 [A Written Exchange with a Japanese Friend (1920)]', in *KYWQJ*, vol. 11, 2007, 118. This is a rare case where Kang employed the neologism, *zongjiao,* without a criticism.

CHAPTER 5

A Celestial Wanderer

> That there is God in heaven is a belief common to all nations and all religions. In China, when the term 'Heaven' is employed, it denotes 'dominion', and he who holds this dominion is God.
>
> KANG YOUWEI[1]

∵

Having taken refuge for seven months in the American legation in Beijing following the debacle of the 1917 Restoration, Kang left it escorted by legation guards, and returned to his residence in Shanghai in early 1918. There he would live for most of his last nine years. He had a magnificent Western-style mansion built in downtown Shanghai near the International Settlement (which he had visited in 1882, a visit that intensified his desire to learn from the West). Such a mansion was required in order to befit his political and social station as well as to house his large family.[2] In the centre of the backyard, he set up a small chapel, which he named *Sanbentang* 三本堂, the Hall of Three Roots. Here he placed the spirit tablets for God the Heaven, Confucius the Sage, and his ancestors. On the occasion of the dedication of the chapel, Kang elaborated on Xunzi's idea that ritual had three roots, namely, in Heaven and Earth, in forefathers and ancestors, and in lords and teachers,[3] a theme that Kang often visited in his later writings:

> Man has his roots. Without Heaven he would not have come into life; without ancestors, he cannot have come forth; without Sages, he would

1 Kang, 'Zhutian jiang', 93.

2 His family at the time included his five wives, six minor children, and their maids. To cover the large expenses, Kang relied on several sources of income, including the selling of his family property in Guangdong which had been formerly confiscated following the failed 1898 Reform but was returned to him after his return to China, the trade of artistic works and the selling of his own works of calligraphy which were popular. See Ma Honglin 马洪林, *Kang Youwei dazhuan* 康有为大传 [*A Comprehensive Biography of Kang Youwei*] (Shenyang: Liaoning renmin chubanshe, 1988), 568–70.

3 John Knoblock, trans., *Xunzi: A Translation and Study of the Complete Works*, vol. III (Stanford: Stanford University Press, 1994), 49.

 | DOI:10.1163/9789004468221_007

> not have been educated. Therefore, Heaven is the root of his life, ancestors are the root of his kinship, and Sages are the root of his education. It follows that one should worship Heaven in association with the Sages and also his ancestors.[4]

Despite the harsh criticisms made against him by the New Culture intellectuals, Kang remained unshaken in his political convictions, and kept abreast of both domestic and international affairs. As the World War drew to a close, Kang called on the embattled factions in North and South China to reach a truce, so as to enable China, a member of the Allied Powers, to unitedly defend her interests at the forthcoming peace negotiations in Paris. His plea was informed by a religious element. 'In the same way that recitation of the names of the Buddha helps,' he wrote in a circular telegram to the regional generals and government leaders, 'the thought of peace, when repeated in your mind for thousands of times, would quench the idea of war, at least temporarily.'[5] His call went unheeded.

China was accorded a very unfair treatment at the Peace Conference, where it was decided that Germany's colonial rights in the Shandong Peninsular, which had recently been seized by the Japanese, be transferred to Japan rather than returned to China. As the news reached Beijing, it touched off a strong student protest on 4 May 1919 against the Japanese demands and the terms of the Versailles Treaty, a protest that soon spread to other cities and developed into what later became known as the May Fourth Movement (which itself has often been considered to be an extension of the New Culture Movement). Although he disagreed with the New Culture intellectuals on many issues, Kang immediately joined with them in supporting the students' movement and demanding that the authorities release the arrested students and punish the 'traitors' within the government who opted for collaboration with Japan.[6] Under pressure, the Beijing government released the students, dismissed the officials, and refused to sign the treaty. Strongly influenced by a rising tide of Chinese nationalism, the Movement shaped the outlooks of a generation of

---

4 Kang Youwei 康有为, 'Sanbentang luocheng jiwen 三本堂落成祭文 [An Oration for the Dedication of the Hall of Three Roots (1922)]', in *KYWQJ*, vol. 11, 2007, 182.

5 Kang Youwei 康有为, 'Cu nanbei suyihe yiying Ouzhou heju dian 促南北速议和以应欧州和局电 [A Telegram Urging a Quick Truce between the South and the North to Better Defend China's Interests at the European Peace Conference (1918)]', in *KYWQJ*, vol. 11, 2007, 93.

6 Kang Youwei 康有为, 'Qing zhu guozei jiu xuesheng dian 请诛国贼救学生电 [A Telegram Demanding the Punishment of the Traitors and the Release of the Arrested Students (1919)]', in *KYWQJ*, vol. 11, 2007, 105–6.

political and cultural leaders. Chen Duxiu, for example, converted to Communism and helped found the Communist Party of China (CPC) in 1921.

However, Kang saw a silver lining in the otherwise dark clouds of the Paris Peace Conference, and he became extremely excited at the proposal of US President Woodrow Wilson (whom he described as 'an old acquaintance') for the creation of a League of Nations to maintain world peace. 'An association of great and small states, all being treated equally and accorded freedom,' he wrote the Chinese delegates to the Peace Conference in early 1919, 'will constitute the solid road to the Universal Peace.'[7] He immediately connected the Wilsonian proposal with the doctrine of the Three Ages, which, for many years, he had hailed and highlighted as the essence of Confucius' teachings. In his view, the proposed League of Nations would usher in the Age of Universal Peace and constitute the beginning of the World of the Great Unity he had earlier envisioned, although it was a century earlier than the timeline he had projected. Accordingly, he arranged for the first two chapters of his *Book of the Great Unity*, the chapters previously published serially in his *Buren* magazine, to be printed in book form and circulated. He commented that when he completed the book in the early 1900s, he had never expected that he would be able to witness during his lifetime the Great Unity coming into being as it soon would, with the establishment of the League of Nations.[8] Although he was subsequently disappointed at how the League, after its establishment in June 1919, was dominated by a few powers, with China being degraded to 'a fourth-class state', and concluded that it had failed to realize the Great Unity to a satisfactory degree, Kang was appreciative of what had already been achieved, and predicted that the world of the Great Unity would be fully accomplished in the fullness of time, but perhaps not until after another world war![9] In this he showed both a prophetic foresight, and an unshakeable certitude in the ultimate attainability of the Great Unity envisioned by the doctrine of the Three Ages.

Kang's primary concern, though, was for contemporary China. Indeed, he was dismayed to observe China's further disunity on military, political and ideological fronts. Successive wars were fought in northern and central China

---

7 Kang Youwei 康有为, 'Zhi yiheweiyuan shu 致议和委员书 [A Letter to the Chinese Delegates to the Peace Conference (1919)]', in *KYWQJ*, vol. 11, 2007, 99; see also Kang Youwei 康有为, 'Ji'nan yanjiangci 济南演讲辞 [A Talk Given at Ji'nan (1923)]', in *KYWQJ*, vol. 11, 2007, 249.

8 The poems and comments in his handwriting were reprinted in Kang, *Datong shu*, [1–11].

9 Kang Youwei 康有为, 'Qing Quanyangyi zhuanda Riben neige chebing jiaohuan dian 请犬养毅转达日本内阁撤兵交还电 [A Telegram Sent via Inukai Tsuyoshi to the Japanese Cabinet on Withdrawing Army and Returning the Shangdong Peninsula (1919)]', in *KYWQJ*, vol. 11, 2007, 108–9; Kang Youwei 康有为, 'Zhi Jin Yunpeng Yan Huiqing shu 致靳云鹏颜惠庆书 [A Letter to Premier Jin and Foreign Minister Yan (1920)]', in *KYWQJ*, vol. 11, 2007, 138.

between regional warlords, who had by then formed several major cliques. As things stood in the early 1920s, the Zhili clique managed to take over the Beijing government following military victories over the other two, but its control over the country remained weak and nominal. North-eastern Manchuria enjoyed *de facto* independence under the still formidable Fengtian clique, who had the backing of the Japanese, while South China was divided among several military factions including the Anhui clique and Sun Yat-sen's self-proclaimed military government in Guangzhou. By that time, Sun had reorganized his Nationalist Party (also known as *Kuomintang,* the KMT), with assistance from the Soviet Union and the Chinese Communists, and was preparing for a northern expedition. Contrary to the popular narrative that Kang withdrew from political activities and was largely forgotten by his countrymen in his later years, the records show that he often expressed his views on state affairs through private and circular telegrams,[10] and he was sometimes consulted by the Beijing government, including President Cao Kun 曹錕 (in office 1923–1924), and by provisional governors and regional generals. While shunning contact with the revolutionary nationalists (and now the Communists as well), Kang actively sought association with certain warlords, in what would prove to be a vain attempt to enlist support for his old proposal of a titular monarchical republic. In particular, he placed his hope in the 'Confucian General' Wu Peifu 吳佩孚 (1874–1939), commander-in-chief of the Zhili armies, with whom he was on friendly terms and had frequent correspondence until the very last year of his life.

1 Science versus Metaphysics

A bitter war was being fought at the same time in the intellectual world, and an unprecedented debate broke out between metaphysicians and defenders of science.[11] The prelude to the debate was Liang Qichao's reflections on the European War and its dire consequences. At the end of 1918, Liang went to Europe on a one-year tour, whose primary purpose was to observe post-war European societies and study their efforts at reconstruction. He was accompanied by six friends, including the New Confucian thinker Zhang Junmai

10 It is estimated that he spent as much as 1000 Chinese dollars, an impressively large sum according to the standard at that time, on telegrams alone each month. Ma, *Kang dazhuan*, 569.

11 The Chinese word for metaphysics, *xuanxue* 玄學, is a protean term that designates the philosophy of life, morality and religion and so on. Here 'metaphysicians' primarily referred to Confucianists.

張君勱 (1886–1969, later known by his English name, Carsun Chang) and Ding Wenjing 丁文江 (V. K. Ting, 1887–1936), a pioneer native geologist trained in Great Britain – these two individuals would later initiate the debate and become its outspoken protagonists and rival participants. While in Europe, they visited battle sites in various countries, and interacted widely with European intellectuals.[12]

Shocked by how scientific and technological advances had contributed to the rise of modern wars and their destructiveness, Liang started to reflect on the 'European dream of omnipotent science'. 'Not only had the advances of modern science [and technology] transformed the external life of Europeans,' he observed, 'but they had also shaken their internal life.' According to him, Christianity and idealist philosophy, which had been the foundations of European internal life, had been dealt a deadly blow by modern science. Everything, including the human mind, was reduced to a mere movement of matter; principles and laws could only be obtained through the method of scientific experiment rather than through philosophical contemplation. As a result, a purely materialist, mechanistic view of life had emerged 'under the shelter' of science, which encouraged pleasure-seeking, self-indulgence and the law of the jungle. The World War, Liang declared, was a 'retribution'.[13] To avoid repeating the same fate of the Westerners, Liang argued that the Chinese people needed to draw on the wisdom of their ancient Sages with respect to the need for harmony between the spiritual and material aspects of life: what distinguished man from the animal kingdom was his capacity for living a life of the spirit; material life, important as it was, should not impair the development of spiritual life, nor should it become the central issue of human life.[14]

Liang's observations exerted a great impact on Chinese intellectual opinion, and his views were shared and promoted by the defenders of Chinese traditional learning, who challenged scientism. In February 1923, Zhang Junmai gave a speech on 'the philosophy of life' at Tsinghua University in which he maintained that traditional Confucianism had contributed to the spiritual civilization of China by offering solutions for the problems of life, and argued that however developed science and technology were, they had no bearing

12 For example, Liang wrote that he had met prominent European politicians, philosophers and writers, including Henry Bergson. He also had an audience with the king of Belgium. He visited the University of Oxford on 24 June 1919. Ding and Zhao, *Nianpu changbian*, 881, 885.

13 Liang Qichao 梁启超, 'Ouyou xinyinglu jielu 欧游心影录节录 [Recorded Reflections During my Travels in Europe: A Selection (1920)]', in *LQCQJ*, 1999, 2973.

14 Liang Qichao 梁启超, 'Xianqin zhengzhi sixiangshi 先秦政治思想史 [A History of the Political Thought in pre-Qin China (1922)]', in *LQCQJ*, 1999, 3694.

on the issue of the philosophy of life.[15] The proponents of a scientific outlook immediately responded. A rationalist and scientific positivist, Ding Wenjiang rebutted Zhang's views in an article entitled 'Metaphysics and Science'. Label ling Zhang as 'a spectre of metaphysics', he asserted that science was all-sufficient, for everything, including the philosophy of life, lay within the province of science and could be examined by the scientific method.[16] Zhang retorted with a lengthy article, arguing that science was limited in its scope and methods, for there was a body of knowledge, such as are to be found in philosophy and religion, that could not be verified by science.[17] As the debate moved into the second round, Liang intervened as an arbitrator who recognized neither the bankruptcy of science nor its omnipotence. He acknowledged that many problems of life could and should be solved with the application of scientific knowledge, but he also stressed that there was a certain portion, indeed the most important part, of human life, such as love, religious experience, moral sentiment, aesthetic feeling, that had to search for solutions in realms that lay beyond the domain of science.[18]

Over the following year, a large number of the leading scholars and thinkers actively participated in the debate. A few, such as Liang Shuming 梁漱溟 (1893–1988) and Zhang Dongsun 張東蓀 (1886–1962), sided with Zhang Junmai, while many others, including Hu Shi and Wu Zhihui 吳稚暉 (1865–1953), joined Ding Wenjiang's camp. All the 'saints' in religion, philosophy and science, be they Eastern or Western, were enlisted to support one side or the other. At a later stage, Marxist thinkers, such as Chen Duxiu and Qu Qiubai 瞿秋白 (1899–1935), also joined in and presented a materialist perspective. Considering the polemics to be a belated repercussion of the debate between

15 Zhang Junmai 张君劢, 'Renshengguan 人生观 [The Philosophy of Life]', in *Kexue yu renshengguan* 科学与人生观 [*Science and the Philosophy of Life*], by Zhang Junmai 张君劢 et al. (Shenyang: Liaoning jiaoyu chubanshe, 1998), 30–37.

16 Ding Wenjiang 丁文江, 'Xuanxue yu kexue 玄学与科学 [Metaphysics and Science]', in *Kexue yu renshengguan* 科学与人生观 [*Science and the Philosophy of Life*], by Zhang Junmai 张君劢 et al. (Shenyang: Liaoning jiaoyu chubanshe, 1998), 38–56. For a critical English biography of Ding, see Charlotte Furth, *Ting Wen-Chiang: Science and China's New Culture* (Cambridge, Mass.: Harvard University Press, 1970).

17 Zhang Junmai 张君劢, 'Zailun renshengguan bingda Ding Zaijun 再论人生观并答丁在君 [Another Discussion on the Philosophy of Life and My Replies to Ding]', in *Kexue yu renshengguan* 科学与人生观 [*Science and the Philosophy of Life*], by Zhang Junmai 张君劢 et al., new (Shenyang: Liaoning jiaoyu chubanshe, 1998), 30–37.

18 Liang Qichao 梁启超, 'Renshengguan yu kexue 人生观与科学 [The Philosophy of Life and Science]', in *Kexue yu renshengguan* 科学与人生观 [*Science and the Philosophy of Life*], by Zhang Junmai 张君劢 et al., new (Shenyang: Liaoning jiaoyu chubanshe, 1998), 125–32.

science and religion in Europe, Chen Duxiu upheld the monistic belief that only objective, material causes could account for social history and determine the philosophy of life.[19] In the end, the 'scientific' faction, or, more exactly, the proponents of scientism, was generally considered to have obtained the upper hand. As D. W. Y. Kwok has observed, science had emerged stronger than before, and its scope was widened, covering aspects of natural and social life, while the scientists claimed to 'have the key to the imponderables of life and the universe'.[20] Coupled with the anti-religious sentiments in the early 1920s, the debate consolidated and even perpetuated the conflict model of science and religion that would prevail in the Chinese intellectual world for decades until the present day.

Kang Youwei did not participate directly in that debate, and it seemed that he was largely ignored by the mainstream intellectuals. However, he was not silent, and he expressed his views on some of the issues under debate through his letters, talks and lectures. He welcomed Liang's observations on post-war Europe and the latter's restored appreciation of Confucius. Kang commented that the disastrous European War had changed the popular views on Western learning. During a public talk he gave in April 1923, he asserted that the War was a 'result of the extravagant spread of Bentham's utilitarianism and Huxley's (representation of the) doctrine of survival of the fittest'. Post-war Europeans had now been alerted to the fact that their doctrines were not perfect. As a result, Kang continued, erudite scholars in many countries had turned to the humanistic teachings of Confucius, and their respect for Confucius had increased. To support this claim, he quoted a passage from Yan Fu, which expressed disappointment in the 'material-oriented European civilization' and a renewed appreciation of the 'revivifying Way of Confucius'.[21] Kang regarded Yan's testimony as being of special significance, not only because Yan had received his education in Europe and had translated seminal Western works, but also because Yan, according to Kang's understanding, had earlier been attracted to Christianity. Furthermore, alluding to his erstwhile student Liang, Kang added that some Chinese scholars, who had once attacked Confucius, had recently changed their position after a trip to Europe.[22]

In condemning indiscriminate Westernization, however, Kang did not blame science for the war. Indeed, he retained a firm belief in science as a powerful positive force for economic and social development. In the same talk,

19 Chen, 'Xuyi'.

20 Kwok, *Scientism in Chinese Thought*, 159.

21 Kang Youwei 康有为, 'Kaifeng yanjiangci 开封演讲辞 [A Talk Given at Kaifeng (1923)]', in *KYWQJ*, vol. 11, 2007, 238–39.

22 Kang, 239.

he extolled the scientific advances recently made in Europe, and emphasized their contribution to Western material development. He lamented that China was weak because science and technology were not sufficiently advanced, but at the same time he maintained that China was a leader in metaphysical studies. He concluded his talk with a discussion on the complementary relationship between Chinese and Western cultures, or, more specifically, between Confucianism and science:

> Therefore, we should learn from Europe and America as far as material development is concerned. Science should be promoted in order to make up for what we are lacking. But in terms of moral education, we have our own teachings, which should be preserved and should never be disregarded.... We could offer better service to our country and the world by externally pursuing European and American sciences while internally preserving and practicing our national essence, the Confucian religion.[23]

Here Kang seems to have advocated a similar version of *zhongti xiyong* 中體西用, the guiding principle of the Self-Strengthening Movement, which regarded Chinese learning (Confucianism) as the internal, spiritual foundation and Western learning (science and technology) as the external, material force for China's development. This view of complementarity can be taken as Kang's basic stance towards the debate between the scientists and the metaphysicians, and it fairly characterizes his outlook on the relationship between Chinese and Western civilisations in general. Nevertheless, while he was reaffirming the importance of Confucianism to contemporary China, a position that resembled the traditionalists' views, Kang had also, in a similar but not exactly the same way as the stance adopted by the proponents of scientism, accorded to science an important role to play in shaping people's views on life, the world and even the universe, as will be seen in the following discussions on the themes of his popular talks given in the North.

2 A Lecture Tour to the North

Throughout 1923, as the debate on science and metaphysics gained momentum, Kang set out on an extensive tour to the North. It started with his visit to the city of Luoyang, where he attended a birthday celebration of General Wu

23 Kang, 239.

Peifu.[24] Then he visited a few other cities such as Kaifeng, Baoding, Ji'nan and Qingdao, where he also delivered public talks. The most important visit was to the northwestern city Xi'an, which had been capital city to many bygone dynasties. Presumably because of Wu's introduction and partly because of his own prestige, Kang was granted an extraordinary welcome when he arrived at the ancient city on 6 November 1923. Seated in an automobile car, he was escorted into the city gate by a brigade of guards consisting of both cavalrymen and infantrymen. As he proceeded, he was greeted by a huge number of spectators standing on both sides of the street, with drums beating and flags waving, and crowds hailing him as 'Sage Kang'.[25] When he reached the Palace of Eight Immortals (*Baxian'an* 八仙庵, a Daoist monastery, where the late Empress Dowager and Emperor Guangxu had temporarily taken refuge during their flight to the city after the capture of Beijing in 1900), he was warmly received by a group of the local elite led by Liu Zhenhua 劉鎮華, the governor-general of Shaanxi.

During his sojourn in Xi'an, Kang was invited to give a dozen talks to audiences from various political, intellectual, religious and journalistic circles. Of particular importance were the talks delivered at the branch of the Confucian Religion Association, the branch of the Universal Morality Society (*Wanguo daodehui* 萬國道德會),[26] the Buddhist Society and the Young Men's Christian Association. His presentations were well-attended, sometimes by thousands of people. Of the various themes that ran through these talks, two recurred and stood out. One was the renewed call to restore Chinese moral character and instil cultural confidence by venerating Confucius and preserving Confucianism, an echo of his previous call for a Confucian religion. The other theme was his repeated emphasis on the importance of material development, and in particular the significance and role of science and technology, on which he had been speaking publicly over the previous decades.

---

24 Kang, however, rarely ever mentioned his attendance at the celebration, possibly because he considered it inconsistent with the preservation of his dignity. See Kang, 'Kang Nanhai shiji', 416; see also Tao Juyin 陶菊隱, *Wu Peifu jiangjun zhuan* 吳佩孚將軍傳 [*A Biography of General Wu Peifu*] (Shanghai: Zhonghua shuju, 1941), 89.

25 Kang, 'Kang Nanhai shiji', 358.

26 Formally established in Shandong in 1921 by a member of the CRA, the Universal Morality Society became one of the most prominent redemptionist movements in China, and Kang would serve as acting president in 1923 before being elected president in the last year of his life. See Kang, 'Zhi Yanshenggong han 致衍圣公函 [A letter to Duke Yansheng]', in *KYWQJ*, vol. 11, 2007, 252; *Wanguo daode zonghui gaikuang 1921–1980* 萬國道德總會概況 *1921–1980* [*An overview of the Head Office of the Universal Morality Society 1921–1980*] (Taibei: Daodehui Press, 1980), 16.

2.1 *The Confucian Way*

More than ever before, during his discourses, Kang placed primacy on religion and Confucianism within a context of comparative religious studies. He started with a reconsideration of the terms, religion, *zongjiao*, and *jiao*. While he had previously denounced *zongjiao* as a mistranslation of the Western concept religion, which he had thought was a broader, generic term, he now acknowledged that *zongjiao* was actually a more accurate and acceptable rendition of the term. He argued that the English word 'religion' and the Japanese term *zongjiao* referred to *shendaojiao* or 'teachings of the way of the spirits', which were primarily concerned with the soul, spirits and deities, and as such, they were inadequate to account for Confucianism, which, although rooted in a Heavenly origin, was oriented to the needs of humanity, and therefore was primarily concerned with *rendao*, or the ways of being human.[27]

There might be two main reasons for this dramatic change in Kang's account of the terms. One is that Kang now had a better grasp of what was meant by 'religion' in English, a term he now said had derived its meaning from a Greek context.[28] Whether in its contemporary use or in its etymological and historical definitions, 'religion' was concerned primarily with reverence for or worship of divine beings. This claim had already been expounded by some of the Chinese participants in the debate over science and metaphysics, who were aware of Max Müller's discussions on the historical meanings of 'religion' three decades earlier.[29]

The second reason is that the semantic concept of *zongjiao* or religion as such had largely been discredited by that time, and to continue to assert that Confucianism was, properly understood, a religion would only do more harm than good to Confucianism itself. In this sense, he had come to accept what Liang had advocated some twenty years before, claiming that Confucianism was different from and exalted above a *zongjiao* or religion in the Western sense. Given the climate of widespread intellectual revulsion towards organized religion, it had become strategically advantageous for Kang to dissociate Confucianism from the term *zongjiao*, and to content himself with classifying it as a *jiao* in general and a *rendaojiao* or 'teachings of the human way' in particular. In other words, Kang had abandoned his attempt to make a universal concept out of the English term 'religion'. However, he maintained that the Chinese concept of *jiao* continued to be a universal phenomenon, but with different manifestations in different cultures.

---

27 Kang, 'Kaifeng yanjiangci', 236; Kang, 'Chang'an yanjianglu', 275.
28 Kang, 'Chang'an yanjianglu', 279.
29 See Müller, *Natural Religion*, 33–43.

During his talk at the Xi'an branch of the Universal Morality Society on 19 November, Kang gave a comprehensive, critical review of the way religions had first appeared and later developed in different regions. He started with a quote from Hong Liangji 洪亮吉, an eighteenth century Chinese scholar and statesman: 'Each of the Nine Continents on the earth had its First Sage.'[30] For Kang, this verse touched upon the question of the origin of religions. 'After the appearance of human beings,' he continued, 'there must have been someone from among them to arise first and proclaim laws or doctrines (*shuofa* 說法), so as to teach his fellow beings to avoid evil, do good and live a moral life.' As the world was not interconnected, each region had its own Sage, and these Sages were the founders of various religions (*jiaozhu*).[31] So those founders were self-driven rather than being instructed to do so by a personal higher power; their religions were founded to regulate the deeds of the people, and the goal of religion was the realization of morality among its practitioners. Therefore, Kang claimed, once born into this world, one stood in 'absolute need' of religion (*jiao*), without which one was 'hardly any better than the beast'.[32]

Having thus put forward his concept of religion (*jiao*) and its origin, Kang discussed each of the major extant religions in the world, often drawing on his own observations of their religious sites, customs and practices during his extensive travels. In doing so, he frequently compared and contrasted them, highlighting both differences and possible connections between them. In the land of India, there had first appeared Brahmanism. When the rival Buddhism arose, Brahmanism came to be considered by Buddhists as 'heterodoxy'. The truth was, Kang argued, that being born within the matrix of Brahmanism, Buddhism had absorbed many of the latter's doctrines. In Europe, the religion of Moses, now known as Judaism, was promulgated before it was eclipsed by the Religion of Jesus. Judaism was disadvantaged, Kang commented, by limiting its God to the Israelites, and by refusing to teach the gentiles. He extolled Jesus for his dramatic reform of the old tradition and his efforts to evangelize all peoples regardless of tribe or race. Echoing his early position, Kang argued that Jesus, being an eclectic, had absorbed Buddhist doctrines. His evidence included the alleged spread of Buddhism to the West following Alexander's conquest of India and the ritualistic resemblance between Catholicism and Buddhism. At the doctrinal level, Kang argued that the Christian notion of Heaven was originally taken from the Buddhist concept of *rentianjiao* 人天教

30 Kang, 'Chang'an yanjianglu', 286. According to Chinese traditional thought, the world consisted of nine continents.

31 Kang, 286.

32 Kang, 282.

(literally, the Human-Heaven teaching), a doctrine that emphasized rebirth in Heaven instead of attaining Buddhahood.[33] In addition, he claimed that a professor at Oxford had proven that over seventy doctrines of Christianity could be traced back to Buddhism.[34] In Arabia, Muhammad founded Islam, which spread to many parts of the world, including Persia, a land that had witnessed the birth of the ancient 'Sun religion' (*Rijiao* 日教) founded by Zoroaster, whose dualistic cosmology, he said, resembled that of Confucius in his view. When it came to Chinese native religions, Kang saved Confucianism for another discussion, but he gave a severe critique of Daoism. As in his earlier writings, he disapproved of the Daoist pursuits of physical immortality and he criticized various occult techniques, saying they were actually not in accordance with what Laozi had taught.[35]

Throughout this historical account, Kang emphasized what he characterised as three common features of all religions: a founder, a book and a community of believers. In other words, a self-motivated teacher brought forth a set of teachings, which were then enshrined in a book (the Vedas, the Bible, Buddhist Sutras and the Qur'an), and were practised by generations of his followers. In addition, he stressed that all religions taught reverence for Heaven and love of humanity (*zuntian airen* 尊天愛人). Moses' teaching, for example, was centred on worship of Heaven, and this was preserved by Jesus in his religion. Muslims were most diligent in worshipping Heaven, which reminded him of the Confucian teaching on being always mindful of 'God' (*Di*) in the *Odes*. The concept of Heaven, as he explicitly claimed during his talk at the YMCA, was the same as that of God in Christianity.[36] As for love of humanity, all the founders of religions loved others and exhorted their followers to do the same, although, Kang commented, there was nevertheless a difference of degree in their teachings. While 'an eye for an eye' should be avoided and injuries should not be returned with revenge, to turn your other cheek when you were slapped on one, as Jesus had taught, was extreme and difficult for common folk to practise. Similarly, to demonstrate Buddhist compassion and love was praiseworthy, but it would be very hard for the followers to emulate the Buddha who is said to have fed an eagle or a tiger with his own flesh in order to rescue their intended prey, according to certain Buddhist legends. In contrast, Kang emphasized that

33 For more details on the doctrine, see 'Rentianjiao 人天教 [Human-Heaven Doctrine]', Digital Dictionary of Buddhism;'Huayan Wujiao 華嚴五教', in *The Princeton Dictionary of Buddhism* (Princeton: Princeton University Press, 2013).

34 I have not been able to locate the source of this claim.

35 Kang, 'Chang'an yanjianglu', 286; see also Kang, 'Woshi', 62.

36 Kang, 'Chang'an yanjianglu', 280, 285–88.

Confucius showed a Middle Way in his teaching on reciprocal love and respect, and Confucianism was therefore more suitable for the majority of humanity.[37]

Given that all religions shared these common teachings, Kang reminded his audience of the necessity for the unification of all religions in the future, a theme that he had, as we have seen, conceived in his youth and later elaborated in his *Book of the Great Unity*.[38] As an initial step, he proposed the union of Confucianism and Buddhism in the context of Chinese culture. For this he had a two-process design. People should practice Confucianism in their prime, honouring parents and giving birth to children; after they reached fifty, when women arrived at their menopause and men were approaching the end of their life, people should practise Buddhism, thereby nourishing their souls.[39]

It might be observed in this context that Kang did not mention any miracles related to those founders, nor did he talk about their claim to divine revelation. This does not mean, however, that Kang was ignorant of the phenomena of miracles and divine revelation claimed in the Abrahamic religions. He chose to overlook them, for in his view, these claims were the mere methods employed by some of the 'self-driven' founders of religions, rather than the universal features of all religions. It seems doubly ironical or paradoxical that Kang was himself occasionally engaged in an effort to point to a divine or heavenly source of inspiration in the teachings of Confucius, as instanced by his not very resolute attempt to deify Confucius as having been the offspring of the 'Black God'. Taking Kang's earlier discussions on the concept of *jiao,* the sages and human nature, it may be argued that Kang also believed in divine inspiration, but that inspiration, instead of being imparted directly by a higher divine source such as the Holy Spirit, came, indirectly, from the Sages' reflections on what Confucian Classics called the Heaven-endowed human nature.

The underlying aim of Kang's discourse on comparative religion was to defend and exalt Confucianism. It was his consistently-held view that religions differed not so much in *what* they taught as in *how* they taught it. He had earlier classified religions according to whether their primary means for instruction was 'the way of the spirits' (*shendao*) or 'the human way' (*rendao*). Kang reiterated his assertion that Confucianism was superior because it encompassed both of those ways. He defended his view by revisiting the beginning verses in the *Mean* on the definition of human nature, the Way and *jiao*. Since *jiao* was about cultivating man's Heaven-endowed nature, Kang started by

37 Kang, 287.

38 Kang, 285–88.

39 Kang, 296.

expounding on the meaning of human nature. He made a categorical claim that the Neo-Confucian master Zhu Xi had been one-sided in his interpretation of human nature as a mere set of moral principles, for human faculties and desires should also be considered part of human nature. If this was so, it followed that a *jiao* would have to tend to both aspects of human nature: moral principles and human desires and faculties. That, in his view, was precisely what Confucius had done: Apart from teaching on the Five Constant Virtues, Confucius recognized such human desires as the sexual impulse and established the institution of marriage to accommodate it. This husband-wife relation provided the foundation for the other four primary relations (father–son, ruler–minister, elder-junior and friend–friend), for each of which Confucius had set up the appropriate rites and norms of ritual propriety. Likewise, dance and music were also designed to please human eyes and ears, and pleasant food and delightful smells were allowed to please tongues and noses. In summary, Confucius' teachings were so comprehensive that they encompassed both the spiritual and physical aspects of human life. That was why it was said that no one could live outside the Way of Confucius, if he chose to live like a human being. On the contrary, the *shendao* religions were focused on the soul and took little care of the body; they even suppressed or denied human desires, such as in the case of the practice of celibacy. In other words, they were outside of the human way and therefore only suitable for certain people rather than for all.[40]

According to Kang, another big difference between Confucianism, a primarily *rendao* religion, and those primarily *shendao* religions consisted in whether gratitude and veneration were duly accorded to parents and ancestors. Kang reasoned that as human nature was imparted to each one by Heaven through one's parents, it followed that for one to come into being, it required the union of three factors: Heaven, a father and a mother. That was why it was said in the Classics, Kang argued, that 'man cannot be born until the union is achieved between the three entities.'[41] For Kang, this notion was one of the most profound doctrines of Confucius and constituted the definitive feature of Confucian teachings in contrast to the tenets of other religions. His reasoning was as follows. Once people realized that they, like everyone else and indeed everything else, were born of Heaven, then they would understand that they were one with their fellow human beings and even with a myriad of things, and it would only be natural for them to show forth benevolence to them. Similarly, once people realized that they were also born of their parents, then it would

40 Kang, 276.
41 Kang, 276.

only be natural for them to honour their parents and ancestors and to show affection towards their family or clan members. This was nothing less than the cardinal virtue of filial duty. The two virtues, benevolence and filial piety, Kang argued, constituted the two mighty pillars of the Way of Confucius. They were ritually manifested in worship of Heaven and veneration of ancestors respectively, both of which were necessary for the maintenance and governance of a country. Evaluating the various faith traditions from this perspective, the disadvantages of other religions soon became obvious – they only revered Heaven (God) without giving due honour and gratitude to deceased ancestors. Thus, Kang concluded, Confucianism was a more balanced, more comprehensive religion and it was better suited to the needs of the people of China and, indeed, the entire world.[42]

One conspicuous feature of Kang's talks and writings in his later years was his downplaying of the dividing line between what he had called the 'false' imperial Confucianism and his own restatement of 'pristine' Confucianism. Whereas he had earlier advocated the New Text school and disregarded many of the Old Text classics as fraudulent, he now rarely stressed the distinction between the two schools. That was quite understandable, for Confucianism itself had been called into question since the 1911 Revolution and in particular during the New Culture Movement. That left Kang in a dilemma. To argue for the continued relevance of Confucianism to modern China, he had to resort to his critique of Neo-Confucianism by means of New Text exegesis, but in doing so he would inevitably weaken the authority of Confucianism and thereby play into the hands its opponents. In the end, he concluded that sectarian disputes must be overcome for the sake of the survival of Confucianism as a whole. Nevertheless, while tactfully restraining himself from public discourse on the Old Text versus New Text controversy, Kang did not entirely ignore the distinction.[43] Both in his private writings and in his letters to foreign (mostly Korean) Confucianists, his attack on the Old Texts remained as strong as his advocacy of his version of Confucianism.[44]

Clearly, Kang's defence of Confucianism was positioned as a response to the contemporary criticisms against religion in general and Confucianism in particular. His strategy of dissociating Confucianism from *zongjiao* as well as

42 Kang, 276; Kang, 'Kaifeng yanjiangci', 236.

43 See the discussion in Hsiao, *K'ang Yu-Wei*, 126.

44 See Kang Youwei 康有为, 'Da Pu jun datixue shu 答朴君大提学书 [A Reply to Mr. Pak-ki-yang, the Korean Commissioner of Education (1924)]', in *KYWQJ*, vol. 11, 2007, 344–46; see also Richard Wilhelm, *The Soul of China*, trans. Arthur Waley (London: Jonathan Cape, 1928), 76–80.

from the Western conception of 'religion' resembled the approach adopted by Liang Qichao in the early 1900s. Nevertheless, Kang maintained a belief in a universal concept of *jiao* with culture-specific forms of manifestation, and he attempted to demonstrate that Confucianism, if properly understood and followed, represented the more advanced stage of such a *jiao*. His views, however, were ignored in mainstream Chinese intellectual thought, which had become deeply anti-traditional and anti-religious in the 1920s.

2.2 *The Power of Science*

The other theme of Kang's talks was the importance of material development, backed by science and technology, for China. Kang asked his audience to extend their love or *ren* to their motherland by making up for what was lacking in China. One of the priorities, in his view, would be to build highways and railways and set up telephone lines following the example of the European nations, as better connections between different cities and regions would lay the foundation for further development. Another priority was to pay greater attention to science and technology, whose power had underpinned European economic and industrial development. This was an echo of the call he made in his *Material Reconstruction* about two decades earlier. It is worth noting that his proposal was not limited to the mere introduction of Western science and technology. He showed both his insight and foresight in emphasizing the need for China to encourage scientific inventions and their application by setting up a mechanism to protect the rights of inventors. As the recent history of nations across the world showed, he asserted, it was the accumulation of inventions that had distinguished between the powerful and the weak nations, the rich and the poor, and even the civilized and the 'barbarian' countries.[45] In an effort to combine the spiritual and the material aspects of life together, Kang stressed that devoting oneself to the material development of one's motherland was nothing short of a manifestation of God's love and compassion towards his fellow countrymen.[46]

Apart from extolling the material application of science, Kang often waxed eloquent in praising the inspirational power of scientific knowledge, which, he argued, would help reorient one's views of man, the earth and even the universe. To begin with, Kang remained fascinated by 'electricity'. Earlier on in his annotation on the *Mencius*, he employed 'electricity', together with 'ether', in his interpretation of the all-important virtue of *ren*, arguing that the former constituted the essence of the latter. He had also considered electricity as the

45 Kang, 'Chang'an yanjianglu', 275.
46 Kang, 282.

medium through which human consciousness or thought was spread, not only within the body but between human beings. In his later years, he dropped the concept of ether, but he went further with his bold claims on electricity.[47] He told his audience that he had written a book entitled *Diantong* 電通, literally, 'Interconnection by the Means of Electricity', in which he had elaborated its universal principles.[48] He distinguished between two types of electricity. The visible was what was commonly known as electricity while the invisible was in fact what was commonly known as spirit. For Kang, the wireless telegraph system then recently developed by Marconi was an extraordinary invention, for it not only made possible the instant transfer of messages between America and Europe, but it also showed a glimpse of the omnipresence as well as the power of the invisible form of electricity.[49] Kang borrowed the idea of a 'transmitter of wireless electricity' (*wuxiantai* 無線台) and attempted to account for the claims and recorded instances of transmission of thought without the means of human sensory channels or physical interaction in Confucian and Buddhist literature.[50] Commenting on Confucius' saying that 'he who is of utmost sincerity is capable of foreseeing', Kang explained that that was possible because 'utmost sincerity' (*zhicheng* 至誠) was in essence a transmitter and whoever possessed the quality was able to send out wireless electricity, through which foreknowledge of things was obtained.[51]

In a similar manner, Kang employed the concept of 'universal connection through electricity' in explaining the doctrine of karma. As briefly noted in the previous chapter, Kang was a firm believer in reincarnation, the doctrine that at physical death the soul would transmigrate to or be reborn in another body, human or animal. For Kang, the truth of reincarnation was incontestable, for it was attested to in the ancient texts of India, Egypt and Babylon as well as in Buddhist sutras and Confucian apocrypha, many instances of which were also reported in Chinese histories and folklore. Although the New Culture intellectuals denounced the concept of reincarnation as sheer superstition, Kang never concealed his belief in it. During his first talk at Xi'an attended by the

47 This indicates that Kang was aware that the concept of ether had been discredited by Einstein's theory of relativity, a fact that he highlighted in a letter penned three years later. See Kang Youwei 康有为, 'Yu Tongfu Tongjian Tongning shu 与同复同铵同凝书 [A Letter to Tongfu, Tongjian and Tongning (1927)]', in *KYWQJ*, vol. 11, 2007, 453.

48 The book was never published, and its manuscript has not been seen.

49 Kang, 'Chang'an yanjianglu', 274. As is shown in the following sentences, Kang used the phrase 'a transmitter of wireless electricity' instead of the standard term 'a radio transmitter'.

50 The popular English term for such phenomena is 'telepathy' coined by Frederic Myers, a founder of the Society for Psychical Research.

51 Kang, 'Chang'an yanjianglu', 274.

local elite in military, civic and literary circles, he recounted an anecdote concerning two brothers in his village. When travelling to another village, they had been greeted by a nine-year boy, who asked knowledgeable questions about their mother and family debts; the boy was believed to be a reincarnation of their deceased father. Kang asserted that it was through the means of electricity that divine reward or retribution were meted out to man, according to which one's welfare and fortune were determined not only by his efforts in this life, but also, and even more importantly, by his deeds in the previous life. Therefore, one should not complain about the adversities in this life, and should always strive to do good so as to have a better next life.[52] Kang's use of electricity to account for divine retribution brings to mind the efforts by scientists such as Oliver Lodge and others, as mentioned in Chapter Two, to use radio waves and electromagnetic waves to try to communicate with the dead.

For Kang, electricity was not only a universal medium that penetrated and connected everything, but it also constituted the essence of human virtues and even the human soul. 'Everyone possesses electricity,' he wrote, 'but its growth requires cultivation on the part of an individual. With diligent nourishment, the scattered dots of electricity in his body can aggregate and form an electrical crystallization, an entity that, if well-preserved, will become bright and lasting.' He argued that it was through the power of his well-gathered and preserved electricity that the Buddha had been able to fathom so many mysteries. He further affirmed that the method for gathering electricity was nothing less than the cultivation of one's virtues, a practice that was advocated in the Confucian tradition. The 'illustrious virtue' mentioned in the *Mean* was meant to refer to electricity, and 'to make the illustrious virtue illustrious' in the *Greater Learning* was in fact an exhortation 'to gather together your electricity'.[53] The best way to gather electricity was to practice the virtue of *ren*, which in Kang's view bore an 'intrinsic resemblance' to electricity. 'The natural attraction between the *yin-negative* and the *yang-positive* electrical charges', he said, 'is in fact the love between two people as conveyed by the written form of the concept of *ren* (仁), which consists of the radicals for "two" and "humans".' He commented that Kropotkin's concept of mutual aid, which was being popularly promoted in Europe and America, corresponded with the benevolent love deeply implied by *ren*. When this love was extended to a family, a nation and the world, *ren* was manifested therein, and electricity was accordingly more prevalent, bringing about interconnected harmony.[54]

52 Kang, 274, 289.

53 Kang, 274.

54 Kang, 275.

'Microscopy' was another modern scientific art that engaged Kang's lasting interest and imagination. Earlier on in his autobiography he had discussed his experience of looking through a microscope in his youth and its inspirational impact on his thought. Now, about twenty-five years later, he resumed the topic and elaborated on microscopy and its religious implications during his talk at the YMCA in Xi'an on 17 November 1923. As in the case of his discussion on electricity, he claimed that he had composed a manuscript on microscopy (*Xianwei* 顯微),[55] from which he quoted: 'I have once observed an ant through a microscope with 300 times magnification,' he said, 'and the ant appeared about five Chinese feet long, with heart, liver, spleen, lungs and kidneys about the same size as a human's.' Based on this observation, Kang began a lengthy process of deduction and speculation. If observed through a microscope with 3,000 times magnification, the ant would appear ten times larger. With increasing magnification, Kang reasoned, the ant would appear larger and larger, conceivably even as large as the earth. Indeed, it could appear infinitely large as the magnification of the observing device infinitely increased. It followed that the ant could, like the earth or even the universe, contain a myriad of entities that were simply too small to be noticed by the naked eye.[56]

For Kang, that process of mental reasoning had powerful spiritual connotations. To start with, it could easily lead one to the realization of the relativity of such common concepts as big and small, manifest and hidden, visible and invisible. Those differences and distinctions could only hold true from the viewpoint of the naked eye, for there were numerous things existing beyond human vision. If one could rise above the limitations of one's physical eye and observe in the mind the vast space that appeared empty, one should be able to know that God (*Shangdi*) and spirits were all around in the invisible realm. They were watching at every moment, and therefore one should always be vigilant even when alone.[57] Here Kang was attempting to offer a 'scientific' rationale for the Confucian doctrine of *shendu* (慎獨 being watchful over oneself when one is alone), drawing his audience's attention to the exhortation in the *Mean* that the superior man did not wait to be cautious and apprehensive until he saw physical indicators with his physical eyes. For Kang, this spiritual awareness and practice was essential to the cultivation of one's illustrious virtue, which he had now unequivocally identified with the human soul. 'The illustrious virtue is the soul in the finer sense,' he claimed. The knowledge that God and spirits were watching all the time would help prevent one from hav-

55 The alleged manuscript was never published and perhaps is not extant any more.

56 Kang, 'Chang'an yanjianglu', 280–81.

57 Kang, 280.

ing evil thoughts or committing acts of evil, which would obscure and defile the soul. As an illustrious soul was needed to match 'the benevolent and loving heart of God', Kang exhorted his audience in these words: 'Remember Heaven's will all day long and remain ever cautious and reverent, and you will be able to bring out your illustrious virtue (*mingde*).'[58]

In addition to knowledge of microscopy, astronomy continued to be a source of inspiration for Kang. Twice during his talks in Xi'an, he dwelled on the theme of how knowledge of astronomy could help relieve one's anxieties and sorrows by reshaping one's views of self, the earth and the heavens. He started with a statement that he had elaborated on in his *Book of the Great Unity* – that everyone, every family, every nation in this world was afflicted with various kinds of worries and sorrows. In a sense, the purpose of this life on the earth was to relieve oneself of worries and become more contented and happy. This was, Kang claimed, the very goal of all religions, whose founders had, like great physicians, offered a number of recommendations for mankind. Each of their prescriptions, however, had its pros and cons. Kang told his audience that he had a better prescription, namely a 'celestial journey or peregrination (*tianyou* 天游), a new doctrine that he had been able to adduce based both on his experience of living through sorrows and on his increased knowledge of the heavens. To experience the joy of celestial journey, one did not have to go to heaven; instead, one simply needed to shift one's perspectives with the help of modern astronomy. First, understanding the Copernican perspective that the earth, rather than being the centre of the universe, was just a small planet in the vast solar system would make it easier for its inhabitants to view their planet as one village, diverse countries as neighbours, and one's own nation as a family. Second, the knowledge that the earth, if viewed from other planets, was high in the heavens would help its inhabitants consider themselves as celestial inhabitants, no longer having to yearn to go to heaven. Third, the realization of how vast and limitless the universe was would help people see how puny and trivial their problems and worries were.[59] In the same vein, he commented that new astronomical discoveries made with the telescope had invalidated the teachings on the heavens in all religions and had also brought great joy to human hearts. Revising the Confucian concept of *gewu qiongli,* he exclaimed, 'Indeed, the investigation of things and the fathoming of principles can bestow upon man limitless power and inestimable joy!'[60] As we shall see, he would dwell on these topics in his last book on the heavens.

58 Kang, 280.

59 Kang, 271–73, 288–89.

60 Kang, 273.

In summary, Kang had in his various talks proposed a twofold programme for saving China – to preserve China's national essence, that is, Confucianism, and to make up for what she was lacking, that is, material development backed by scientific inventions. This position represented his eclectic view on the complementary relationship between Chinese and Western civilizations, and can be seen as his response to the iconoclastic New Culture–May Fourth intellectuals who had advocated an indiscriminate Westernization. As for individual salvation, Kang also put forward a two-fold recommendation – to revere God and practice *ren*, and to heed the spiritual implications of science, in particular electrical studies, the art of microscopy and astronomy. Ironically, Kang's stress on and appreciation of the enormous impact of scientific knowledge upon one's views of life and of the world resembles the views of the advocates of scientism in the ongoing debate over science and metaphysics. But there was a noticeable difference – he always bore in mind the spiritual dimension of life and maintained a belief in a spiritual Heaven. 'It is my hope,' Kang exhorted his audience at the end of his first talk in Xi'an, 'that all of you will always heed Heaven's will and bear in mind the universal connection of electricity, so that you will be able, under the guiding principle of *ren*, to fulfil your responsibilities, determine your destiny and secure boundless and eternal blessings.'[61]

3 Celestial Peregrination

> All under Heaven are one family, and the Middle Empire one person.
> To fathom out a myriad of things, and to nourish all with benevolence.[62]

Kang Youwei composed this couplet in early 1926 for his newly-established Academy of Celestial Peregrination (*Tianyou xueyuan* 天游學院). Located within his own residence in Shanghai, the academy was the third that Kang had established in China in his lifetime, the first two being the *Wanmu caotang* in Guangzhou and the *Shengxuehui* in Guilin. He made himself president of the academy, and employed several tutors. Its mission statement echoed the universal ideals of the couplet – 'to investigate the principles of heaven, earth, humans and things for the benefit of the mind, the body, the family, the state and the world (*tianxia jiaguo shenxin* 天下家國身心)'. Its curriculum covered five disciplines – classical studies, philosophy, literature, politics and foreign

61 Kang, 275.

62 Kang, 'Kang Nanhai shiji', 416.

languages. Under philosophy were listed such subjects as astronomy, geography, electricity, biology, science of man (*renxue* 人學), the human way (*rendao*), Chinese philosophy, Japanese philosophy, Western philosophy, psychology, ethics, sociology, the way of the spirits (*shendao*), and the doctrine of the Great Unity.[63] The curriculum demonstrated Kang's usual eclectic approach to teaching and his emphasis on the vision of the Great Unity. Compared with his earlier curricula, it also revealed the increased importance Kang had now come to attach to the study of supernatural and metaphysical subjects such as souls, ghosts and spirits.

To many of his contemporaries and some of his future Chinese biographers, however, the mixture of 'idealist and materialist' subjects did not make much sense and was a sign of 'a confused mind'.[64] This may partially explain why the academy was able to recruit only around twenty students for the first term. The small number also indicated the dramatic decline of Kang's prestige among young students. Kang was disappointed, but he did not lose hope. 'Jesus had only a dozen disciples,' he commented, 'but his religion has now been promulgated across the world. So, numbers do not matter much.'[65] The academy lasted only a year till the end of 1926. Kang closed it because he decided to leave Shanghai, which was likely to be overrun by the southern Nationalist forces, backed by the communists, which had recently defeated General Wu Peifu in their northern expedition. As always, Kang was determined to shun the revolutionaries, nationalists or communists.

3.1 *Lectures on the Heavens*

Kang took on the task of lecturing on astronomy, which, as the academy's name suggested, occupied a central place in its curriculum. He claimed that his lecture notes had been drafted some forty-two years earlier, though he added that they had been improved through additions and revisions.[66] Although there has been suspicion that Kang may have back-dated the origins of these notes, it is beyond doubt that, as discussed in Chapter One, he had already become enormously interested in astronomy in the 1880s. He had turned himself into a well-informed amateur in astronomy through diligent reading of translations

---

63 Cited in Wu Tianren 吳天任, ed., *Kang Youwei xiansheng nianpu* 康有為先生年譜 [*A Chronological Biography of Mr. Kang Youwei*] (Taipei: Yiwen yinshuguan, 1994), 768.

64 Ma, *Kang dazhuan*, 610.

65 Ren Qisheng 任启圣, 'Kang Youwei wannian jiangxue jiqi shishi zhi jingguo 康有为晚年讲学及其逝世之经过 [My Recollection of Kang's Lectures in His Later Years and His Passing Away]', in *Wenshi ziliao xuanji* 文史资料选辑 [*A Selection of Literary and Historical Materials*], vol. 31, 1962, 241.

66 Kang, 'Zhutian jiang', 13.

of Western books, a learning process that was enriched by his observations through telescopes. His interest in astronomy was sustained throughout his life. During his tours around the world following the stormy reform years, he took every opportunity available to him to look through the telescopes at a number of observatories. 'When observed through the telescope,' he wrote concerning fixed stars being all suns, 'their rays and brilliance differ from those of planets and resemble those of the sun.'[67] In his last years, when the political and military situations had frustrated each of his efforts to 'save China', he resumed his studies of astronomy, finding solace in 'celestial peregrination'.

As he was giving lectures on astronomy, he further revised his lecture notes in preparation for their publication as a book. Although he wrote that he had completed the manuscript in the summer of 1926, its revision did not seem to end there. In a letter to his children dated 10 February 1927, less than two months before his death, he instructed that two lines should be added to the chapter on Einstein. 'Einstein has surpassed Newton,' he wrote, 'in his abandonment of the concept of ether and in this discovery of the conic curvature of light concerning universal gravity.'[68] Clearly Kang was referring to Einstein's theory of relativity. This indicates that despite his advanced age, he tried to keep abreast of new scientific knowledge. By the early 1920s, both Einstein and his theory of relativity had become well-known among Chinese intellectuals in China through an increase of publications on the topic in Chinese prompted by such events as Bertrand Russell's lectures in China in 1921 and Einstein's planned visit in 1922.[69] As a result, the theory of relativity was quickly and widely accepted by the New Culture intellectuals, many of whom found its revolutionary nature and implications particularly attractive.[70]

Although Kang wanted that book to be finalized with all due haste, it was not published by his students niuntil 1930, three years after his death, under the title of *Zhutian jiang* or *Lectures on the Heavens*. For the next three decades, however, the book was almost completely ignored, and even his admirers and students attached little importance to it. It was Hsiao who first seriously examined the book in the 1960s; he offered original insights on Kang's scientific attitude, which permeated those lectures.[71] The book consisted of fifteen chapters of various lengths, twelve of which dealt primarily with astronomical

67 Kang, 55.

68 Kang, 'Yu Tongfu'.

69 Einstein only stopped in Shanghai on his way to Japan; it had been arranged for him to pay a formal visit to China afterwards, but that did not materialize due to disagreement on the amount payable to him.

70 See Danian Hu, 'The Reception of Relativity in China', *Isis* 98, no. 3 (2007): 539–57.

71 Hsiao, 'Excursion into Science', 375.

knowledge in the West, while the remaining three, Chapters 10, 11 and 15, were largely concerned with content of a religious, philosophical or cosmological nature. With those chapters on the moon and the Milky Way being the longest ones, the lengths of the sections were decided less by the intrinsic importance of the matter under discussion than by the author's personal interest or the amount of information available to him in translated works.[72]

Kang began the first chapter with an account of the study of astronomy in the history of China but considered it 'like the prattle of children' from the perspective of the knowledge available during his time.[73] In the next nine chapters, he dwelt respectively on the earth, the moon, the sun, planets, comets, meteors, the Galaxy, and the extragalactic regions which he called 'Heavens of Rosy Clouds', covering such subjects as their origin, composition, form and movements. A survey of these parts shows that his knowledge of astronomy, although not free from inaccuracies, was exceptionally broad and, as Hsiao points out, was quite up-to-date at that time, a remarkable achievement for a man without special training who was dependent on translations for his astronomical knowledge.[74] When discussing the origin of the earth, for instance, Kang introduced the 'New Nebular Hypothesis' (Planetesimal Hypothesis) of T. C. Chamberlin (1843–1928) and F. R. Moulton (1872–1952) as well as the 'Nebular Hypothesis' of Immanuel Kant (1724–1804) and Pierre-Simon Laplace (1749–1829), and pointed out that the new theory was extremely popular.[75]

In the tenth chapter, 'on the extragalactic heavens', Kang flew from astronomical facts into sheer cosmic imagination. Although Europeans 'have stopped with the extragalactic regions' in fathoming the heavens, he asserted that there must be numerous heavens beyond the extragalactic sphere.[76] Since those limitless heavens transcended understanding, he claimed he was 'only' able to speculate and name 240 of them. Having thus established the infinity of the universe based on what he believed were astronomical facts, Kang marched into the transcendental realm of religion in the next chapter, asserting the existence of God above all the heavens. In the following chapter on Buddhist literature on heavens, he concluded that the Buddhist views were fanciful and baseless. This, however, was understandable because the Buddha, like the other founders of a religion, was born in ancient times without the use

72 Hsiao, 282.
73 Kang, 'Zhutian jiang', 17.
74 Hsiao, 'Excursion into Science', 380.
75 Kang, 'Zhutian jiang', 20.
76 Kang, 83–84.

of a telescope.[77] Armed with an invincible belief in the infinity of the universe, Kang categorically dismissed as erroneous Einstein's theory of a finite universe in the ending chapter.[78]

One distinguishing feature of the *Lectures* is that Kang's belief that scientific facts obtained through the telescope were the only valid criterion of astronomical knowledge. He disowned any time-honoured tradition or belief concerning the earth and the heavens if it failed to conform to the scientifically proven facts. This was reminiscent of his earlier position in the *Shili gongfa* that all principles, if and when possible, should be subjected to scientific verification. He repudiated China's native astronomy as 'only half intelligible' and being 'of little worth' due to the absence of fine telescopes.[79] He deeply appreciated the introduction of Western astronomy by the Jesuits in the late Ming dynasty, which had advanced China's astronomy, especially in calendrical reform, but he regarded it as a pity that the Copernican heliocentric theory and the Newtonian notion of universal gravitation had not been made known earlier to the Chinese.[80] Kang further repudiated the practice of Chinese astrology: 'It can be readily seen that', he explained, 'as our earth is but one of the sun's planets, and a small one at that, it is impossible to expect any correspondence between the stars and the divination of the officials of a single state.'[81] Convinced that the earth was but a tiny planet in the vast universe, he did not hesitate to point out the fallacy of honouring the earth as the mate of heaven, a seminal concept of Confucianism.[82] In a similar manner, he condemned as erroneous the views on the heavens held in Buddhism, Islam and Christianity, which in his opinion had risen due to lack of telescopes.[83] While astronomy had been advanced in the West through the use of telescope and mathematics, Kang promoted a method for pursuing astronomy that centred on the use of the telescope, without taking into consideration the crucial role of mathematics.[84] This inadequacy, however, should not lead one to undervalue his appreciation and promotion of scientific attitudes in general.

As was mentioned earlier, Kang's motivation for engaging in prolonged astronomical studies was not merely the acquisition of scientific knowledge. As in his Xi'an talks, he emphasized the metaphysical implications of astronomical

77 Kang, 101.
78 Kang, 116.
79 Kang, 17.
80 Kang, 18.
81 Kang, 18; translated in Hsiao, 'Excursion into Science', 378.
82 Kang, 'Zhutian jiang', 19.
83 Kang, 101.
84 Hsiao, 'Excursion into Science', 394.

knowledge for this mundane life. In the traditional Chinese worldview, Kang wrote, the earth and the heavens constituted the universe. The Chinese often envied the life of the denizens of the heavens, and Kang himself was a firm believer in the existence of life on other planets. As astronomy advanced, however, it had become demonstrated that the earth turned out to be merely one of the planets circling the sun, which itself was only one of the countless stars in the universe. Kang wholeheartedly welcomed this overthrowing of the traditional geocentric and anthropocentric view of the universe, which had proven so disheartening to the ecclesiastical elites in Europe.

Kang earnestly, and somewhat fancifully too, believed that his theory of 'celestial peregrination' could help his fellow human beings to get rid of worries and to achieve happiness through intellectual thought and celestial reflection. By reflecting on the infinity and vastness of the heavens, and the minuscule size and insignificance of the earth in the cosmos, they could see how microscopically infinitesimal and trifling their sufferings and vexations were. Once aware of the fact that they were the inhabitants of a tiny planet in the vast immensity of space, and thus coming to the realization that they were actually denizens of the heavens too, people would rationally transcend the pain-inducing limits of families, tribes and nation states in the realms of their knowledge, thought and actions, thus achieving utmost happiness. It was due to ignorance of those facts, Kang argued, that the founders of religions invented such concepts as paradise and hell in their compassionate efforts to bring joy to the anguished human hearts.[85] Here one observes the progression of Kang from the path-finder Confucian theologian to proponent of mystic studies in astronomy.

3.2 *A Treatise on God*

It was no surprise that Kang devoted a whole chapter of his book on the heavens to a discourse on God (*Shangdi*).[86] For one thing, Heaven (*Tian*) and God (*Shangdi*) were interchangeable terms in ancient Chinese texts. He pointed out in the Chinese context when the term 'Heaven' is used, it denotes God, the lord of dominion.[87] He also cited for his support the examples where the two terms were used together as one concept in Chinese classics, such as *Ditian*

85 Kang, 'Zhutian jiang', 11–12.

86 Kang's treatise on God has been rendered into English by me and Patrick Swire, with reference being made to the partial translation in Hsiao, *K'ang Yu-Wei*, 183–84. Our translation has recently been published in a Chinese-English translation magazine. Kang Youwei, 'Treatise on God', trans. Zhaoyuan Wan and Patrick Swire, *Renditions* 93, no. Spring (2020): 43–48.

87 Kang, 'Zhutian jiang', (Chapter 11), 93.

帝天 or God the Heaven.[88] It was based on this understanding that he often used Heaven to refer to 'God' in other religions, such as his repeated comment that 'reverence for *Tian* (Heaven)' was taught in Christianity. For another thing, it was commonly taught in various religions that heaven was the abode of God. It was also Kang's declared belief that 'high above the inscrutable primal primary heavens (*yuanyuantian* 元元天) exists God.'[89] So it was only natural for Kang to talk of these two terms together. This recalls his earlier attempt in the 1880s to establish the universality of the concept of God by striving to find the most appropriate term for the concept of the ultimate Reality. Now in his last book, not only did he assert the existence of God, but he also attempted to prove it, against the broader context of the spread of strong anti-religion sentiments and the rise of scientific materialism in the 1920s.

Kang's chapter on God consisted of a presentation of utterances concerning God by European philosophers followed by a forceful argument for the existence of God. He started by stating that ancient European discourse regarding the heavens and deities was highly erroneous. He cited two illustrations. One was the Aristotelian model of the universe, described as being comprised of 'concentric crystalline spheres', which was held in place and put into motion by an external God. The other was the Ptolemaic model, a revision of the Aristotelian model, which divided the heavens into two parts, the sublunary and the superlunary. They were like 'the layers of an onion' with circular and cycloid movements, high above which 'there was the abode of God'. Kang's representation of these models was hardly intelligible to a Chinese reader, but he made his point clear: 'In the millennium of the Middle Ages, Ptolemy's theories were universally upheld; today, they are seen as quite laughable.'[90]

Kang then entered a discussion on theism and atheism in European philosophy. He stated that European and American philosophers generally subscribed to either one or the other of the two doctrines. 'Atheism is untenable,' he claimed, 'since perfection exists', which denotes the existence of God. With atheism being thus summarily dismissed, he presented five 'current arguments' for the doctrine of theism: 1) the ontological argument, which took God's perfection as proof of his existence; 2) the psychological argument, as advanced by Descartes, which claimed that in men's hearts there existed an awareness of God, and that this must be because God's existence had made it so; 3) the cosmological argument, which held that the existence of the universe must have

88 Kang, 'Chang'an yanjianglu', 273.

89 Kang, 273.

90 Kang, 'Zhutian jiang', 92. For a concise introduction to the Ptolemaic system, see E. Edson and E. Savage-Smith, *Medieval Views of the Cosmos* (Oxford: Bodleian Library, 2004).

had an origin, and that this origin lay in God; 4) the argument from design, which maintained that the universe was such a work of fine art that its good order presupposed a maker who must have been its architect, if not its Creator; and 5) the moral argument, which stated that man's willingness to sacrifice himself for the good of the group must have come from a source in the unseen world, God.[91]

Although Kang's presentation of these well-known proofs was occasionally obscure,[92] his message was clearly conveyed: the five proofs fell short of proving the existence of God, which according to Kant was an existential judgment or an *a posteriori* statement. Kang reasoned that 'from experience we cannot know for sure that there is no God, nor can we say with confidence that there is.'[93] Nevertheless, he stressed that the ontological argument should be sufficient to demonstrate the existence of God from at least one aspect. That people always hesitated to acknowledge the argument was due to dualistic thinking, for which there were three explanations: The first held that both perfection and imperfection were derived from God. The second maintained that God represented perfection, and imperfection, which was embodied in matter, coexisted with but did not come from God, the encounter of the two producing various forms of suffering. The third doctrine claimed that God and the Devil both existed in the world, the former being light and the latter darkness.[94]

Having discussed the existence of God, Kang moved to the subject of how God existed. He stated that there were two theories in this regard. One was monotheism, which envisaged that 'God has an innate noumenal substance', and that 'he is the highest entelechy', from which all the worldly phenomena were derived or emanated. The other was pantheism, which held that 'God does not possess a noumenal substance, for his substance exists in a state of change and variation, and thus the universe is in a changing process of non-stop daily renewal'. The two theories held different views on the possibility of attaining ultimate knowledge of the universe. Monotheists would argue that once the nature of God was known, the phenomena of the universe could be deduced therefrom. The pantheist thesis, however, held that despite our knowledge of God, the phenomena of the universe could not be fathomed as they were changing all the time.[95]

91 Kang, 'Zhutian jiang', 92; Kang Youwei, 'Treatise on God', 45.

92 See the discussion in Hsiao, *K'ang Yu-Wei*, 182.

93 Kang, 'Zhutian jiang', 92.

94 Kang, 92–93.

95 Kang, 93; Kang Youwei, 'Treatise on God', 46.

Although Kang tried to appear impartial, his implicit preference for pantheism could be detected. He commented that the concept of pantheism held by 'us [the Chinese]' was slightly different from the view proposed by those like Spinoza and Goethe. 'They believe that God is everywhere, hence the term pantheism. We stress that God does not have a substance but exists in constant change and variation, which is in agreement with Bergson's thesis.'[96] By 'constant change and variation' Kang seems to be alluding to the Confucian concept of change as best illustrated in the title and content of the *Changes*. He made an insightful observation here about the difference between European and Chinese pantheism – the former stressed God's omnipresence while the latter emphasized the character of being in constant change.

Kang went on to discuss monotheism, which could be divided into two doctrines: emanation and creation. Emanationists, according to Kang, considered God as being of the same substance as the universe through which he manifested himself. Creationists, however, held that this world had a Creator who existed outside and beyond it, and who, having created the world, existed along with it. In Kang's view, Neo-Platonism, Indian philosophy and the theories of Spinoza, Schelling, and Hegel, all belonged to the former, while Judaism, Christianity and Islam subscribed to the latter. Kang highlighted Schelling's attempt to mediate between the two by arguing that 'the world as it is has resulted from the substance of God', but 'it lies solely in God's freedom whether to create or not'. That was, in Kang's view, a compromise between the fixed limits of God having a substance and the freedom consequent upon his being without substance.[97]

Having given a relatively comprehensive account of European utterances on God, Kang started to put forth his views on why God necessarily existed. 'That there is God high in heaven', he asserted, 'is a belief common to all nations and all religions'.[98] Then he took great pains to illustrate this 'universal truth'. In ancient Chinese literature, to start with, references to God and Heaven were present in many places. The *Odes,* for example, spoke of 'serving God with entire intelligence', and also said that 'God is with you, have no doubts in your heart.'[99] It was on this account, Kang argued, that grand ritual sacrifices were made by the state to God outside the city. Similarly, Kang continued, Mencius also said, 'Though a man may be wicked, yet if he adjust his thoughts, fast, and bathe, he

96 Kang, 'Zhutian jiang', 93.

97 Kang, 93.

98 Kang, 93.

99 *Shi* 3:1:2, Legge, *Chinese Classics IV*, 433, 436, with minor adaptations.

may sacrifice to God.'[100] Kang went on to show how the same God was worshipped but was called by different names in different religions: God was called Brahma in Hinduism, and Yahweh in Christianity. In Islam God was worshipped five times and bowed to seventy times a day, which Kang felt was particularly austere. Only the Buddha did not revere God and was said to have fought with God, and God, being defeated, became a disciple of the Buddha.[101]

Kang upheld the unity of God, that is, he claimed that there was but one God; but it would be wrong to assume that his notion of God was completely congruent with Western monotheism. As was mentioned above, although he maintained a firm belief in the existence of God, he held a somewhat pantheist view of the ultimate Reality. Moreover, he thought there were a number of questions left unanswered by monotheism, and in particular, by the doctrine of creation.

> First, if one posits that the universe came into being by means of God's creation, then when did the creation begin? Second, if man is regarded as free, that would affect God's omniscience and omnipotence, for the two ideas are incompatible. Third, if we follow the doctrine of emanation, then the day when man is emancipated should also be the day when God is emancipated; according to the doctrine of creation, however, God exists outside this world, that means even if man is completely redeemed, that should be nothing more than the redemption of the things he has created. What would then be the position of God?[102]

Whether these questions constitute an adequate criticism of monotheism is irrelevant for our purposes. What is significant is that they clearly reveal Kang's resistance to the idea of God as Creator, an entity existing beyond the world. Rather, his God was the source and sustainer of everything; without having a substance, he existed in constant change, a change which permeated a myriad of things; he preordained human affairs and exercised his decrees through reward and punishment. In fact, Kang had professed unwavering belief in this concept of God since he was a young man. Therefore, it is hard for us to agree

100 *Mengzi*, 4(2):25, in Legge, *Chinese Classics II*, 330.

101 Kang, 'Zhutian jiang', 93. I have not been able to find the source for Kang's statement on this peculiar 'Buddhist' concept of God. A similar account in Chinese folk mythology relates that the God of the Northern Heaven (*Xuantian Shangdi* 玄天上帝), also a major Deity in Daoism, was said to have become a disciple of Dīpankara Buddha, a Buddha of the Past.

102 Kang, 93; Kang Youwei, 'Treatise on God', 47.

with Hsiao's statement that Kang 'remained an agnostic (or an atheist)' before he entered his old age.[103]

Kang agreed that both God and the universe were unknowable in their essence, and he argued that their ultimate comprehension was beyond the limited capabilities of men. In this regard, he offered a critique of certain limited concepts of God. 'In the older religion of Christ [Judaism] God is called Yahweh,' he commented, 'but to say Yahweh is of the Israeli people would immediately make it an unbefitting name of God.'[104] Furthermore, he was aware of the attacks on the belief in God from mechanistic scientists and materialists of modern times. Newton had, in Kang's view, held that the heavens were formed by the power of attraction and repulsion while Laplace's theory of mechanistic materialism developed on Newton's theories, going so far as to claim that there was no God. In recent times, Kang continued, most of the adherents of materialism professed atheism, while there were those such as Bergson who still stuck to pantheism. Kant revered God as the mysterious, but his view was attacked by the school of Newton and Laplace by means of their theories of attraction and repulsion, so much so that belief in God was almost shaken among Europeans.[105]

Kang's basic position on the issue can be summarized as follows: science can modify religious beliefs, but scientists cannot disprove God. Kang did not insist on the infallibility of the founders of religions, who in his view were just 'a bit wiser' than the ordinary. Nor did he believe in the absolute truth of religion. He considered religious beliefs to stand in need of adjustment wherever established scientific evidence made it imperative to correct previous misconceptions of reality. Kang was ready to disown any traditional views, religious or otherwise, when they were found contradictory to established scientific facts. A typical example was his critique of the various discussions on the physical heavens in religious teachings, especially in the Buddhist Sutras: 'The Buddha's descriptions of the various worlds, which increased in number by thousand folds, were mere random speculations rather than factual presentation of the universe.'[106]

Referring to Confucius' remarks on the limits of his own knowledge, Kang implied that the Sage's humility should also be maintained by scientists in their attitude towards the existence of God, for he was too exalted to be measured,

103 Hsiao, *K'ang Yu-Wei*, 141.

104 Kang, 'Chang'an yanjianglu', 273.

105 Kang, 'Zhutian jiang', 93.

106 Kang, 101.

too subtle to fathom.[107] Therefore, it was inappropriate for a scientist to assert the non-existence of God. Kang did not believe that by relying purely on the physical senses, all facts and reasons could be revealed. In this sense, he set out to rebut Newton, Laplace, Darwin and the like, whom he otherwise praised in glowing terms, for their impudent and arrogant denial of God:

> These people, although famous for their study of matter, could not have known of these very material things. In the future, there will be great inventions related to magnetism, and other inventions similar to electromagnetism will come along as well. Today's material inventions are just the germs of greater inventions in the future. The above-named were not able to know of the existence of electricity [in every entity], so how much less could they pass judgment on the existence of God or know about all the mysterious things in the world?[108]

Quoting Zhuangzi's saying that 'a man's life is limited while knowledge is limitless', Kang concluded his discussion on the subject by a series of forceful questions: 'Their knowledge being so limited, how could Newton, Laplace, and Darwin hope to comprehend all under heaven? How could they assert there is no God? In so doing they only showed their ignorance of their own limitations.'[109]

In summary, as a work on astronomy, Kang's *Lectures on the Heavens,* being a mixture of a recitation of Western philosophers and his own speculations, neither presented new findings nor advanced scientific theories.[110] Its significance, however, lies more in the intellectual sphere than in the scientific field. Kang demonstrated an appreciation of scientific method and a reliance on science, notably astronomy, in advancing his cosmological understanding. For this reason, Kang has been regarded by some historians as the first thinker of modern (or pre-modern) China to enrich Chinese philosophy by injecting

---

107 Kang, 40.

108 Kang, 93–94; Kang Youwei, 'Treatise on God', 48.

109 Kang, 'Zhutian jiang', 94; Kang Youwei, 'Treatise on God', 49. It should be pointed out, however, that in purporting to refute Newton's denial of God's existence, Kang was mistaken, for not only did Newton believe in God and regard his scientific work a way of knowing the mind of God, he also shared Kang's fundamental position: 'concerning things unknown nothing is rashly to be affirmed or denied'. See J. E. McGuire, 'Newton on Place, Time and God: An Unpublished Source', *The British Journal for the History of Science* 11, no. 2 (July 1978): 121. See also Rob Iliffe, *Priest of Nature: The Religious Worlds of Isaac Newton* (Oxford; New York: Oxford University Press, 2017).

110 See also Hsiao, 'Excursion into Science', 394.

Western elements into it.[111] Indeed, Kang had unknowingly helped to usher in another new phase of China's response to the West by the appropriation of natural science as a means of enriching Chinese intellectual life.[112] However, one of Kang's contributions, often neglected, was his effort to construct a universal concept of God that was in harmony with science. He spoke of science in glowing terms while affirming the existence of God, emphasizing his view that no matter how far sciences may advance, scientists could never disprove God, whose nature lay beyond the comprehension of the limited faculties of men. Driven by a messianic self-image as sage for his age, Kang Youwei had set forth his own *jiao* or doctrine of 'celestial peregrination' in his final book, and had become, at least in his own self-estimation and within his own private mental universe, a *jiaozhu* or founder of a teaching/religion.

4 The Fate of Kang's Skull

> To come into being and to abide, to assemble and to disperse –
> These are the natural course[s] of man and nature.
>
> KANG YOUWEI[113]

Early on the morning of 31 March 1927, Kang Youwei passed away at his residence in Qingdao, Shandong, which he had named the House of Celestial Peregrination.[114] It has been suggested that he might have died of food poisoning from something ingested on the eve of his death, but that has never been verified. Kang may have had a premonition of his imminent death a couple of weeks earlier when he was celebrating his seventieth birthday in Shanghai. On that occasion, he received a congratulatory tablet and a jade sceptre from the abdicated emperor Puyi, to whom he had maintained his allegiance; in response, he wrote a 'memorial of thanks'. In an elegant style of prose, which he had rarely adopted in the vast body of his writing, Kang penned a summary review of his eventful life since the Reform of 1898. Completing the letter, he threw down his brush and declared, 'Here are the affairs of my mind during my life!' Just before leaving Shanghai for Qingdao after the birthday party, he

111 Hsiao, *K'ang Yu-Wei*, 140.
112 Hsiao, 'Excursion into Science', 395.
113 Quoted in Lo, 'Sequel to K'ang's Autobiography', 249. In Chinese, 成住聚散, 人天之常.
114 In addition to Shanghai, Kang also had a residence in Qingdao, a city that had become the preferred domicile for members of the former royal family, including Puyi, after they were expelled from the Forbidden City in 1923 on the grounds that they had participated in the failed Restoration attempts.

walked around his garden, and said, 'This is the end of my fated stay in Shanghai!'[115] According to his attendant, Kang observed the stars on the eve of his death, and made his usual remarks – 'Alas, all is lost!' Then he added, 'There is no foothold left for me in China, but I simply cannot die on foreign soil.'[116]

By the declaration of 'All is lost' and saying there was 'no foothold left' for him in China, Kang was alluding to the triumphant advances, which he by no means wished to see, and must have dreaded, of the northern expedition of the southern Revolutionary Army (*gemingjun* 革命軍), led by Sun Yat-sen's successor, Generalissimo Chiang Kai-shek 蔣介石 (1887–1975). The nationalist forces had wrested control over large territories, including Shanghai and Nanjing, in South and East China, from the control of the Zhili clique headed by Wu Peifu and the Anhui clique. They would forge ahead and in a year's time would reunify China under the control of their government (whose seat had been moved to Nanjing), thereby ushering in what is called the Nanjing era.[117] Kang had correctly foreseen the victory of the nationalists, but he had never backed off from his disapproval of the revolutionary approach and discourse adopted by the nationalists and later the communists; Kang loathed their political views so much that, if at all possible, he would avoid living or dying at a place controlled by the revolutionaries and their ilk.

Kang died a disappointed man. He had not accomplished his mission to save China. Politically, his proposal of a titular monarchy for China was rejected outright, and his efforts at restoration failed disastrously. So did his religious cause. The Confucian Religion Association, which he and Chen had established to institutionalize his *Kongjiao*, had lost its vigour and appeal.[118] His pleas on behalf a Confucian religion were, in the words of Hsiao, 'not so much opposed as ridiculed or ignored' by the younger generations of Chinese intellectuals.[119]

---

115 Lo, 'Sequel to K'ang's Autobiography', 249–50.

116 Li Keliang 李可良, 'Wo yinxiangzhong zhi Kang Youwei 我印象中之康有为 [My Recollections of Kang Youwei]', in *ZYKYW*, 1996, 482–83.

117 For more on the northern expedition, see C. Martin Wilbur, 'The Nationalist Revolution: From Canton to Nanking, 1923–28', in *The Cambridge History of China*, ed. John K. Fairbank, vol. 12 (Cambridge: Cambridge University Press, 1983), 527–720.

118 The Confucian Religion Association officially ceased to exist in 1937 when the nationalist government ordered it to change its name to Confucian Studies Association (*Kongxuehui* 孔學會). Chen Huanzhang, however, continued his pursuit of Confucian Religion in Hong Kong under British rule, where he established the Confucian (Religion) Academy (*Kongjiao xueyuan* 孔教學院) in 1930, an institute that has continued to thrive till this day. It has been officially recognized as one of the major religious communities of Hong Kong. For the history of the institute, see http://www.confucianacademy.com/load.php?link_id=27660.

119 Hsiao, *K'ang Yu-Wei*, 126.

He had also been criticized for his character deficiencies. 'Kang had attempted to promote Confucianism as the state religion,' wrote a magazine review on Kang's work and life published a year after his death, 'but regrettably his deeds and conduct had failed to conform to the teachings of Confucius. Under a moral scrutiny, he could well be said to be a transgressor of the Confucian ethical code (*mingjiao zuiren* 名教罪人).'[120]

Kang was buried at the foot of Xiang'er Hill in the suburbs of Qingdao, at a site that he had chosen himself according to the art of Chinese geomancy (*feng shui* 風水). His death received little attention within the political and cultural circles, partly because the gaze of the whole nation was fixed on the ongoing northern expedition, and partly because Kang had lost the high esteem he had once enjoyed among the mainstream Chinese intellectuals. That stood in stark contrast to the treatment accorded to Liang Qichao, who died two years later and was mourned nationwide, with hundreds of elegiac couplets written by the political and cultural elite as eulogies to honour him. While Liang shared Kang's disapproval of revolution, he managed to remain on good terms with the revolutionaries. In contrast, Kang's long-standing anti-revolutionary position and his insistent advocacy of constitutional monarchy had caused him to be regarded as an enemy of republicans, nationalists and communists, the last two groups having dominated China's political and intellectual stage during Kang's last years. Naturally, he was viewed negatively during the Nanjing era, a two-decade period that witnessed the rise of the Communist movement, the Sino-Japanese War, and the KMT-CPC conflicts which finally culminated in the defeat of the KMT forces (who withdrew to Taiwan) and the establishment of the People's Republic (PRC) on the mainland in 1949.

A keynote communist perspective on Kang was articulated by Mao Zedong 毛澤東 (1893–1976) on the eve of the founding of the PRC. An admirer of Kang during his own adolescence, Mao regarded Kang as one of the Chinese progressives who had gone through hardships in their quest for truth from the West; unfortunately yet inevitably, Mao said, their proposed remedies for China had failed one after another. In particular, he commented that Kang, despite that fact he had written the *Book of the Great Unity*, could not find the road to attain that vision.[121] Mao believed that the CPC had found that road in Marxism-Leninism, and about ten years later he would attempt to realize

120 Huang Yanyu 黃延毓, 'Kang Youwei de shengping jiqi xueshu gailue 康有為的生平及其學術概畧 [A Short Account of Kang's Life and His Scholarship]', *Nanfeng* 南風 3, no. 1 (1928): 45.

121 Tse-Tung Mao, *Selected Works of Mao Tse-Tung*, vol. IV (Peking: Foreign Languages Press, 1961), 412, 414.

that vision in the communes of the Great Leap Forward. Inspired by Mao's argument and in a state of socialist ferment, PRC scholars started a debate about the legacy of Kang, in particular on his utopian thought, around 1958, the sixtieth anniversary of the Reform of 1898. Most of the participants in the debate adopted the Marxist approach of analysing his work through the prism of class struggle, claiming that Kang's scholarship had purely served his bourgeois political agenda, although among them there was also the more balanced voice of a young Li Zehou 李澤厚 (who would later become a prominent Chinese philosopher) who defended Kang's utopianism against those harsh critiques.[122]

Posthumously, Kang's reputation deteriorated. Being the chief proponent of constitutional monarchy, he was denounced as a reactionary defender of the old tradition, a view that had dramatic consequences when it led to the destruction of Kang's tomb in August 1966 at the height of the campaign to 'do away with the Four Olds' (*po sijiu* 破四舊) during the initial stage of Mao's Cultural Revolution.[123] Kang's remains were dug up by a group of Red Guards consisting of students and teachers from Qingdao No. 5 Middle School. His bones were burned, while his skull was placed in a wheelbarrow and paraded through the streets, amidst shouts of the slogan, 'Down with the founder of China's Royalist Faction!' By that time, the term 'royalist faction' had taken on a new political connotation, denoting those who defended the more liberal political incumbents who were then coming under attack. When the parade ended, Kang's skull was left on the ground, and it was picked up and hidden away by a curator of the municipal museum.[124] In hindsight, Kang would probably not have been too upset about the fate of his skull, for its treatment had put him among the rank of those religious (in particular Buddhist and Catholic) and scientific saints in history, whose remains or brains were preserved,

---

122 For a good discussion, see Brusadelli, 'A Tale of Two Utopias'; see also Xu Guangren 徐光仁, 'Ping Luo Mengwei xiansheng du renmin minzhu zhuanzheng duiyu Kang Youwei de pingjia 评罗孟韦先生"读人民民主专政"对于康有为的评价 [A Review of Mr Luo's Essay Entitled "Reading Mao's Comment on Kang"]', *Huanan shiyuan xuebao 华南师院学报*, no. 2 (1959): 78–84.

123 The 'Four Olds' was a pejorative reference to 'old ideas, old culture, old customs and old habits'. For more on the campaign to destroy the Four Olds, see Guo Jian, Yongyi Song, and Yuan Zhou, *The A to Z of the Chinese Cultural Revolution* (Lanham, Md.: Scarecrow Press, 2009), 70–72.

124 Lin Zhiye 林治业, 'Yuan Kang Youwei mu shi zheyang bei wahuide 原康有为墓是这样被挖毁的 [An Account of the Destruction of Kang Youwei's Original Tomb]', *Laonian shenghuo bao 老年生活报* [*Daily for the elderly*], 12 December 2012, sec. 6, http://epaper.qingdaonews.com/html/lnshb/20121212/lnshb514317.html.

worshipped or studied.[125] That episode was a prelude to even fiercer attacks on Kang's historic role at the height of the Cultural Revolution. Among other things, he was condemned as an anti-revolutionary reformist, whose legacy had been allegedly inherited by Liu Shaoqi 劉少奇, the recently disgraced *de jure* head of the PRC, who had been labelled a traitor to the revolution due to his 'capitalist leanings'.[126] Kang was not alone in receiving this treatment, for the historic Confucian Cemetery in Qufu was also destroyed, and Confucius himself also became a subject of criticism during the politically motivated anti-Confucian campaign in the 1970s, with the Confucian legacy being considered to be a hindrance to the progress of socialist China.

By contrast, Kang received a much more lenient and positive treatment across the Pacific in US academia. The year 1958, which also marked his hundredth birthday anniversary, witnessed an appropriate celebration by way of the publication of Laurence G. Thompson's translation of his *Datong shu*, the first book-length translation of his works in English.[127] That was followed by a well-attended symposium on Kang Youwei, whose contributions were edited and published as a book by Kang's grandson Jung-pang Lo, then associate professor of history at the University of California, who also added to the volume his translations of Kang's *Chronological Autobiography* and of its *Sequel* edited by Kang's daughter Tongbi (Lo's mother).[128] One of the key participants at the symposium was Kung-Chuan Hsiao (often quoted in this book), whose careful examination of Kang's works and manuscripts resulted in the publication of a monograph on Kang's thought, which has remained the definitive work of scholarship on the subject in the English language. As was mentioned in the introduction, Hsiao spoke highly of Kang as a 'patriotic' revisionist who occupied a seminal position in the Confucian tradition.[129]

When Hsiao's work was introduced to post-Mao China, it was welcomed – but not without reservation – by Chinese intellectuals. Among other things, they criticized the author for being too sympathetic to Kang.[130] In the 1980s,

125 For a recent volume on 'savant relics', see Marco Beretta, Maria Conforti, and Paolo Mazzarello, eds., *Savant Relics: Brains and Remains of Scientists* (Sagamore Beach, MA: Science History Publications, 2016).

126 Ma, 'Kang Youwei yanjiu', 88.

127 See Thompson, *Ta T'ung Shu*.

128 Jung-pang Lo, ed., *K'ang Yu-Wei: A Biography and a Symposium* (Tucson: University of Arizona Press, 1967).

129 See Hsiao, *K'ang Yu-Wei*, 122, 125.

130 Wong Young-Tsu 汪荣祖, 'Yizhe xuyan 译者序言 [Preface by the Translator]', in *Jindai Zhongguo yu xinshijie* 近代中国与新世界 [*A Modern China and a New World*] (Nanjing: Jiangsu renmin chubanshe, 1997), 1.

FIGURE 5.1 Kang's Restored Tomb in Qingdao.
PHOTO COURTESY OF PAN WEIJUN (KANG'S GREAT GRANDDAUGHTER), 2018

Kang was rehabilitated, but only partially. His thought was subject to an analysis based on dialectic materialism – he was recognized for his 'progressive' role in the 1898 Reform, a movement whose own nature was being heatedly debated, but his thought and political activities after the Reform continued to be denounced as anti-revolutionary. Subsequently, as a material manifestation of Kang's political rehabilitation, his skull was finally interred in a dignified new grave built by order of the local government of Qingdao in 1985, seventeen years after it had been dug out and vilified by the Red Guards. Meanwhile, his old house in Nanhai had been restored to its former state, and was made a site for patriotic education.

Conclusion

Unlike the image of Kang Youwei as he is depicted in conventional accounts, Kang is primarily presented in this book as a religious figure. From his childhood, by reason of his having been immersed in Confucian education at home, he had developed a reverence for *Shangdi*, the Supreme Ruler in Heaven spoken of in ancient Confucian Texts, to whom a grand annual ceremony of sacrifices was still being performed by the Qing emperor. At the same time, he was familiar with the practices of ancestor worship within his family and was exposed to popular local beliefs based on a folkloric admixture of Buddhist and Daoist ideas. His early religious leanings were further nurtured through his visits to nearby temples and holy places and by his reading of Buddhist literature. His short periods of intense immersion in the practice of Buddhist meditation in seclusion led him to experience several mystical states, analogous to those canvassed by William James, during which he sensed that he was becoming one with the universe while suffused with brilliant light. He also records a vision of his being born to be a sage with a bodhisattva's mission to deliver from suffering not only his compatriots in China but the whole of humanity. This sense of having been bestowed with a mission, accompanied and buttressed by a belief in the Mandate of Heaven, persisted as a strong inner driving force throughout his life and sustained his tenacious commitment to his reform cause.

To carry out his redemptive mission, Kang concurrently embarked on two interrelated realms of reformation: political and religious. His political aim was to transform the imperial government into a constitutional monarchy through gradual reforms so as to better prepare China for survival and dignified prosperity among the competing nations. His religious goal was to recast Confucianism as a religion and in particular, to enshrine it as a state religion. A firm believer in the religious origin of morality and in the moral basis of society, he considered politics and religion as 'two wheels' of the chariot of state. Religion would serve as a foundation for good political order and national solidarity, while politics would provide assistance and support to religious institutions for the common good of all. In this light, Kang's political reform and his religious cause can be regarded as different aspects of a larger project. Despite his repeated efforts, Kang's political reform failed disastrously, with dire consequences for his personal life. His own analysis and interpretation of the 'eleven' instances of his narrow escape from capture and death, however, only served to strengthen his faith in his mission and his belief in the will and protection of Heaven.

His religious reform, on the other hand, achieved a degree of success. His determination to reform the Confucian tradition came from several

 | DOI:10.1163/9789004468221_008

sources, including a personal intellectual dissatisfaction with orthodox Neo-Confucianism, its perceived inadequacy to help China counter external colonial aggression, and the stimulus and challenge from an advancing and intruding Christianity. His Confucian reformation was carried out in two main areas. At the institutional level, he proposed to the Throne the creation of a national Confucian Church modelled after its Christian counterpart, with branches set up in every locality across the country, each being assigned a Confucian pastor to offer weekly sermons to its congregation. By means of this Christianity-like design, Kang had hoped to bring Confucianism out of the exclusive province of the educated elite and to connect it with the general populace. His vision of a proper Confucian Church, however, differed from the Christian model in an important way: it should always be attached to or accompanied by a school or an academy (or a university at the national level), an idea that could be traced back to the practice in ancient China, where a state school had often served as a Confucian temple or shrine. His proposal was only partially realized through the establishment (with the assistance of his young disciple, Columbia graduate Chen Huanzhang) of the Confucian Religion Association and its various branches during the early Republican period; but, in later years, the Association slid into a decline and eventually lost its appeal to the public. Kang's agenda to make Confucianism entrenched as a state religion in the national Constitution was rejected by the National Assembly, and was severely criticized by the New Culture intellectuals who took the opportunity to mount a concerted campaign against religion in general. The failure of his initiative to institutionalize Confucianism as a religion, it can be seen, was perhaps due not so much to the imperfection of his programme as to the adversities he faced and the exigencies of the times. The forces of political turmoil and revolutionary zeal, along with the forces of anti-traditionalism, the emergence and rise of scientism and the urgent demands for national survival and enlightenment, all combined to frustrate Kang's vision of a Confucian religion revitalized to harness the power of science and technology to propel China onward and upwards towards spiritual as well as material prosperity. Herein lies the germ of an idea that posterity might well ponder.

At the doctrinal level, Kang sought to recast the Confucian tradition through a radical reinterpretation of the Confucian Canon drawing on the New Text exegesis of the Gongyang school. Breaking from convention, he presented Confucius as a founder of a religion whose teachings also sanctioned political and social reforms. On the one hand, he attempted to revitalize ancient Chinese belief in *Shangdi* found in classical Confucianism and he defended the religious nature of Confucianism by demonstrating that Confucian Texts contained numerous discussions on such subjects as Heaven (God), spirits and the human soul. Appropriating apocryphal accounts and Zhuangzi's statements, he also attempted to ascribe divinity to Confucius – by extolling the Sage as a divine teacher, born after a miraculous conception, who was charged with the Mandate of Heaven to carry forward an ever-advancing Way. Kang thereby placed Confucius in the same rank as the founders of the great religions. On the other hand, he emphasized that, although rooted in Heaven, the religion of Confucius was human-oriented. Unlike other founders of religions, Confucius had focused his attention on the human world by setting forth teachings to nurture proper human relations so as to ensure social order and the well-being of the people. Beyond that, Confucius had a vision for the future of mankind, as manifested in the progressive doctrines of the Three Ages (*sanshi*) and the Great Unity (*datong*), concepts cherished by Kang as being among the most profound teachings of Confucius whose true meaning had been inaccurately understood (in the case of the former) or even forgotten (in the case of the latter) largely due to the prevailing influence but corrupt content the Old Texts. Conceived of this way, Confucianism represented a highly advanced form of religion originally founded by Confucius and now being restored by his exegesis, a religion which called for profound change both at the level of the individual and in the structure of society.

The concept of *jiao* or 'religion' occupies a pivotal position in Kang's religious discourse. Kang had developed a progressive view of religion as an instrument for the education and development of humanity. To begin with, religion was born when a self-driven man, who had become enlightened ahead of others, arose and expounded on a set of instructions to teach his fellow beings to avoid evil and do good. Naturally there appeared multiple religions in the earlier periods of time, when individual groups of people who lived in isolated regions would each have developed its own religion in accordance with the teachings of its founder. Despite their multiplicity, all religions shared the same goal – the realization of a moral life for their followers, but they might, necessarily, differ in their means of instruction. Even within one geographical region, later religions might have adopted ways of teaching that were different from the older ones. But there were regularities to follow in analysing them, according to Kang. At an earlier stage or in a certain region, if people had been ignorant, and their intellect had not been fully developed, the founder of their religion may have resorted to *shendao* (the way of the spirits), making use of, for example, notions of paradise and hell, clear-cut doctrines on the retributive powers of ghosts and spirits and the ritual specification of various practices, so as to make people avoid evil and do good. But, he argued, to a more civilized audience, whose intellect had been developed to a higher degree, the founder of their religion may have employed primarily *rendao* (the human way), by coming away from superstitious concepts and focusing his teachings on nurturing proper human relations and establishing higher forms of social order through the practice of moral/heavenly principles as well as satisfying human desires in a seemly and moderate manner. According to Kang's reinterpretation, which dramatically differed from the Neo-Confucian account, human desires, like moral principles, were also part of human nature as endowed by Heaven. Therefore, a proper 'religion' or *jiao*, Kang argued, was one that, as defined in the *Means*, cultivated the way for nourishing human nature without suppressing the natural human desires. The Sages, or the founders of religions, each constituted in his own self the highest exemplar that his followers could model themselves on in their quest for a fuller development of their Heaven-endowed nature.

One can hardly miss the end-point of Kang's theory concerning the history of religion. It followed that when peoples in different regions became more connected by virtue of improved transportation and communication, their religions also came into contact, as evidenced by what had happened in history. Thereupon Kang entered a comparative analysis of the major religions of his times, an analysis often adorned with anecdotal observations from his extensive world travels. Earlier in his youth, he had, following the traditional *yin-yang* dualistic philosophy, attempted to classify all religions into two groups according to their attitudes towards human desires. The *yang* religions, represented by Confucianism, were more oriented to meeting human needs, while the *yin* religions, as typified by Buddhism and others, were more oriented to suppressing them, especially the human sexual urge. Kang later gave up his intellectual construct of the *yin-yang* classification, and instead opted for the *shendao-rendao* categorization, but that was more of a change of terminology than of substantive content. He argued that authentic Confucianism encompasses both ways, based on *rendao* yet retaining *shendao*, while Buddhism, Christianity and Islam were primarily confined to *shendao*. Thus, Kang asserted, Confucianism was not only a religion, but, as a religion of superior wisdom, represented a more advanced stage in cultural progress and therefore, he believed, was better suited to the modern world, and more specifically, to the immediate needs of the Chinese people. That was his clear and stern response to the emerging view that Confucianism was not a religion or *zongjiao* (the Japanese neologism for 'religion') on the grounds that Confucius had not discoursed much on such key elements of religion as spirits, the soul and the afterlife. That view was first raised by his erstwhile disciple Liang Qichao and soon gained momentum, thereby triggering the first debate over the religious nature of Confucianism among Chinese intellectuals at the beginning of the twentieth century.

In this connection, Kang began to tackle the question of the term *zongjiao*/religion. He argued that the Japanese neologism denoted only the *shendao* doctrines, and therefore was a mistranslation of the English term 'religion'. For in his understanding, the original concept of *religion* [*lilijin*] was a broader, generic term that encompassed both *shendao* and *rendao*, although it had long been narrowed down by institutionalized Christianity. He maintained that *jiao* and 'religion', if properly understood, were equivalent terms. He continued to employ this etymological approach in his later defence of Confucianism when propagating the Confucian Religion movement, arguing that the Confucian tradition, properly understood, was a religion, even in the narrow sense of *zongjiao*. In the post New Culture era, however, when the concept and practice of religion had become intellectually and politically discredited, Kang pragmatically switched his strategy. He now acknowledged that *zongjiao* was an accurate translation of 'religion', both of which resorted to *shendao* teachings, and as such, both these terms were inadequate to encompass Confucianism, which enjoyed the distinction and advantage of being primarily a *rendao* teaching while retaining the *shendao* teaching as well. He stressed one unique teaching of Confucianism, the veneration of one's ancestors, something that had been rejected by some religionists, especially the Christian missionaries, as being incompatible with worship of God. But for Kang, the two practices were in perfect harmony and honoured a fundamental verity that humans came into being through the agency of God and one's parents, and therefore, one should show gratitude to both one's parents and Heaven (God). Kang's revised position seems very close to the now widely-accepted view that Confucianism is primarily an ethical and humanistic body of teachings that contained shades of religious belief. But considering his understanding of *jiao* as the cultivation of the Way by which the Heaven-endowed human nature can progress towards sagehood, Kang's Confucianism can be more accurately described as a religious system of knowledge that primarily manifests its society-transforming effects by extolling a moral code based on attainable human standards.

With an eye vigilantly looking for signs of an ingathering unity despite the diversity and complexity plainly evident in the world of phenomena, Kang had pondered the theme of unity of religion throughout his life. He stressed that all the major religions of the world, despite their differences, were united in a number of fundamental teachings. To begin with, all religions taught reverence for *Shangdi* or 'God', a concept that he believed was universal, though given different designations in different cultures. In his youth he had attempted to find a more suitable common term for naming the ultimate Reality, and in his later years he had explicitly sought to prove the existence of God against the spread of strong anti-religion sentiments and the rise of scientific materialism in the 1920s. Kang himself favoured the conception of a pantheist God who did not have a visible substance but who existed with and within this world that was in constant change. Second, it was his conviction that all religions without exception taught love of one's fellow beings, which was best captured by Jesus's counsel to 'love your neighbour', the teaching of compassion exemplified by the Buddha and the Confucian virtue of *ren*. Thirdly, all religions taught the doctrine of the existence of the soul, although the human soul, like the concept of God, had been given different names in the different religions. He believed in the immortality of the soul, but that immortality, he held, was manifested in repeated reincarnations within this world rather than in its continued life in another world. For Kang, there was only one world, with two realms (visible and invisible). It was his conviction that unity of religion was both necessary and inevitable. Indeed, his progressive view of religion would lead to the logical conclusion that as peoples around the world become increasingly connected, diverse religions would also be unified, and there would appear a new, universal religion, which would help usher humankind into the ideal world of the Great Unity.

An earnest self-taught student of Western science by means of an arduous reading of Chinese translations, the young Kang turned himself into one of the most learned Chinese scholars in Western learning in the late nineteenth century. Having experienced the intellectual exhilaration of encountering the most recent developments in science, he became one of the pioneer Confucian teachers who lectured on Western learning, and he further called for science education to be included in the reform of the national educational system. He recognized the potential of science and technology to transform the material conditions of existence and argued that China could be saved through the vigorous promotion of technoscience. His observations from visits to industrially advanced powers such as Great Britain, the United States and Germany convinced him of the urgent need for China to encourage scientific inventions by setting up a mechanism to protect patents so as to give China a cutting-edge advantage in the fierce competition with other countries. Furthermore, it was clear to Kang that his ideal world of the Great Unity could not come into practical realization without a high degree of scientific and technological development, for the universally desired increase of human happiness and decrease of suffering also depended on material progress. He spoke of the value of science as well as its utility in such glowing terms that he can indeed be regarded as one of the pioneers who helped usher in a new phase of China's response to the West. Unintentionally, he contributed to the rise of scientism as a new faith in China, for which he naturally declined to take credit. Unlike the New Culture-May Fourth intellectuals, however, his passionate appreciation of science did not convert him to fervid materialistic scientism or cause him to advocate indiscriminate Westernization.

In the intellectual realm of his holistic, associative thinking, Kang maintained an unfractured concept of knowledge, by holding firmly the view that scientific knowledge and Confucian verities can be integrated by virtue of their as being complementary aspects of one body of knowledge. He attempted to synthesize his Confucian religion and Western science, both of which, in his view, were essential to individual intellectual development as well as to national prosperity. He saw no conflict between them, for he believed that the Confucian Classics had already contained the origins of modern science, which, as it turned out, had been recently developed into finer stages of scientific discovery in the West. So, embracing Western science was entirely legitimate and consonant with his religious cause, which called for going back to 'pristine' Confucianism. Epistemologically speaking and in terms of the Confucian concept of *gezhi* that was infused with a moral orientation, the principles of nature discovered through natural science are expected to conform with or lead to the higher principles enunciated in the Confucian Classics. Conflict with science might occur with other 'inferior' religions like Christianity, but rarely if ever with the more advanced Confucian religion. The Darwinian principle of evolution by natural selection, for example, contradicted the Christian doctrine of creation, but it fitted in well with the Confucian notion of evolution through constant change and transformation. Further, he did not insist on the infallibility of the founders of religions and he considered religious doctrines as standing in need of adjustment when scientific evidence made it imperative to correct previous misconceptions of reality. But there was a limit to the sovereignty of scientific reasoning. No matter how far science advanced, it could never disprove the existence of God, whose infinite nature Kang believed lay beyond the grasp and measure of finite scientific minds, even those as brilliant as Newton's or Einstein's.

Nowhere is Kang's view of a complementary relationship between science and religion more clearly manifest than in his appropriation of scientific notions in the process of his Confucian exegesis. Like his reformist ally and disciple in spirit, Tan Sitong, Kang related the all-important Confucian virtue *ren* to 'ether' and 'electricity', arguing that the latter constituted the essence of the former. This attempt has been recognized by some later scholars as a new paradigm for the Confucian tradition. Upon learning that the concept of 'ether' had been shattered by Einstein's theory of relativity, however, he willingly stopped using the term himself. In doing so, he demonstrated one aspect of scientific spirit: abiding by verified facts and proven theories. At the same time, he waxed more eloquent about 'electricity'. He distinguished between two types of electricity, a subtle yet visible material power and a more refined invisible power. He claimed that the latter was in fact what was commonly known as 'spirit', and was the very medium through which human consciousness was spread within the body and between human beings, and through which Heaven and man were connected. He applied the notion of electricity to his explanation of the nature of the soul and the mysterious mechanism of divine retribution. All of this brings to mind the efforts of some of the eminent scientists in late nineteenth century Europe to reconcile their religious convictions with contemporary scientific theories through the mediating role of 'ether' or 'electricity'.

As a self-proclaimed sage for his age, Kang did not content himself with the role of a mere reformer of Confucianism. He desired to establish his own *jiao* or religion, and he indeed embarked upon this enterprise in his later years. He began in earnest by expounding his doctrine of 'celestial peregrination', a doctrine that mixed, or, synthesized, his astronomical knowledge with his metaphysical speculations. He taught that knowledge of the heavens, coupled with observation by the naked eye or through a powerful telescope, could provide a paradigm shift in thinking. As an offer of supernal solace to humanity, Kang stated a theoretical proposition that the realization of this truth would assist his fellow human beings in getting rid of worries and achieving happiness – the ultimate goal of all religions in his understanding. For the realization of the vastness of the universe would make them look beyond and thereby transcend their trifling sufferings and vexations. The awareness of the fact that if observed from other planets they were actually 'denizens of the heavens' would help them transcend the limits of families, tribes and states and eventually achieve utmost happiness. His intellectual convictions as to the metaphysical implications of astronomical knowledge for this mundane life, however, had not shaken his belief in *Shangdi,* the Supreme Ruler in Heaven. In the final analysis, Kang had eventually become, at least within his own private mental universe, a *jiaozhu* or founder of a *jiao*/teaching/religion, believing that his doctrine would be practised in the future world of Great Unity, a world that Confucius had prepared humanity for.

Seen from a broader perspective, Kang's construction of a universal concept of 'God', his vision of the future unification of religions, and his conceptualization of a compatible and complementary relationship between science and religion, all exemplified the rise of transnational, trans-religious spirituality. This intellectual activity took place in the context of major events in the formation of imperial and colonial modernity. In these processes, the cultural and religious encounters between the Christian occidental and the Confucian-Hindu oriental worlds led many to a search for a common essence between different traditions as well as between different domains of human thought. The sometimes seemingly Quixotic yet farsighted concepts enunciated by Kang beckon an increasing number of scholars to a reconsideration of his hopes, dreams, ideals, principles, systems, institutions and his agenda for the alliance of science and religion working hand in hand to address the crying issues of humanity on the path to build 'a community of shared future'.

This book is focused on Kang Youwei. To fully illustrate the intellectual richness and complexity of the period of transition from Imperial to Republican China, other prominent contemporary thinkers have been included, in particular, Kang's allies Liang Qichao, Tan Sitong and Yan Fu, and Kang's political detractor, the revolutionary thinker Zhang Binglin. They have all been discussed or touched upon in this book in terms of their relation to Kang, but each of them, being a gigantic intellect, deserves more substantial treatment on his own terms by way of concentrated focus on the views they articulated in their discourse on science and religion. This may be one direction for systematic research into the intellectual ferment generated by this exceptional and stellar group of towering intellects who were pondering the great questions of science, religion and development during that turbulent turning point in Chinese history.

Another direction for further research would be a more comprehensive study devoted to Kang's religious thought. His voluminous writings contain elaborate discussions on various religious themes, many of which have only begun to be revisited with vigour and increasing frequency by today's scholars in relevant fields. In particular, his reconstruction of the concept of 'religion' in light of the Chinese traditional notion of *jiao* has of late begun to resonate with present-day thinking, as illustrated by the recent proposal from a number of leading Chinese scholars to reconsider the generic concept of 'religion' in a consciously Chinese context rather than from a Eurocentric perspective as has been the practice for the past century and a half, so as to provide a new paradigm for the re-evaluation of Chinese religion, and in that sense, of Chinese culture as well (see the discussion in the Introduction).

Having completed the research to produce this book, I have come to perceive the acute foresight and profound insightfulness of Liang Qichao's pronouncement in his summation of the significance of his teacher's life and work – a statement that was made a century ago but which has hitherto been largely ignored – that Kang's mind was dominated by religious thought, and it is his religious cause that would in due course be recognized as paramount among all his contributions to the progress and development of the Chinese nation.[1] I humbly suggest that Kang is not reduced as a result of the foregoing unconventional presentation of him as a religious thinker. He only fails, to borrow Joshua A. Fogel's comment on Liang, when he is judged 'ahistorically' by later dichotomous standards.[2]

1 Liang, 'Nanhai Kang', 488.

2 Joshua A. Fogel, ed., *The Role of Japan in Liang Qichao's Introduction of Modern Western Civilization to China* (Berkeley, Calif.: Institute of East Asian Studies, University of California Berkeley, 2004), 12.

Bibliography

'A State Religion for China?' *The North-China Herald*, 30 August 1913.

'America in China.' *New York Times*, 29 March 1868.

Asad, Talal. *Genealogies of Religion: Discipline and Reasons of Power in Christianity and Islam*. Baltimore & London: Johns Hopkins University Press, 1993.

Baldrian-Hussein, Farzeen. 'Hun and Po'. In *The Encyclopedia of Taoism*, edited by Fabrizio Pregadio, 521–23. Routledge, 2008.

Barbour, Ian G. *Religion in an Age of Science*. San Francisco: Harper & Row, 1990.

Barrows, John. *The World's Parliament of Religions*. Vol. 1. 2 vols. Chicago: The Parliament Publishing, 1893.

Bentley, Jerry H. *Old World Encounters: Cross-Cultural Contacts and Exchanges in Pre-Modern Times*. Oxford: Oxford University Press, 1993.

Beretta, Marco, Maria Conforti, and Paolo Mazzarello, eds. *Savant Relics: Brains and Remains of Scientists*. Sagamore Beach, MA: Science History Publications, 2016.

Billioud, Sébastien. *Thinking Through Confucian Modernity*. Leiden; Boston: Brill, 2011.

Bodde, Derk. *Chinese Thought, Society, and Science: The Intellectual and Social Background of Science and Technology in Pre-Modern China*. Honolulu: University of Hawaii Press, 1991.

Boorman, H. L., ed. *Biographical Dictionary of Republican China*. Vol. 2. New York: Columbia University Press, 1968.

Borrell, Ari. 'Ko-Wu or Kung-an? Practice, Realization, and Teaching in the Thought of Chang Chiu-Ch'eng'. In *Buddhism in the Sung*, edited by Peter Gregory and Daniel A. Getz, 62–108. Honolulu: University of Hawai'i Press, 1999.

'Brenan to Sir C. MacDonald'. In *Correspondence Respecting the Affairs of China*, 307–8. London: Harrison and Sons, 1899.

Brière, O. *Fifty Years of Chinese Philosophy, 1898–1950*. Westport, Conn.: Greenwood Press, 1979.

Brockey, Liam M. *Journey to the East: The Jesuit Mission to China, 1579–1724*. Cambridge, MA.: Harvard University Press, 2007.

Brooke, John H. *Science and Religion: Some Historical Perspectives*. Cambridge: Cambridge University Press, 1991.

Brooke, John H., and Ronald L. Numbers, eds. *Science and Religion around the World*. Oxford: Oxford University Press, 2011.

Brown, Robert H. *The Wisdom of Science: Its Relevance to Culture and Religion*. New York: Cambridge University Press, 1986.

Brusadelli, Federico. 'A Tale of Two Utopias: Kang Youwei's Communism, Mao Zedong's Classicism and the "Accommodating Look" of the Marxist Li Zehou'. *Asian Studies* 5, no. 1 (2017): 103–22.

Bryce, James B. *The American Commonwealth*. 2 vols. New York: Macmillan, 1915.

Buck, Peter. *American Science and Modern China, 1876–1936*. Cambridge: Cambridge University Press, 1980.

Butterfield, Herbert. *The Origins of Modern Science: 1300–1800*. New ed. London: G. Bell and Sons Ltd, 1957.

Butterfield, Herbert. *The Whig Interpretation of History*. New ed. New York: W. W. Norton & Company, 1965.

Cai Yuanpei 蔡元培. 'Fu Ren Hongjun han 复任鸿隽函 [A Letter in Reply to Ren Hongjun (1915)]'. In *Cai Yuanpei quanji 蔡元培全集* [*Complete Works of Cai Yuanpei*], 10:249–50. Zhejiang jiaoyu chubanshe, 1998.

Cai Yuanpei 蔡元培. 'Tongxin 通信 [Correspondence]'. *Xinqingnian 新青年* [*New Youth Journal*] 3, no. 1 (1917).

Cai Yuanpei 蔡元培. 'Yi meiyu dai zongjiao shuo 以美育代宗教說 [On Substituting Aestheticism for Religion]'. *Xinqingnian 新青年* [*New Youth Journal*] 3, no. 6 (1917): 2nd article.

Campany, Robert F. *Strange Writing: Anomaly Accounts in Early Medieval China*. Albany: State University of New York Press, 1996.

Cantor, G. N. and Hodge, M. J. S., eds. *Conceptions of Ether: Studies in the History of Ether Theories, 1740–1900*. Cambridge: Cambridge University Press, 1981.

Cantor, Geoffrey. 'Science and Religion: From the Historian's Perspective'. In *The Routledge Companion to Religion and Science*, edited by James W. Haag, Gregory R Peterson, and Michael L. Spezio, 24–33. London: Routledge, 2012.

Chalmers, A. F. *What Is This Thing Called Science?* 4th ed. Queensland: University of Queensland Press, 2013.

Chan, Sin-wai, trans. *An Exposition of Benevolence: The Jen-Hsüeh of T'an Ssu-t'ung*. Hong Kong: The Chinese University Press, 1984.

Chan, Sin-wai. *Buddhism in Late Ching Political Thought*. Hong Kong: Chinese University Press, 1985.

Chang, Hao. *Chinese Intellectuals in Crisis : Search for Order and Meaning (1890–1911)*. Berkeley: University of California Press, 1987.

Chang, Hao. 'Intellectual Change and the Reform Movement, 1890–8'. In *The Cambridge History of China*, edited by John K. Fairbank and Kwang-Ching Liu, 11:274–338. Cambridge: Cambridge University Press, 1980.

Chang, Hao. *Liang Ch'i-Ch'ao and Intellectual Transition in China, 1890–1907*. Cambridge, MA: Harvard University Press, 1971.

Chang, Hao. 'The Intellectual Heritage of the Confucian Ideal of Ching-Shih'. In *Confucian Traditions in East Asian Modernity*, edited by Wei-ming Tu, 72–91. Cambridge, Mass.: Harvard University Press, 1996.

Chen Baozhen 陳寶徵. 'Kongjiao yi dingwei guojiao lun 孔教宜定為國教論 [On Confucianism Being Qualified for Being the State Religion]'. *Kongjiaohui zazhi 孔教會雜誌* [*The Confucian Association Monthly*], no. 1 (1913): 27–30.

Chen Duxiu 陈独秀. 'Xuyi 序一 [Preface One]'. In *Kexue yu renshengguan 科学与人生观* [*Science and the Philosophy of Life*], by Zhang Junmai 张君劢 et al., 1–7. Shenyang: Liaoning jiaoyu chubanshe, 1998.

Chen Duxiu 陳獨秀. 'Benzhi zuian zhi dabianshu 本誌罪案之答辯書 [A Reply to the Charges Against Our Magazine]'. *Xinqingnian 新青年* [*New Youth Journal*] 6, no. 1 (1919): 10–11.

Chen Duxiu 陳獨秀. 'Bo Kang Youwei zhi Zongtong Zongli shu 駁康有為致信總統總理書 [A Refutation of Kang Youwei's Letter to the President and the Premier]'. *Xinqingnian 新青年* [*New Youth Journal*] 2, no. 2 (1916): 3rd article.

Chen Duxiu 陳獨秀. 'Falanxi renmin yu jinshi wenming 法蘭西人民與近世文明 [The French People and Modern Civilization]'. *Xinqingnian 新青年* [*New Youth Journal*] 1, no. 1 (1915): 2nd article.

Chen Duxiu 陳獨秀. 'Jinggao qingnian 敬告青年 [My Solemn Plea to Youth]'. *Xinqingnian 新青年* [*New Youth Journal*] 1, no. 1 (1915): 1–6.

Chen Duxiu 陳獨秀. 'Ouxiang pohuai lun 偶像破壞論 [On the Destruction of Idols]'. *Xinqingnian 新青年* [*New Youth Journal*] 5, no. 2 (1918): 89.

Chen Duxiu 陳獨秀. 'Zailun Kongjiao wenti 再論孔教問題 [My Second Discussion on the Question of Confucianism as a Religion] [1917]'. *Xinqingnian 新青年* [*New Youth Journal*] 2, no. 5 (1917): 1st article.

Chen, Hsi-yuan. 'Confucianism Encounters Religion: The Formation of Religious Discourse and the Confucian Movement in Modern China'. Ph.D., Harvard University, 1999.

Chen Huaiyu 陈怀宇. *Jindai chuanjiaoshi lun Zhongguo zongjiao: yi Muweilian Wujiao tongkao wei zhongxin 近代传教士论中国宗教: 以慕维廉五教通考为中心* [*Missionary Writings on Chinese Religions: William Muirhead and His Introduction to Five Religions in China*]. Shanghai: Shanghai renmin chubanshe, 2012.

Chen, Huan-Chang. *The Economic Principles of Confucius and His School*. 2 vols. New York: Longmans, Green & Co., 1911.

Chen Huanzhang 陈焕章. 'Gaoyao Yanzhou Yingchuan xuetang xiaoge 高要砚洲颖川学堂校歌 [Songs for the School of Yingchuan Village, Xianzhou, Gaoyao (1929)]'. In *Chen Huanzhang wenlu 陈焕章文录* [*Collected Writings of Chen Huanzhang*], 453–54. Changsha: Yuelu shushe, 2015.

Chen Huanzhang 陈焕章. 'Jiamu Li taifuren...shilue 家母李太夫人...事略 [A Biographical Sketch of My Mother, Mrs. Chen, née Li (1929)]'. In *Chen Huanzhang wenlu 陈焕章文录* [*Collected Writings of Chen Huanzhang*], 419–23. Changsha: Yuelu shushe, 2015.

Chen Huanzhang 陈焕章. 'Meizhou huaqiao chuangli changjiaohui xuyan 美洲华侨创立昌教会叙言 [Introduction to the Changjiaohui Established Among the Chinese in America]'. In *Chen Huanzhang wenlu 陈焕章文录* [*Collected Writings of Chen Huanzhang*], 363–66. Changsha: Yuelu shushe, 2015.

Chen Huanzhang 陈焕章. 'Shen Yi'an xiansheng zhuan 沈乙盦先生传 [A Short Biography of Mr. Shen Yi'an]'. In *Chen Huanzhang wenlu 陈焕章文录* [*Collected Writings of Chen Huanzhang*], 391–93. Changsha: Yuelu shushe, 2015.

Chen Huanzhang 陈焕章. 'Zhi yinbing xuezhang shu 致饮冰学长书 [Letter to Liang Qichao]'. In *Chen Huanzhang wenlu 陈焕章文录* [*Collected Writings of Chen Huanzhang*], 2. Changsha: Yuelu shushe, 2015.

Chen Huanzhang 陳煥章. 'Guohui xianfa huiyi duiyu Kongjiao dageming jinggao guoren shu 國會憲法會議對於孔教大革命敬告國人書 [A Letter to Compatriots on How the Constitutional Session of the National Assembly Has Slain the Life of the Confucian Religion (1917)]'. In *Kongjiao shinian dashi 孔教十年大事* [*Events on Confucian Religion during a Decade*], edited by Ke Huang 柯璜, 34a–36b. Taiyuan: Zongshenghui, 1924.

Chen Huanzhang 陳煥章. 'Kongjiaohui guojiao yijianshu 孔教會國教意見書 [The Views of the Confucian Religion Association on State Religion]'. *Ching Shih Pao 經世報* [*World Managing Monthly*] 2, no. 3 (1913): [jiaoshi] 9–15.

Chen Huanzhang 陳煥章. 'Kongjiaohui qingyuanshu 孔教會請願書 [A Petition of the Confucian Religion Association]'. *Kongjiaohui zazhi 孔教會雜誌* [*The Confucian Association Monthly*], no. 6 (1913): 13–21.

Chen Huanzhang 陳煥章. 'Kongjiaohui wei xianfa qicao weiyuanhui foujue guojiao jinggao quanguo tongbao shu 孔教會為憲法起草委員會否決國教警告全國同胞書 [Letter to Compatriots on the Rejection of the State Religion Proposal by the Constitution-Drafting Committee (1917)]'. In *Kongjiao shinian dashi 孔教十年大事* [*Events on Confucian Religion during a Decade*], edited by Ke Huang 柯璜, 83a–87b. Taiyuan: Zongshenghui, 1924.

Chen Huanzhang 陳煥章. *Kongjiao lun 孔教論* [*On the Confucian Religion*]. Shanghai: Shangwu Press, 1912.

Chen Huanzhang 陳煥章. 'Mingding yuanyou zhi guojiao bingbu aiyu xinjiao ziyou zhi xinmingci 明定原有之國教並不礙於信教自由之新名詞 [Specifying the Previous State Religion in Clear Terms not Contradicting the Neologism for Religious Freedom]'. *Kongjiaohui zazhi 孔教會雜誌* [*The Confucian Association Monthly*], no. 1 (1913): 1–12.

Chen, Zhi-yun. 'Liang Ch'i-Ch'ao's Missionary Education: A Case Study of Missionary Influence on the Reformers'. In *Papers on China*, 16:66–125. Cambridge, MA.: East Asian Research Center, Harvard University, 1962.

Chen, Zhongping. 'Kang Youwei's Activities in Canada and the Reformist Movement Among the Global Chinese Diaspora, 1899–1909'. *Twentieth-Century China* 39, no. 1 (2014): 3–23.

Chow, Tse-tsung. 'The Anti-Confucian Movement in Early Republican China'. In *The Confucian Persuasion*, edited by Arthur F. Wright, 288–312. Stanford: Stanford University Press, 1960.

Chu, Pingyi. 'Scientific Dispute in the Imperial Court: The 1664 Calendar Case'. *Chinese Science*, no. 14 (1997): 7–34.

'Chunqiu Guliangzhuan 春秋穀梁傳 [The Guliang Commentary on the Annals]'. In *Baiwen shisanjing 白文十三經 [The Thirteen Classics: An Unannotated Edition]*, 1–107. Shanghai: Shanghai guji chubanshe, 1983.

Cohen, Paul A. 'Christian Missions and Their Impact to 1900'. In *The Cambridge History of China*, edited by John K. Fairbank and Kwang-Ching Liu, 11:543–91. Cambridge: Cambridge University Press, 1980.

Craig, Edward. *Philosophy: A Very Short Introduction*. Oxford: Oxford University Press, 2002.

Crowther, J. G. *The Social Relations of Science*. Revised ed. London: Cresset Press, 1967.

De Groot, J. J. M. *The Religious System of China*. Vol. VI. Leyden: E. J. Brill, 1910.

DeBold, Richard C., and Russell C. Leaf, eds. *LSD, Man & Society*. Middletown, Ct.: Wesleyan University Press, 1967.

Dessein, Bart. 'Yearning for the Lost Paradise: The "Great Unity" (Datong) and Its Philosophical Interpretations'. *Azijske Studije* 5, no. 1 (2017): 83–102.

Ding Weiliang 丁韙良 [William A. P. Martin]. *Xixue kaolue 西學考略 [A Brief Review of Western Learning]*. Beijing: Zongli yamen, 1883.

Ding Wenjiang 丁文江. 'Xuanxue yu kexue 玄学与科学 [Metaphysics and Science]'. In *Kexue yu renshengguan 科学与人生观 [Science and the Philosophy of Life]*, by Zhang Junmai 张君劢 et al., 38–56. Shenyang: Liaoning jiaoyu chubanshe, 1998.

Ding Wenjiang 丁文江, and Zhao Fengtian 赵丰田, eds. *Liang Qichao nianpu changbian 梁启超年谱长编 [A Chronological Biography of Liang Qichao]*. Shanghai: Shanghai renmin chubanshe, 1983.

Dixon, Thomas. *Science and Religion: A Very Short Introduction*. Oxford: Oxford University Press, 2008.

Draper, J. William. *History of the Conflict between Religion and Science*. London: Henry S. King & Co., 1875.

Eber, Irene. *The Jewish Bishop and the Chinese Bible: S.I.J. Schereschewsky (1831–1906)*. Leiden: Brill, 2016.

Edkins, Joseph. *Religion in China; Containing a Brief Account of the Three Religions of the Chinese*. 2nd ed. London: Trübner & Co. Ltd, 1893.

Edson, E., and E. Savage-Smith. *Medieval Views of the Cosmos*. Oxford: Bodleian Library, 2004.

Elman, Benjamin. *A Cultural History of Modern Science in China*. Cambridge, MA.: Harvard University Press, 2006.

Elman, Benjamin. *Classicism, Politics, and Kinship: The Ch'ang-Chou School of New Text Confucianism in Late Imperial China*. Berkeley: University of California Press, 1990.

Elman, Benjamin. *On Their Own Terms: Science in China, 1550–1900*. Cambridge, MA: Harvard University Press, 2005.

Elshakry, Marwa S. 'Knowledge in Motion: The Cultural Politics of Modern Science Translations in Arabic'. *Isis* 99, no. 4 (2008): 701–30.

Engels, Friedrich. 'Dialectics of Nature'. In *Marx & Engels Collected Works*, translated by Clemens Dutt, 25:313–588. London: Lawrence & Wishart, 2010.

Fairbank, John K. 'Introduction: The Old Order'. In *The Cambridge History of China*, edited by John K. Fairbank, 10:1–34. Cambridge: Cambridge University Press, 1978.

Fairbank, John K., and Kwang-Ching Liu, eds. *The Cambridge History of China*. Vol. 11. Cambridge: Cambridge University Press, 1980.

Fairbank, John King, and Merle Goldman. *China: A New History, Second Enlarged Edition*. Harvard University Press, 2006.

Fan Hongye 樊洪业. 'Cong gezhi dao kexue 从"格致"到"科学" [From "gezhi" to "Science"]'. *Ziran bianzhengfa tongxun 自然辩证法通讯 [Journal of Dialectics of Nature]*, no. 3 (1998): 39–50.

Fang Min 方敏. 'Kexue yu zongjiao guanxi de xueshu guanzhu 科学与宗教关系的学术关注 [Scholarly Attention to the Relationship between Science and Religion]'. *Ziran bianzhengfa tongxun 自然辩证法通讯 [Journal of Dialectics of Nature]*, no. 3 (2008).

Ferngren, Gary B. *The History of Science and Religion in the Western Tradition: An Encyclopedia*. New York & London: Garland Publishing, 2000.

Feyerabend, Paul K. *Against Method: Outline of an Anarchistic Theory of Knowledge*. London: New Left Books, 1975.

Fogel, Joshua A., ed. *The Role of Japan in Liang Qichao's Introduction of Modern Western Civilization to China*. Berkeley, Calif.: Institute of East Asian Studies, University of California Berkeley, 2004.

Force, James E., and Richard H. Popkin, eds. *The Books of Nature and Scripture: Recent Essays on Natural Philosophy, Theology, and Biblical Criticism in the Netherlands of Spinoza's Time and the British Isles of Newton's Time*. Luxemburg: Springer Science+Business Media Dordrecht, 1994.

Fryer, John. 'An Account of the Department for the Translation of Foreign Books at the Kiangnan Arsenal, Shanghai'. *The North-China Herald*. 29 January 1880.

Fu Lanya 傅蘭雅 [John Fryer]. 'Gezhi luelun 格致略論 [Sketches of Science]'. *Gezhi huibian 格致彙編 [The Chinese Scientific Magazine]* I, no. 1 (1876): 1–6.

Fu Lanya 傅蘭雅 [John Fryer]. 'Hundun shuo 混沌說 [On Chaos]'. *Gezhi huibian 格致彙編 [The Chinese Scientific Magazine]*, no. 7 (1877): 6–8.

Fung, Yu-lan. *A History of Chinese Philosophy*. Translated by Derk Bodde. 2nd ed. Vol. 1. Princeton: Princeton University Press, 1983.

Fung, Yu-lan. *A History of Chinese Philosophy*. Translated by Derk Bodde. Vol. 2. Princeton: Princeton University Press, 1983.

Fung, Yu-Lan. 'Why China Has No Science - An Interpretation of the History and Consequences of Chinese Philosophy'. *International Journal of Ethics* 32, no. 3 (1922): 237–63.

Furth, Charlotte. 'Intellectual Change: From the Reform Movement to the May Fourth Movement, 1895–1920'. In *An Intellectual History of Modern China*, edited by Merle Goldman and Leo Ou-fan Lee, 13–96. Cambridge: Cambridge University Press, 2002.

Furth, Charlotte. *Ting Wen-Chiang: Science and China's New Culture*. Cambridge, Mass.: Harvard University Press, 1970.

Gan Chunsong 干春松. *Baojiao liguo: Kang Youwei de xiandai fanglue 保教立国：康有为的现代方略* [*To Establish the State through Preserving the Faith: Kang's Strategies for Modernization*]. Beijing: Sanlian shudian, 2015.

Gan Chunsong 干春松. *Kang Youwei yu Ruxue de xinshi 康有为与儒学的新世* [*Kang Youwei and the New Epoch of Confucianism*]. Shanghai: Huadong shifan daxue chubanshe, 2015.

Gan Chunsong 干春松. 'Kang Youwei yu xiandai Ruxue sichao de guanxi bianxi 康有为与现代儒学思潮的关系辨析 [On the Relationship between Kang Youwwei and Contemporary Schools of Confucianist Thought]'. *Zhongguo renmin daxue xuebao 中国人民大学学报* [*Journal of China Renmin University*], no. 5 (2015): 22–30.

Gao Like 高力克. 'Yi kexue dai zongjiao: Chen Duxiu de kexuezhuyi zongjiaoguan 以科学代宗教: 陈独秀的科学主义宗教观 [Substituting Science for Religion: Chen Duxiu's Scientism-Oriented Concept of Religion]'. *Shixue yuekan 史学月刊* [*Journal of Historical Science*], no. 1 (2017): 89–97.

Gasster, Michael. 'The Republican Revolutionary Movement'. In *The Cambridge History of China*, edited by John K. Fairbank and Kwang-Ching Liu, 11:463–534. Cambridge: Cambridge University Press, 1980.

Ge Zhaoguang 葛兆光. 'Yixiangtiankai: jinnianlai dalu Xinruxue de zhengzhi suqiu 异想天开: 近年来大陆新儒学的政治诉求 [A Fantasy: On Recent Political Appeals of the New Confucianists on the Chinese Mainland]'. www.aisixiang.com 爱思想, 1 July 2017. http://www.aisixiang.com/data/104951.html.

Geng Yunzhi. *An Introductory Study on China's Cultural Transformation in Recent Times*. Translated by Wang Huimin. Beijing & Heidelberg: Foreign Language Teaching and Research Press & Springer-Verlag, 2015.

Geng Yunzhi 耿云志. 'Kang Youwei de shengren qingjie jiqi yi Kongjiao wei guojiao shuo 康有为的"圣人"情结及其以孔教为国教说 [Kang Youwei's Sage Complex and His Views on Confucianism as a State Religion]'. In *Kang Youwei yu jindai wenhua 康有为与近代文化* [*Kang Youwei and Modern Culture*], edited by Fang Zhixin 方志钦 and Wang Jie 王杰. Kaifeng: Henan daxue chubanshe, 2006.

Giles, Herbert A. *Religions of Ancient China*. London: Archibald Constable & Co Ltd, 1905.

Giles, Herbert A., and Charles Aylmer. 'The Memoirs of H. A. Giles'. *East Asian History*, no. 13/14 (1997): 1–90.

Godart, G. Clinton. '"Philosophy" or "Religion"? The Confrontation with Foreign Categories in Late Nineteenth Century Japan'. *Journal of the History of Ideas* 69, no. 1 (2008): 71–91.

Gongyang Shou 公羊壽, He Xiu 何休, and Xu Yan 徐彥. *Chunqiu Gongyangzhuan zhushu 春秋公羊傳注疏* [*The Gongyang Commentary on the Annals with Annotations and Commentaries*]. Beijing: Beijing daxue chubanshe, 2000.

Goossaert, Vincent. 'Republican Church Engineering: The National Religious Associations in 1912 China'. In *Chinese Religiosities: Afflictions of Modernity and State Formation*, edited by Mayfair Mei-hui Yang, 209–32. Berkeley: University of California Press, 2008.

Goossaert, Vincent. 'The Concept of Religion in China and the West'. *Diogenes* 52 (2005): 13–20.

Goossaert, Vincent. 'The Social Organization of Religious Communities in the Twentieth Century'. In *Chinese Religious Life*, edited by David A. Palmer, Glenn Landes Shive, and Philip L. Wickeri, 172–90. Oxford: Oxford University Press, 2011.

Goossaert, Vincent, and David A. Palmer. *The Religious Question in Modern China*. Chicago: University of Chicago Press, 2011.

Gou Wenzeng 勾文增. 'Dangdai kexue yu zongjiao yanjiu de yanjin 当代科学与宗教研究的演进 [A Review of Contemporary Research on Science and Religion in China]'. *Nanjing ligong daxue xuebao 南京理工大学学报 (社科版)* [*Journal of Nanjing University of Science and Technology (Social Sciences)*], no. 2 (2012): 44–51.

Gower, Barry. *Scientific Method: A Historical and Philosophical Introduction*. London and New York: Routledge, 1997.

Grant, Edward. *A History of Natural Philosophy: From the Ancient World to the Nineteenth Century*. Cambridge; New York: Cambridge University Press, 2007.

Guan Xiang 关祥, ed. *Xiqiaoshan zhi 西樵山志* [*A Gazette of the Xiqiao Mountain*]. Guangzhou: Guangdong renmin chubanshe, 1982.

'Guangdong Chen dudu ji gesheng dudu jian 廣東陳都督及各省都督鑒 [Cable to Governor Chen of Guangdong and Governors of Other Provinces]'. *Linshi zhengfu gongbao 臨時政府公報* [*Provisional Government Gazette*], *No. 1*, 29 January 1912.

Gupta, Krishna Prakash. 'Religious Evolution in Premodern Asia: A Comparative Study of K'ang Yu-Wei and Vivekananda'. Ph.D., Harvard University, 1976.

Han Qi 韩琦. *Tongtian zhixue: Yesuhui shi he tianwenxue zai Zhongguo de chuanbo 通天之学: 耶稣会士和天文学在中国的传播* [*Learning of Heaven: The Jesuits and the Spread of Astronomy in China*]. Beijing: Sanlian shudian, 2018.

Harrell, Stevan. 'The Concept of Soul in Chinese Folk Religion'. *The Journal of Asian Studies* 38, no. 3 (May 1979): 519.

Harrison, Peter. *The Territories of Science and Religion*. Chicago: University of Chicago Press, 2015.

Henry, John. *The Scientific Revolution and the Origins of Modern Science*. 3rd ed. New York: Palgrave Macmillan, 2008.

Hexuli 赫胥黎 [Huxley]. *Tianyan lun 天演論* [*On Cosmic Evolution (Evolution and Ethics)*]. Translated by Yan Fu 嚴復. 2 vols. Gaiyang: Shenshijizhai 慎始基斋, 1898.

Hopkins, L. C. 'The Shaman or Chinese Wu: His Inspired Dancing and Versatile Character'. *The Journal of the Royal Asiatic Society of Great Britain and Ireland*, no. 1 (1945): 3–16.

Howard, Richard C. 'K'ang Yu-Wei (1858–1927): His Intellectual Background and Early Thought'. In *Confucian Personalities*, by Arthur F. Wright and Denis Twitchett, 294–316. Stanford: Stanford University Press, 1962.

Howard, Richard C. 'The Early Life and Thought of K'ang Yu-Wei, 1858–1895'. Ph.D., Columbia University, 1972.

Hsiao, Kung-ch'uan. *A Modern China and a New World: K'ang Yu-Wei, Reformer and Utopian*. Seattle: University of Washington Press, 1975.

Hsiao, Kung-ch'uan. 'K'ang Yu-Wei's Excursion into Science'. In *K'ang Yu-Wei: A Biography and a Symposium*, edited by Jung-pang Lo, 375–407. Tucson: University of Arizona Press, 1967.

Hu, Danian. 'The Reception of Relativity in China'. *Isis* 98, no. 3 (2007): 539–57.

Hu Mingfu 胡明復. 'Kexue fangfa lun yi 科學方法論一 [On the Method of Science: I]'. *Kexue 科學 [Science]* 2, no. 1 (1916): 719–28.

Huang Jing 黄晶. *Kang Youwei zhuan 康有为传 [A Biography of Kang Youwei]*. Beijing: Beijing lianhe chuban gongsi, 2013.

Huang Kan 黃侃, ed. 'Liji 禮記 [The Book of Rites]'. In *Baiwen shisanjing 白文十三經 [The Thirteen Classics: An Unannotated Edition]*, 1–245. Shanghai: Shanghai guji chubanshe, 1983.

Huang Yanyu 黄延毓. 'Kang Youwei de shengping jiqi xueshu gailue 康有為的生平及其學術概畧 [A Short Account of Kang's Life and His Scholarship]'. *Nanfeng 南風* 3, no. 1 (1928): 40–45.

'Huayan Wujiao 華嚴五教'. In *The Princeton Dictionary of Buddhism*, 361. Princeton: Princeton University Press, 2013.

Hummel, Arthur W., ed. *Eminent Chinese of the Ch'ing Period, 1644–1912*. 2 vols. Folkestone: Global Oriental, 2010.

Ichiko, Chuzo. 'Political and Institutional Reform, 1901–11'. In *The Cambridge History of China*, edited by John K. Fairbank and Kwang-Ching Liu, 11:375–415. Cambridge: Cambridge University Press, 1980.

Iliffe, Rob. *Newton: A Very Short Introduction*. Oxford: Oxford University Press, 2007.

Iliffe, Rob. *Priest of Nature: The Religious Worlds of Isaac Newton*. Oxford; New York: Oxford University Press, 2017.

James, William. *Varieties of Religious Experience*. London: Longmans, Green, and Co., 1902.

Jami, Catherine. 'Revisiting the Calendar Case (1664–1669): Science, Religion, and Politics in Early Qing Beijing'. *Korean Journal of History of Science* 27, no. 2 (2015): 459–77.

Jami, Catherine, Peter M Engelfriet, and Gregory Blue, eds. *Statecraft and Intellectual Renewal in Late Ming China: The Cross-Cultural Synthesis of Xu Guangqi (1562–1633)*. Leiden: Brill, 2001.

Jastrow, Morris. *The Study of Religion*. London: Walter Scott, 1901.

Jensen, Lionel M. *Manufacturing Confucianism: Chinese Traditions & Universal Civilization*. Durham: Duke University Press, 1997.

Jian Chaoliang 简朝亮. 'Nianpu 年谱 [A Chronological Biography]'. In *Zhu Jiujiang (Ciqi) xiansheng ji 朱九江（次琦）先生集* [*Collected Works of Mr. Zhu Ciqi*], 9–91. Taipei: Wenhai chubanshe, 1967.

Jian, Guo, Yongyi Song, and Yuan Zhou. *The A to Z of the Chinese Cultural Revolution*. Lanham, Md.: Scarecrow Press, 2009.

Jiang Guangxue 蒋广学. *Liang Qichao pingzhuan* [*A Critical Biography of Liang Qichao*]. Nanjing: Nanjing University Press, 2005.

Jiang Guiling 蒋貴麟, ed. *Kang Nanhai xiansheng yizhu huikan 康南海先生遗著汇刊* [*Collected Works of the Late Mr. Kang of Nanhai*]. 22 vols. Taipei: Hongye shuju, 1976.

Jiang Qing 蒋庆, Chen Ming 陈明, Kang Xiaogang 康晓光, Yu Donghai 余东海, and Qiufeng 秋风. *Zhongguo bixu zairuhua 中国必须再儒化* [*China Must Be Confucianized Again*]. Singapore: World Scientific, 2016.

Jin, Xiaoxing. 'The Evolution of Evolutionism in China, 1870–1930'. *Isis* 111, no. 1 (2020): 46–66.

Kang, Phee Seng. 'Bridging Science and Religion in China: Emerging Opportunities for Global Dialogue'. *Theology and Science* 4, no. 2 (2006).

Kang Youwei. 'Treatise on God'. Translated by Zhaoyuan Wan and Patrick Swire. *Renditions* 93, no. Spring (2020): 43–48.

Kang Youwei 康有为. 'Baoguohui zhangcheng 保国会章程 [The Charter of the Society for Preserving China (1898)]'. In *KYWQJ*, 4:54–55, 2007.

Kang Youwei 康有为. 'Canyiyuan tiyi liguo zhi jingshen yishu hou 参议院提议立国之精神议书后 [On the Senate's Proposal to Establish the Moral Spirit to Sustain China (1914)]'. In *KYWQJ*, 10:203–6, 2007.

Kang Youwei 康有为. 'Chang'an yanjianglu 长安演讲录 [Recorded Talks (by Kang) in Xi'an (1923)]'. In *KYWQJ*, 11:271–99, 2007.

Kang Youwei 康有为. 'Changxi xueji 长兴学记 [The Curriculum for Changxing Academy (1891)]'. In *KYWQJ*, 1:340–51, 2007.

Kang Youwei 康有为. 'Chunqiu bixiao dayi weiyan kao 春秋笔削大义微言考 [A Study of the Profound Principles and Hidden Meanings of the Emended Annals (1901)]'. In *KYWQJ*, 6:1–310, 2007.

Kang Youwei 康有为. 'Cu nanbei suyihe yiying Ouzhou heju dian 促南北速议和以应欧州和局电 [A Telegram Urging a Quick Truce between the South and the North to Better Defend China's Interests at the European Peace Conference (1918)]'. In *KYWQJ*, 11:92–93, 2007.

Kang Youwei 康有为. 'Da Pu jun datixue shu 答朴君大提学书 [A Reply to Mr. Pakki-yang, the Korean Commissioner of Education (1924)]'. In *KYWQJ*, 11:344–46, 2007.

Kang Youwei 康有为. *Datong shu 大同書* [*The Book of the Great Unity*]. Shanghai: Zhonghua shuju, 1935.

Kang Youwei 康有为. 'Gonghe pingyi 共和平议 [A Dispassionate Treatise on Republicanism (1917)]'. In *KYWQJ*, 11:2–76, 2007.

Kang Youwei 康有为. 'Gonghe zhengti lun 共和政体论 [On the Polity of a Republic (1911)]'. In *KYWQJ*, 9:241–50, 2007.

Kang Youwei 康有为. 'Guixue wenda 桂学问答 [Answers to Questions on Learning in Guiling (1894)]'. In *KYWQJ*, 2:18–25, 2007.

Kang Youwei 康有为. 'Jiaoxue tongyi 教学通义 [The General Import of Teaching and Learning (1885)]'. In *KYWQJ*, 1:17–58, 2007.

Kang Youwei 康有为. 'Ji'nan yanjiangci 济南演讲辞 [A Talk Given at Ji'nan (1923)]'. In *KYWQJ*, 11:248–49, 2007.

Kang Youwei 康有为. 'Jiuwang lun 救亡论 [A Treatise on Saving China (1911)]'. In *KYWQJ*, 9:222–39, 2007.

Kang Youwei 康有为. 'Kaifeng yanjiangci 开封演讲辞 [A Talk Given at Kaifeng (1923)]'. In *KYWQJ*, 11:236–39, 2007.

Kang Youwei 康有为. 'Kang Nanhai xiansheng jiangxue ji 康南海先生讲学记 [Notes of Mr. Kang's Lectures (1896)]'. In *KYWQJ*, 2:103–24, 2007.

Kang Youwei 康有为. 'Kang Nanhai xiansheng shiji 康南海先生诗集 [Collected Poems of Mr. Kang of Nanhai (1888–1927)]'. In *KYWQJ*, 12:133–421, 2007.

Kang Youwei 康有为. 'Kanggong shizhuang 康公事状 [A Eulogy for His Excellency Mr. Kang (1884)]'. In *KYWQJ*, 1:5–13, 2007.

Kang Youwei 康有为. 'Kangshi jiamiao bei 康氏家庙碑 [An Inscription on the Stone Tablet of the Kang Shrine (1917)]'. In *KYWQJ*, 10:438–40, 2007.

Kang Youwei 康有为. 'Kangzi neiwai pian 康子内外篇 [The Esoteric and Exoteric Essays of Master Kang (1886)]'. In *KYWQJ*, 1:95–112, 2007.

Kang Youwei 康有为. 'Kongjiaohui xu (er) 孔教会序(二) [(A Second) Introduction to the Confucian Religion Association (1912)]'. In *KYWQJ*, 9:343–46, 2007.

Kang Youwei 康有为. 'Kongjiaohui xu 孔教会序 [Introduction to the Confucian Religion Association (1912)]'. In *KYWQJ*, 9:341, 2007.

Kang Youwei 康有为. 'Kongzi gaizhi kao 孔子改制考 [A Treatise on Confucius as a Reformer (1897)]'. In *KYWQJ*, 3:1–260, 2007.

Kang Youwei 康有为. 'Lun shiwu 论时务 [On Current Affairs (1888)]'. In *KYWQJ*, 1:164–67, 2007.

Kang Youwei 康有为. 'Lunyu zhu 论语注 [The Analects Annotated (1902)]'. In *KYWQJ*, 6:375–540, 2007.

Kang Youwei 康有为. 'Mengzi wei 孟子微 [The Esoteric Meanings of the Mencius (1901)]'. In *KYWQJ*, 5:409–505, 2007.

Kang Youwei 康有为. 'Mingong pian 民功篇 [On Civil Feats (1886)]'. In *KYWQJ*, 5:63–89, 2007.

Kang Youwei 康有为. 'Minzhengbu zhun Diguotongyidang zhuce lun 民政部准帝国统一党注册论 [On the Registration of the Imperial Unification Party by the Ministry of the Interior (1911)]'. In *KYWQJ*, 9:192–93, 2007.

Kang Youwei 康有为. 'Nanhai Zhu xiansheng mubiao 南海朱先生墓表 [Epitaph for Mr. Zhu of Nanhai (1882)]'. In *KYWQJ*, 1:1, 2007.

Kang Youwei 康有为. 'Ni Zhonghua minguo xianfa cao'an 拟中华民国宪法草案 [A Proposed Draft Constitution for the Republic of China (1913)]'. In *KYWQJ*, 10:38–86, 2007.

Kang Youwei 康有为. 'Ouzhou shiyiguo youji xu 欧洲十一国游记序 [Preface to Accounts of My Travels to Eleven European Countries (1904)]'. In *KYWQJ*, 7:344–45, 2007.

Kang Youwei 康有为. 'Qing Quanyangyi zhuanda Riben neige chebing jiaohuan dian 请犬养毅转达日本内阁撤兵交还电 [A Telegram Sent via Inukai Tsuyoshi to the Japanese Cabinet on Withdrawing Army and Returning the Shangdong Peninsula (1919)]'. In *KYWQJ*, 11:108–9, 2007.

Kang Youwei 康有为. 'Qing shangding jiaoan falv ... yi zun shengshi er bao dajiao zhe 请商定教案法律......以尊圣师而保大教折 [A Memorial to the Throne Requesting the Making of Laws on Religious Disputes ... so as to Revere the Sacred Teacher and Preserve Our Great Faith (1898)]'. In *KYWQJ*, 4:92–95, 2007.

Kang Youwei 康有为. 'Qing zhu guozei jiu xuesheng dian 请诛国贼救学生电 [A Telegram Demanding the Punishment of the Traitors and the Release of the Arrested Students (1919)]'. In *KYWQJ*, 11:105–6, 2007.

Kang Youwei 康有为. 'Qingfei bagushitie kaifa shishi gaiyong celun zhe 请废八股试贴楷法试士改用策论折 [A Memorial to the Throne Requesting the Replacement of Eight-Legged Essays with Policy Papers in the Civil Service Examinations (1898)]'. In *KYWQJ*, 4:78–79, 2007.

Kang Youwei 康有为. 'Qinggai zhisheng shuyuan wei zhongxuetang xiangyi yinci wei xiaoxuetang 请改直省书院为中学堂乡邑淫祠为小学堂 ... [A Memorial to the Throne Requesting the Transformation of Provincial Academies into Middle Schools and Improper Temples in Towns and Villages into Primary Schools ...(1898)]'. In *KYWQJ*, 4:317–19, 2007.

Kang Youwei 康有为. 'Qingzhao jingjiteke li tuixing shengtong suikeshi pian 请照经济特科例推行生童岁科试片 [A Request for the Application of the Practices of the Special Examinations to the Civil Service Examinations at Various Levels (1898)]'. In *KYWQJ*, 4:50–51, 2007.

Kang Youwei 康有为. 'Qingzun Kongsheng wei guojiao li jiaohui jiaobu yi Kongzi jinian er fei yinci zhe 请尊孔圣为国教立教会教部以孔子纪年而废淫祠折 [A Memorial Requesting that Confucius Be Worshiped as the Founder of the State Religion, that a Ministry of Religion and a Church Be Established, that Calendar Years be Reckoned from the Birth of Confucius, and that Improper Temples Be Suppressed (1898)]'. In *KYWQJ*, 4:96–98, 2007.

Kang Youwei 康有为. 'Riben shumu zhi 日本书目志 [An Annotated Bibliography of Japanese Titles (1898)]'. In *KYWQJ*, 3:261–524, 2007.

Kang Youwei 康有为. 'Sanbentang luocheng jiwen 三本堂落成祭文 [An Oration for the Dedication of the Hall of Three Roots (1922)]'. In *KYWQJ*, 11:182, 2007.

Kang Youwei 康有为. 'Shang Qingdi disanshu 上清帝第三书 [My Third Memorial to the Throne (1895)]'. In *KYWQJ*, 2:68–80, 2007.

Kang Youwei 康有为. 'Shang Qingdi diyishu 上清帝第一书 [My First Memorial to the Throne (1888)]'. In *KYWQJ*, 1:180–84, 2007.

Kang Youwei 康有为. 'Shengxuehui houxu 圣学会后序 [An Afterword to the Society of Sacred Learning (1897)]'. In *KYWQJ*, 2:265–66, 2007.

Kang Youwei 康有为. 'Shengxuehui yuanqi 圣学会缘起 [An Introduction to the Society of Sacred Learning (1897)]'. In *KYWQJ*, 2:268–69, 2007.

Kang Youwei 康有为. 'Shili gongfa quanshu 实理公法全书 [A Compendium of True Principles and Universal Laws (c. 1885)]'. In *KYWQJ*, 1:145–60, 2007.

Kang Youwei 康有为. 'Woshi 我史 [My History (1899)]'. In *KYWQJ*, 5:58–106, 2007.

Kang Youwei 康有为. 'Wuxu lunzhou zhong juebishu ji Wuwu bahou 戊戌轮舟中绝笔书及戊午跋后 [My "Last Note" Penned in a Ship in 1898, with Postscripts Added in 1918]'. In *KYWQJ*, 5:4–6, 2007.

Kang Youwei 康有为. 'Wuzhi jiuguo lun 物质救国论 [On Saving China Through Material Reconstruction (1904)]'. In *KYWQJ*, 8:63–101, 2007.

Kang Youwei 康有为. 'Xinxue weijing kao 新学伪经考 [An Investigation into the Forged Classics of the Xin Dynasty (1891)]'. In *KYWQJ*, 1:353–558, 2007.

Kang Youwei 康有为. 'Yidali youji 意大利游记 [An Account of My Travels in Italy (1904)]'. In *KYWQJ*, 7:346–406, 2007.

Kang Youwei 康有为. 'Yingguo Jianbuliezhu daxue huawen zongjiaoxi Zhailushi huijianji 英国监布列住大学华文总教习斋路士会见记 [An Interview with Herbert Allen Giles, Professor of Chinese at Cambridge University of England (1904)]'. In *KYWQJ*, 8:27–57, 2007.

Kang Youwei 康有为. 'Yingguo youji 英国游记 [An Account of My Travels in Great Britain (1904)]'. In *KYWQJ*, 8:1–26, 2007.

Kang Youwei 康有为. 'Yu Chen Huanzhang shu 与陈焕章书 [Letter to Chen Huanzhang (1912)]'. In *KYWQJ*, 9:337, 2007.

Kang Youwei 康有为. 'Yu Hong Youcheng jijian lun zhongxi yixue shu 与洪右臣给谏论中西异学书 [A Discussion with Hong Youchen, the Supervising Censor, on the Differences between Chinese and Western Learning (1891)]'. In *KYWQJ*, 1:336–37, 2007.

Kang Youwei 康有为. 'Yu Huang Zhongtao bianxiu shu 与黄仲弢编修书 [A Letter to Huang Zhongtao, the Hanlin Compiler (1888)]'. In *KYWQJ*, 1:188, 2007.

Kang Youwei 康有为. 'Yu Li Yuanhong, Huang Xing, Tang Hualong shu 与黎元洪、黄兴、汤化龙书 [A Letter to Li Yuanhong, Huang Xing and Tang Hualong (1911)]'. In *KYWQJ*, 9:210–17, 2007.

Kang Youwei 康有为. 'Yu Pan Wenqin shu 与潘文勤书 [A Letter to Pan Wenqin (1888)]'. In *KYWQJ*, 1:168–70, 2007.

Kang Youwei 康有为. 'Yu riren moujun bitan 与日人某君笔谈 [A Written Exchange with a Japanese Friend (1920)]'. In *KYWQJ*, 11:118, 2007.

Kang Youwei 康有为. 'Yu Shen Xingbu Zipei shu 与沈刑部子培书 [A Letter to Shen Zipei (1889)]'. In *KYWQJ*, 1:236–39, 2007.

Kang Youwei 康有为. 'Yu Tongfu Tongjian Tongning shu 与同复同钱同凝书 [A Letter to Tongfu, Tongjian and Tongning (1927)]'. In *KYWQJ*, 11:453, 2007.

Kang Youwei 康有为. 'Yu Yiqin xueshi shu 与峄琴学士书 [A Letter to Pan Yiqin, a Member of the Hanlin (1888)]'. In *KYWQJ*, 1:190–91, 2007.

Kang Youwei 康有为. 'Yu Youbo shu 与幼博书 [A Letter to Youbo (1888)]'. In *KYWQJ*, 1:175, 2007.

Kang Youwei 康有为. 'Yu Zhu Yixin lunxue shudu 与朱一新论学书牍 [Letters to Zhu Yixin on Learning (1891)]'. In *KYWQJ*, 1:314–37, 2007.

Kang Youwei 康有为. 'Zhi Jin Yunpeng Yan Huiqing shu 致靳云鹏颜惠庆书 [A Letter to Premier Jin and Foreign Minister Yan (1920)]'. In *KYWQJ*, 11:137–38, 2007.

Kang Youwei 康有为. 'Zhi Li Yuanhong Duan Qirui shu 致黎元洪段祺瑞书 [A Letter to the President and the Premier (1916)]'. In *KYWQJ*, 10:316–17, 2007.

Kang Youwei 康有为. 'Zhi Yanshenggong han 致衍圣公函 [A Letter to Duke Yansheng (1923)]'. In *KYWQJ*, 11:252, 2007.

Kang Youwei 康有为. 'Zhi yiheweiyuan shu 致议和委员书 [A Letter to the Chinese Delegates to the Peace Conference (1919)]'. In *KYWQJ*, 11:98–101, 2007.

Kang Youwei 康有为. 'Zhi Zhang Yishan shu 致章一山书 [A letter to Mr. Zhang Yishan] [1918]'. In *KYWQJ*, 11:82, 2007.

Kang Youwei 康有为. 'Zhongguo shanhou yi 中国善后议 [On the Reconstruction of Post-Revolution China][1916]'. In *KYWQJ*, 10:269–80, 2007.

Kang Youwei 康有为. 'Zhonghua jiuguo lun 中华救国论 [On Saving China (1912)]'. In *KYWQJ*, 9:209–39, 2007.

Kang Youwei 康有为. 'Zhonghua Xianzhenghui qiaomin gongshang qingyuanshu 中华宪政会侨民公上请愿书 [A Petition of the Chinese Constitutional Association and Overseas Chinese (1908)]'. In *KYWQJ*, 8:410–21, 2007.

Kang Youwei 康有为. 'Zhongyong zhu 中庸注 [The Doctrine of the Mean Annotated (1901)]'. In *KYWQJ*, 5:367–92, 2007.

Kang Youwei 康有为. 'Zhutian jiang 诸天讲 [Lectures on the Heavens (1927)]'. In *KYWQJ*, 12:1–132, 2007.

'Kang Yu-Wei's Story'. *The North-China Herald*, 17 October 1898.

Katz, Paul R. *Religion in China and Its Modern Fate*. Waltham, MA.: Brandeis University Press, 2014.

Keynes, John M. 'Review: The Economic Principles of Confucius and His School'. *The Economic Journal* 22, no. 88 (1912): 584–88.

Knoblock, John, trans. *Xunzi: A Translation and Study of the Complete Works*. Vol. III. Stanford: Stanford University Press, 1994.

Kong Xiangji 孔祥吉. *Kang Youwei bianfa zouzhang jikao 康有为变法奏章辑考* [*A Critical Collection of Kang Youwei's Memorials on Reforms*]. Beijing: Beijing tushuguan chubanshe, 2008.

Küng, Hans, and Julia Ching. *Christianity and Chinese Religions*. New York: Doubleday Religious Publishing Group, 1989.

Kuo, Y. -P. '"Christian Civilization" and the Confucian Church: The Origin of Secularist Politics in Modern China'. *Past & Present* 218, no. 1 (2013): 235–64.

Kwok, Danny W. Y. *Scientism in Chinese Thought: 1900–1950*. New Haven: Yale University Press, 1965.

Legge, James, trans. *The Chinese Classics I: Confucian Analects, The Great Learning, and The Doctrine of the Mean*. 2nd ed. Oxford: The Clarendon Press, 1893.

Legge, James, trans. *The Chinese Classics II: The Works of Mencius*. 2nd ed. Oxford: The Clarendon Press, 1895.

Legge, James, trans. *The Chinese Classics III: The Shoo King, or The Book of Historical Documents*. Hong Kong: At The Author's, 1865.

Legge, James, trans. *The Chinese Classics IV: The She King, or The Book of Poetry*. Hong Kong: Lane, Crawford & Co., 1871.

Legge, James. *The Notions of the Chinese Concerning God and Spirits*. Hong Kong: The Hongkong Register Office, 1852.

Legge, James, trans. *The Texts of Confucianism II: The Yi King*. 2nd ed. The Sacred Books of the East, XVI. Oxford: Clarendon Press, 1899.

Legge, James, trans. *The Texts of Confucianism IV: The Li Ki, XI-XLVI*. Oxford: The Clarendon Press, 1885.

Legge, James, trans. *The Texts of Taoism II: The Writings of Kwang-Tze, XVIII-XXXIII, The Thai-Shang Tractate of Actions and Their Retributions*. Oxford: The Clarendon Press, 1891.

Lei Libai 雷立柏 [Leopold Leeb]. *Zhang Heng, kexue yu zongjiao 张衡，科学与宗教* [*Zhang Heng, Science and Religion*]. Beijing: Shehui kexue wenxian chubanshe, 2000.

Leixiaer 雷侠兒 [Lyell]. *Dixue qianshi 地學淺釋* [*Elements of Geology*]. Translated by Magaowen 瑪高溫 [MacGowan] and Hua Hengfang 華蘅芳. New ed. Vol. 13. Shanghai: Kiangnan Arsenal, 1873.

Li Chichang 李炽昌. *Shenghao lunheng 圣号论衡* [*Discussions on the Chinese Translations of the Divine Name*]. Shanghai: Shanghai guji chubanshe, 2008.

Li Gui 李圭. *Huanyou diqiu xinlu 环游地球新录* [*A New Account of a Trip Around the Globe*]. New edn. Changsha: Hunan renmin chubanshe, 1985.

Li Keliang 李可良. 'Wo yinxiangzhong zhi Kang Youwei 我印象中之康有为 [My Recollections of Kang Youwei]'. In *ZYKYW*, 478–83, 1996.

Li, San-pao. 'K'ang Yu-Wei's Iconoclasm: Interpretation and Translation of His Earliest Writings, 1884–1887'. Ph.D., University of California, Davis, 1978.

Li Shen 李申, ed. *Gaokeji yu zongjiao 高科技与宗教* [*Advanced Science-Technology and Religion*]. Tianjin: Kexue jishu chubanshe, 2000.

Li Shen 李申, ed. *Zhongguo Rujiao shi 中国儒教史 A History of the Confucian Religion of China*. 2 vols. Shanghai: Shanghai renmin chubanshe, 1999.

Li Tiangang 李天纲. *Jinze: Jiangnan minjian jisi tanyuan 金泽: 江南民间祭祀探源* [*Jinze Town: Exploring the Origins of Folk Sacrificial Rituals in Jiangnan*]. Beijing: Sanlian shudian, 2017.

Li Zehou 李泽厚. *Kang Youwei Tan Sitong sixiang yanjiu 康有为谭嗣同研究* [*A Treatise on the Thought of Kang Youwei and Tan Sitong*]. Shanghai: Renmin chubanshe, 1958.

Liang, Ch'i-ch'ao. *Intellectual Trends in the Ch'ing Period*. Translated by Immanuel C. Y. Hsü. Cambridge, MA: Harvard University Press, 1959.

Liang Dingfen 梁鼎芬. 'Bo panfan Kang Youwei nishu 駁叛犯康有為逆書 [Refuting the Blasphemous Letter from the Fugitive Kang Youwei]'. *Shen Pao 申報* [*Shanghai News*], no. 9173 (27 October 1898): 3.

Liang Dingfen 梁鼎芬. 'Kang Youwei shishi 康有为事实 [Facts about Kang Youwei]'. In *Cong wuxu dao xinhai 从戊戌到辛亥* [*From 1898 to 1911*], edited by Tang Zhijun 汤志钧, 65. Nanjing: Jiangsu guji chubanshe, 1990.

Liang Qichao 梁启超. 'Baojiao feisuoyi zunkong lun 保教非所以尊孔论 [Preserving the Faith Is Not the Way to Exalt Confucius (1902)]'. In *LQCQJ*, 3:756–70, 1999.

Liang Qichao 梁启超. 'Du xixueshu fa 读西学书法 [On How to Read Books on Western Learning (1896)]'. In *Yinbingshi heji jiwaiwen 饮冰室合集集外文* [*Essays not Included in the Collected Works from the Ice-Drinker's Studio*], edited by Xia Xiaohong 夏晓红, 2:1159–70. Beijing: Beijing daxue chubanshe, 2005.

Liang Qichao 梁启超. 'Dushu fenyue kecheng 读书分月课程 [A Monthly Reading Schedule (1892)]'. In *LQCQJ*, 1:1–9, 1999.

Liang Qichao 梁启超. 'Jinshi wenming chuzu erdajia zhi xueshuo 近世文明初祖二大家之学说 [On the Thought of Bacon and Descartes, the Two Pioneers of Modern Civilization (1902)]'. In *LQCQJ*, 4:1030–35, 1999.

Liang Qichao 梁启超. 'Lun Fojiao yu qunzhi zhi guanxi 论佛教与群治之关系 [On the Relationship between Buddhism and Social Governance (1902)]'. In *LQCQJ*, 4:906–10, 1999.

Liang Qichao 梁启超. 'Lun Zhina zongjiao gaige 论支那宗教改革 [On the Religious Reformation in China (1899)]'. In *LQCQJ*, 2:263–66, 1999.

Liang Qichao 梁启超. 'Lun zongjiaojia yu zhexuejia zhi duanchang 论宗教家与哲学家之短长得失 [On the Advantages and Disadvantages, and the Gains and Losses of Religious Leaders and Philosophers (1902)]'. In *LQCQJ*, 3:762–64, 1999.

Liang Qichao 梁启超. 'Nanhai Kangxiansheng zhuan 南海康先生传 [A Biography of Mr Kang of Nanhai (1901)]'. In *LQCQJ*, 2:481–99, 1999.

Liang Qichao 梁启超. 'Ouyou xinyinglu jielu 欧游心影录节录 [Recorded Reflections During my Travels in Europe: A Selection (1920)]'. In LQCQJ, 2968–87, 1999.

Liang Qichao 梁启超. 'Qingdai xueshu gailun 清代学术概论 [Intellectual Trends in the Qing Period (1920)]'. In LQCQJ, 10:3066–3122, 1999.

Liang Qichao 梁启超. 'Renshengguan yu kexue 人生观与科学 [The Philosophy of Life and Science]'. In *Kexue yu renshengguan 科学与人生观* [*Science and the Philosophy of Life*], by Zhang Junmai 张君劢 et al., 125–32, New. Shenyang: Liaoning jiaoyu chubanshe, 1998.

Liang Qichao 梁启超. 'Sanshi zishu 三十自述 [An Account of Myself at 30 (1902)]'. In LQCQJ, 1:957–60, 1999.

Liang Qichao 梁启超. 'Shihua 诗话 [Literary Criticisms]'. In LQCQJ, 18:5295–5413, 1999.

Liang Qichao 梁启超. 'Tan Sitong zhuan 谭嗣同传 [A Biography of Tan Sitong (1898)]'. In LQCQJ, 2:231–34, 1999.

Liang Qichao 梁启超. 'Xianqin zhengzhi sixiangshi 先秦政治思想史 [A History of the Political Thought in pre-Qin China (1922)]'. In LQCQJ, 3603–95, 1999.

Liang Qichao 梁启超. 'Xin Zhongguo jianshe wenti 新中国建设问题 [On the Reconstruction of New China (1911)]'. In LQCQJ, 8:2433–43, 1999.

Liang Qichao 梁启超. 'Xixue shumu biao xuli 西学书目表序例 [Preface to A Bibliographical Table of Books on Western Learning (1896)]'. In LQCQJ, 1:82–84, 1999.

Liang Qichao 梁启超. 'Xixue shumu biao 西学书目表 [A Bibliographical Table of Books on Western Learning (1896)]'. In *Yinbingshi heji jiwaiwen 饮冰室合集集外文* [*Essays not Included in the Collected Works from the Ice-Drinker's Studio*], edited by Xia Xiaohong 夏晓红, 2:1121–58. Beijing: Beijing daxue chubanshe, 2005.

Liang Qichao 梁启超. 'Yu Yan Youling shu 与严又陵书 [A Letter to Yan Fu (1897)]'. In LQCQJ, 1:71–73, 1999.

Liang Qichao 梁啟超. 'Wuxu zhengbian ji 戊戌政變記 [An Account of the Coup D'état of 1898]'. In *Xuxiu Sikuquanshu 續修四庫全書* [*A Sequel to the Complete Library in Four Sections*], 446:181–258. Shanghai: Shanghai guji chubanshe, 1995.

Liao Daochuan 廖道传. 'Qing zun Kongjiao wei guojiao shang dazongtong dengshu 请尊孔教为国教上大总统等书 [A Petition to President (et al.) to Enshrine Confucianism as State Religion]'. In *Zhonghua minguoshi dang'an ziliao huibian 3: wenhua 中华民国史档案资料汇编3: 文化* [*Compilation of Archival Materials Pertaining to the History of the Republic of China 3: Culture*], edited by ZDLDG 中国第二历史档案馆 [China Second Historical Archives], 47–50. Nanjing: Jiangsu guji chubanshe, 1991.

Liezi. *Liezi* [*in Chinese-English Version*]. Translated by Liang Xiaopeng. Beijing: Zhonghua shuju, 2005.

Lin Zhiye 林治业. 'Yuan Kang Youwei mu shi zheyang bei wahuide 原康有为墓是这样被挖毁的 [An Account of the Destruction of Kang Youwei's Original Tomb]'. *Laonian shenghuo bao 老年生活报* [*Daily for the elderly*], 12 December 2012, sec. 6. http://epaper.qingdaonews.com/html/lnshb/20121212/lnshb514317.html.

Lindberg, David C. *The Beginnings of Western Science: The European Scientific Tradition in Philosophical, Religious, and Institutional Context, Prehistory to A.D. 1450*. 2nd ed. Chicago and London: The University of Chicago Press, 2007.

Liu Dachun 刘大椿. *Xixue dongjian 西学东渐* [*Western Learning Spreading to the East*]. Beijing: Zhongguo renmin daxue chubanshe, 2018.

Liu Wei 刘巍. 'Jiaoxue tongyi yu Kang Youwei de zaoqi jingxue luxiang jiqi zhuanbian 教学通义与康有为的早期经学路向及其转变 [The Jiaoxue tongyi and Kang's Approach to Classical Studies and Its Ensuing Change]'. *历史研究* [*Historical Research*], no. 4 (2005): 190–49.

'Liyan 例言 [The Preamble]'. *Kexue 科學* [*Science*], no. 1 (1915): 1–2.

Lloyd, G. E. R. *Early Greek Science: Thales to Aristotle*. London: Chatto & Windus, 1970.

Lo, Jung-pang, ed. 'Chronological Autobiography'. In *K'ang Yu-Wei: A Biography and a Symposium*, 17–174. Tucson: University of Arizona Press, 1967.

Lo, Jung-pang, ed. *K'ang Yu-Wei: A Biography and a Symposium*. Tucson: University of Arizona Press, 1967.

Lo, Jung-pang, ed. 'Sequel to the Chronological Autobiography of K'ang Yu-Wei'. In *K'ang Yu-Wei: A Biography and a Symposium*, 175–278. Tucson: University of Arizona Press, 1967.

Loewe, Michael, ed. *Early Chinese Texts: A Bibliographical Guide*. Berkeley: Institute of East Asian Studies, University of California, 1993.

Lu Naixiang 陆乃翔, and Lu Dunkui 陆敦骙. 'Nanhai xiansheng zhuan 南海先生传 [A Biography of Mr. Kang of Nanhai]'. In *KYWQJ*, 12:440–71, 2007.

Ma Honglin 马洪林. *Kang Youwei dazhuan 康有为大传* [*A Comprehensive Biography of Kang Youwei*]. Shenyang: Liaoning renmin chubanshe, 1988.

Ma Honglin 马洪林. *Kang Youwei pingzhuan 康有为评传* [*A Critical Biography of Kang Youwei*]. Nanjing: Nanjing University Press, 2009.

Ma Honglin 马洪林. 'Kang Youwei yanjiu bainian huigu yu zhanwang 康有为研究百年回顾与展望 [A Review of Studies of Kang Youwei over the Past Century and Prospects for Future Research]'. *Dongfang luntan 东方论坛*, no. 5 (2008): 83–92.

Ma Jishi 瑪吉士 [Martinho José Marques]. *Xinshi dili beikao quanshu 新釋地理備考全書*. Guangzhou: Haishan xianguan, 1847.

Makeham, John. *New Confucianism: A Critical Examination*. New York: Palgrave Macmillan, 2003.

Mao Haijian 茅海建. *Cong jiawu dao wuxu: Kang Youwei Woshi jianzhu 从甲午到戊戌：康有为我史鉴注* [*Kang's 'My History' Verified and Annotated: 1894–1998*]. Beijing: Sanlian shudian, 2009.

Mao Haijian 茅海建. 'Kang Youwei zixie nianpu shougaoben yuedu baogao "康有为自写年谱手稿本"阅读报告 [A Report on Reading Kang's Manuscript of His Autobiography]'. *Jindaishi yanjiu 近代史研究* [*Journal of Modern Chinese History*], no. 4 (2007).

Mao, Tse-Tung. *Selected Works of Mao Tse-Tung*. Vol. IV. Peking: Foreign Languages Press, 1961.

McGrath, Alister E. *Science and Religion: A New Introduction*. 2nd ed. Malden, MA: Wiley-Blackwell, 2010.

McGuire, J. E. 'Newton on Place, Time and God: An Unpublished Source'. *The British Journal for the History of Science* 11, no. 2 (July 1978): 114–29.

McNamara, Patrick. *The Neuroscience of Religious Experience*. Cambridge: Cambridge University Press, 2009.

Minamiki, George. *The Chinese Rites Controversy: From Its Beginning to Modern Times*. Chicago: Loyola University Press, 1985.

Mou Zhongjian 牟钟鉴, and Zhang Jian 张践. *Zhongguo zongjiao tongshi 中国宗教通史* [*A General History of Chinese Religions*]. Revised ed. Beijing: Zhongguo shehui kexue chubanshe, 2007.

Müller, Max. *Natural Religion*. London: Longmans, Green, & Co., 1889.

Nedostup, Rebecca. *Superstitious Regimes: Religion and the Politics of Chinese Modernity*. Cambridge, MA.: Harvard University Asia Center, 2009.

Needham, Joseph. *Science and Civilisation in China: History of Scientific Thought*. Vol. 2. Cambridge: Cambridge University Press, 1956.

Noakes, Richard. 'Ethers, Religion and Politics in Late-Victorian Physics: Beyond the Wynne Thesis'. *History of Science* 43, no. 4 (2005): 415–55.

Noriko, Mori. 'Liang Qichao, Late-Qing Buddhism, and Modern Japan'. In *The Role of Japan in Liang Qichao's Introduction of Modern Western Civilization to China*, edited by Joshua A. Fogel, 222–46. Berkeley: Institute of East Asian Studies, University of California Berkeley, 2004.

Notovitch, Nicolas. *The Unknown Life of Jesus Christ*. Translated by Alexina Loranger. Chicago and New York: Rand, McNally & Company, 1894.

Oppenheim, Janet. *The Other World: Spiritualism and Psychical Research in England, 1850–1914*. Cambridge: Cambridge University Press, 1988.

Osterhammel, Jürgen. *The Transformation of the World: A Global History of the Nineteenth Century*. Translated by Patrick Camiller. Princeton and Oxford: Princeton University Press, 2014.

Pan, Wei-tung. *The Chinese Constitution: A Study of Forty Years of Constitution-Making in China*. Washington, D.C.: The Institute of Chinese Culture, 1945.

Paramore, Kiri. '"Civil Religion" and Confucianism: Japan's Past, China's Present, and the Current Boom in Scholarship on Confucianism'. *Journal of Asian Studies* 72, no. 2 (2015): 269–82.

Peng Chunling 彭春凌. *Ruxue zhuanxing yu wenhua xinming 儒学转型与文化新命* [*The Transformation of Confucianism and the New Cultural Mission*]. Beijing: Bejing daxue chubanshe, 2014.

Peterson, Willard. 'Learning from Heaven: The Introduction of Christianity and Other Western Ideas into Late Ming China'. In *The Cambridge History of China*, edited by Denis Twitchett and Frederick Mote, 8:789–839. Cambridge: Cambridge University Press, 1998.

Pfister, Lauren. 'A Modern Ruist Religious Vision of a Global Unity: Kang Youwei's Utopian Vision and Its Humanistic Religious Refraction in European Sinology'. In *Globalization and the Making of Religious Modernity in China*. Leiden and Boston: Brill, 2014.

'Putong jiaoyu zanxing banfa 普通教育暫行辦法 [Provisional Regulations for General Education]'. *Linshi zhengfu gongbao 臨時政府公報 [Provisional Government Gazette], No. 4*, 1 February 1912.

Qian Mu钱穆. 'Liu Xiang Xin fuzi nianpu 刘向歆父子年谱 [Chronicles of Liu Xiang and Liu Xin, Father and Son]'. *Yenching Journal*, no. 7 (1930).

Reilly, Thomas H. *The Taiping Heavenly Kingdom: Rebellion and the Blasphemy of Empire*. Seattle: University of Washington Press, 2004.

Ren Hongjun 任鴻雋. 'Fakanci 發刊詞 [Foreword]'. *Kexue 科學 [Science]* 1, no. 1 (1915): 3–7.

Ren Hongjun 任鴻雋. 'Kexue yu gongye 科學與工業 [Science and Industry]'. *Kexue 科學 [Science]* 1, no. 10 (1915): 1089–99.

Ren Hongjun 任鴻雋. 'Kexue yu jiaoyu 科學與教育 [Science and Education]'. *Kexue 科學 [Science]* 1, no. 12 (1915): 1343–52.

Ren Hongjun 任鴻雋. 'Shuo Zhongguo wu kexue zhi yuanyin 說中國無科學之原因 [On the Reasons Why China Did Not Have Science]'. *Kexue 科學 [Science]* 1, no. 1 (1915): 6–12.

Ren Jiyu 任继愈. 'Rujiao de xingcheng 儒教的形成 [On the Formation of Confucian Religion]'. *Zhongguo shehui kexue 中国社会科学 [Chinese Social Sciences]*, no. 1 (1980): 61–74.

Ren Jiyu 任继愈, ed. *Rujiao wenti zhenglunji 儒教问题争论集 [Collected Essays on the Debate over Ruism/Confucianism as a Religion]*. Beijing: Zongjiao wenhua chubanshe, 2000.

Ren Qisheng 任启圣. 'Kang Youwei wannian jiangxue jiqi shishi zhi jingguo 康有为晚年讲学及其逝世之经过 [My Recollection of Kang's Lectures in His Later Years and His Passing Away]'. In *Wenshi ziliao xuanji 文史资料选辑 [A Selection of Literary and Historical Materials]*, 31:237–45, 1962.

'Review: A Compendious Description of the World (Lit. the Oceanic Circle) by Seu Sung Lung in 6 Vols, 1849'. *The North-China Herald*. 28 September 1850.

Ricci, Matteo. *The True Meaning of the Lord of Heaven [Tianzhu Shiyi]*. Translated by Douglas Lancashire and Peter Kuo-chen Hu. St. Louis: Ricci Institute, 1985.

Richard, Timothy. *Forty-Five Years in China: Reminiscences*. New York: Frederick A. Stokes Company, 1916. translated into Chinese in 2005.

Scalapino, Robert, and George T. Yu. *The Chinese Anarchist Movement*. Berkeley: Center for Chinese Studies, Institute of International Studies, University of California, 1961.

Schmitt, Carl. *Political Theology: Four Chapters on the Concept of Sovereignty*. University of Chicago Press, 2005.

'Scientific Method'. In *Oxford Dictionaries Online*. Oxford University Press. Accessed 28 January 2014. http://www.oxforddictionaries.com/.

Shan, Paul. 'Science and Faith in China Today'. In *Western Learning and Christianity in China: The Contribution and Impact of Johann Adam Schall von Bell, S.J. (1592–1666)*, edited by Roman Malek, Vol. 11. Nettetal: Steyler Verlag, 1998.

Shapin, Steven. *The Scientific Revolution*. Chicago: University of Chicago Press, 1998.

Shen Jie 沈洁. 'Fanmixin huayu jiqi xiandai qiyuan 反迷信话语及其现代起源 [The Discourse against Superstition and Its Modern Origins in China]'. *Shilin* 史林 [*Historical Review*], no. 2 (2006): 30–42.

Sivin, Nathan. *Medicine, Philosophy and Religion in Ancient China: Researches and Reflections*. Aldershot: Variorum, 1995.

Smith, W. Cantwell. *The Meaning and End of Religion*. Minneapolis, Minnesota: Fortress Press., 1991.

Solé-Farràs, Jesús. *New Confucianism in Twenty-First Century China: The Construction of a Discourse*. Routledge, 2013.

Stenmark, Mikael. 'Ways of Relating Science and Religion'. In *The Cambridge Companion to Science and Religion*, edited by Peter Harrison, 278–95. Cambridge: Cambridge University Press, 2010.

Stewart, Balfour, and Peter G. Tait. *The Unseen Universe: Or, Physical Speculations on a Future State*. Macmillan and Company, 1875.

Stewart, Melville Y. *Science and Religion in Dialogue*. John Wiley & Sons, 2009.

Streng, Frederick J. *Understanding Religious Life*. Belmont: Wadsworth Publishing Company, 1985.

Sun, Anna. *Confucianism as a World Religion: Contested Histories and Contemporary Realities*. Princeton, N.J.: Princeton University Press, 2013.

Sun Chunzai 孫春在. *Qingmo de Gongyang sixiang* 清末的公羊思想 [*On the Thought of the Gongyang School in Late Qing*]. Taibei: Taiwan shangwu yinshuguan, 1985.

Sun Zhongshan 孙中山. *Sun Zhongshan quanji* 孙中山全集 [*Complete Works of Sun Zhongshan*]. Vol. 2. 11 vols. Beijing: Zhonghua shuju, 1982.

Takeuchi Hiroyuki 竹内弘行. *Chūgoku no jukyō kindaika ron* 中国の儒教近代化論 [*The Modernization of Chinese Confucianism*]. Tokyo: Ken bunshuppan, 1995.

Tan Sitong 譚嗣同. 'Renxue 仁學 [The Science of Benevolence]'. In *TSTQJ*, 289–374, 1981.

Tan Sitong 譚嗣同. 'Shang Ouyang Zhonghu 上歐陽中鵠 [Letters to Ouyang Zhonghu]'. In *TSTQJ*, 449–78, 1981.

Tan Sitong 譚嗣同. 'Zhishi shipian 治事十篇 [Ten Essays on Managing Affairs]'. In *TSTQJ*, 435–45, 1981.

Tang Wenming 唐文明. *Fujiao zaikuan: Kang Youwei Kongjiao sixiang shenlun* 敷教在宽: 康有为孔教思想申论 [*Teaching with Tolerance: A Study of Kang's Thought on Confucian Religion*]. Beijing: Zhongguo remin daxue chubanshe, 2012.

Tang Zhijun汤志钧. *Kang Youwei yu Wuxu bianfa* 康有为与戊戌变法 [*Kang Youwei and the Reform of 1898*]. Beijing: Zhonghua shuju, 1984.

Tao Juyin 陶菊隱. *Wu Peifu jiangjun zhuan* 吳佩孚將軍傳 [*A Biography of General Wu Peifu*]. Shanghai: Zhonghua shuju, 1941.

Tay, Wei Leong. 'Kang Youwei, The Martin Luther of Confucianism and His Vision of Confucian Modernity and Nation'. In *Secularization, Religion and the State*, edited by Haneda Masashi, 97–109. Tokyo: The University of Tokyo Centre for Philosophy, 2010.

Tay, Wei Leong. 'Secularization as Sacralization: Religion and the Formation of Modern Chinese Nationalism and Nation-State, 1840–1939'. The University of Oxford, 2019.

Taylor, Rodney L. *The Religious Dimensions of Confucianism*. Albany: State University of New York Press, 1990.

The Science Council. 'Our Definition of Science', 2009. https://sciencecouncil.org/about-science/our-definition-of-science/.

Thompson, Laurence G. *Chinese Religion: An Introduction*. 5th ed. Belmont: Wadsworth Publishing Company, 1996.

Thompson, Laurence G. *Ta T'ung Shu: The One-World Philosophy of K'ang Yu-Wei*. London: George Allen & Unwin Ltd, 1958.

Thompson, Laurence G., and O. Briere, trans. *Fifty Years of Chinese Philosophy 1898–1950*. London: George Allen and Unwin, 1956.

'Treaty of Tientsin [Tianjin], 1858', 1860. http://www.chinaforeignrelations.net/node/162.

Tu Wei-ming. *Centrality and Commonality: An Essay on Confucian Religiousness*. New York: State University of New York Press, 1989.

Tu Wei-ming. *Confucian Thought: Selfhood as Creative Transformation*. Albany, N.Y.: SUNY Press, 1985.

Tu Wei-Ming. *Way, Learning, and Politics: Essays on the Confucian Intellectual*. Albany: State University of New York Press, 1993.

Tu Wei-ming. *Zhongyong dongjian* 中庸洞见 / *An Insight of Chung-yung*. Bilingual ed. Beijing: People's Publishing House, 2008.

Udias, A. 'Jesuit Astronomers in Beijing, 1601–1805'. *Quarterly Journal of The Royal Astronomical Society* 35, no. 4 (1994): 463–78.

Veer, Peter van der. *The Modern Spirit of Asia: The Spiritual and the Secular in China and India*. Princeton: Princeton University Press, 2013.

Wan, Sze-kar. 'The Viability of Confucian Transcendence: Grappling with Tu Weiming's Interpretation of the Zhongyong'. *Dao: A Journal of Comparative Philosophy* 7, no. 4 (2008): 407–28.

Wan, Zhaoyuan, and David A. Palmer. 'The Cosmopolitan Moment in Colonial Modernity: The Bahá'í Faith, Spiritual Networks and Universalist Movements in Early Twentieth Century China'. *Modern Asian Studies* 54, no. 6 (2020): 1787–1827. https://doi.org/10.1017/S0026749X19000210.

Wang, Ban, ed. *Chinese Visions of World Order: Tianxia, Culture, and World Politics*. Durham: Duke University Press, 2017.

Wang, Ban, ed. 'Confucianism and Nature: Ecological Motifs in Kang Youwei's Great Community'. *Telos* 2018, no. 183 (2018): 47–67.

Wang Fansen 王汎森. *Zhiniu de diyin: yixie lishi sikao fangshi de fansi* 执拗的低音 -- 一些历史思考方式的反思 [*An Insistent Low Voice: Some Reflections on the Modes of Historical Thinking*]. Beijing: Sanlian shudian, 2014.

Wang Fansen 王汎森. *Zhongguo jindai sixiangshi de zhuanxing shidai* 中國近代思想史的轉型時代 [*The Transitional Era of Modern Chinese Intellectual History*]. Taipei: Linking Publishing, 2007.

Wang Hui 汪晖. *Xiandai Zhongguo sixiang de xingqi* 现代中国思想的兴起 [*The Rise of Modern Chinese Thought*]. 4 vols. Beijing: Sanlian shudian, 2004.

Wang Qiyuan 王啟元. *Qingshu jingtan* 清署经谈 [*Discussions on Scripture at Qingshu*]. Shanghai: Shanghai guji chubanshe, 2017.

Wang, Zuoyue. 'Saving China through Science: The Science Society of China, Scientific Nationalism, and Civil Society in Republican China'. *Science and Civil Society*, Osiris, 17 (2002): 291–322.

Wanguo daode zonghui gaikuang 1921–1980 萬國道德總會概況 *1921–1980* [*An Overview of the Head Office of the Universal Morality Society 1921–1980*]. Taibei: Daodehui Press, 1980.

Weale, B. L. Putnam. *The Fight for the Republic in China*. New York: Dodd, Mead and Company, 1917.

Wei Lianchen 韋廉臣 [Williamson]. *Gewu tanyuan* 格物探原 [*Natural Theology and the Method of Divine Government*]. 2nd ed. 6 vols. Shanghai: Christian Literature Society, 1880.

Wei Lianchen 韋廉臣 [Williamson]. *Gujiao huican* 古教彙參 [*Ancient Religions and Philosophies*]. Translated by Dong Shutong 董樹堂. 3 vols. Shanghai: Yizhu shuhui, 1882.

Wei Lianchen 韋廉臣 [Williamson]. *Jidu shilu* 基督實錄 [*The Light and Life of the World*]. 3 vols. Shanghai: Meihua shuguan [The Presbyterian Press], 1879.

Wei Yuan 魏源. *Haiguo tuzhi* 海國圖志 [*An Illustrated Gazetteer of Maritime Countries*]. Yangzhou: Guweitang, 1852.

Welch, Holmes. *The Buddhist Revival in China*. Harvard University Press, 1968.

Wenshizu 文史组, ed. *Nanhai wenshi ziliao: Xiqiaoshan zhuanji* 南海文史资料: 西樵山专辑 [*Historical Materials of Nanhai: On Xiqiao Mountain*]. Nanhai, 1983.

Weston, Timothy B. *The Power of Position: Beijing University, Intellectuals, and Chinese Political Culture, 1898–1929*. Berkeley: University of California Press, 2004.

White, A. Dickson. *A History of the Warfare of Science with Theology in Christendom*. 2 vols. New York: D. Appleton and Company, 1896.

Whitehead, Alfred N. *Science and the Modern World*. New York: The New American Library, 1925.

Whittaker, Edmund. *A History of the Theories of Aether and Electricity*. Thomas Nelson and Sons Ltd, 1951.

Wilbur, C. Martin. 'The Nationalist Revolution: From Canton to Nanking, 1923–28'. In *The Cambridge History of China*, edited by John K. Fairbank, 12:527–720. Cambridge: Cambridge University Press, 1983.

Wilhelm, Richard. *The I Ching or Book of Changes*. Translated by Cary F. Baynes. 3rd ed. London and Henley: Routledge & Kegan Paul, 1985.

Wilhelm, Richard. *The Soul of China*. Translated by Arthur Waley. London: Jonathan Cape, 1928.

Williams, S. Wells. *Controversy among the Protestant Missionaries on the Proper Translation of the Words God and Spirit into Chinese*. Andover: Warren F. Draper, 1878.

Wong, Aida Yuen. *The Other Kang Youwei: Calligrapher, Art Activist, and Aesthetic Reformer in Modern China*. Leiden: Brill, 2016.

Wong, David. 'Confucian Political Philosophy'. In *The Oxford Handbook of the History of Political Philosophy*, edited by George Klosko, 771–88. Oxford: Oxford University Press, 2011.

Wong, Man-Kong Timothy. 'The Rendering of God in Chinese by the Chinese: Chinese Responses to the Term Question in the Wanguo Gongbao.' In *Mapping Meanings: The Field of New Learning in Late Qing China*, edited by Michael Lackner and Natascha Vittinghoff, 589–614. Leiden; Boston: Brill, 2004.

Wong Young-Tsu 汪荣祖. 'Yizhe xuyan 译者序言 [Preface by the Translator]'. In *Jindai Zhongguo yu xinshijie* 近代中国与新世界 [*A Modern China and a New World*], 1–3. Nanjing: Jiangsu renmin chubanshe, 1997.

Wood, Henry. *Ideal Suggestion through Mental Photography: A Restorative System for Home and Private Use, Preceded by a Study of the Laws of Mental Healing*. Boston: Lee and Shepard Publishers, 1893.

Worden, Robert Leo. 'A Chinese Reformer in Exile: The North American Phase of the Travels of K'ang Yu-Wei, 1899–1909'. Ph.D., Georgetown University, 1972.

Wright, David. 'Tan Sitong and the Ether Reconsidered'. *Bulletin of the School of Oriental and African Studies* 57, no. 3 (1994).

Wright, David. *Translating Science: The Transmission of Western Chemistry into Late Imperial China, 1840–1900*. Leiden: Brill, 2000.

Wu Guosheng 吴国盛. *Shenme shi kexue* 什么是科学 [*What Is Science?*]. Guangzhou: Guangdong renmin chubanshe, 2016.

Wu Tianren 吳天任, ed. *Kang Youwei xiansheng nianpu* 康有為先生年譜 [*A Chronological Biography of Mr. Kang Youwei*]. 2 vols. Taipei: Yiwen yinshuguan, 1994.

Wu Zongci 吳宗慈. *Zhonghua minguo xianfashi* 中華民國憲法史 [*A History of the Constitution of the Republic of China*]. Shanghai & Beijing, 1924.

Wute Hengli 烏特亨利 [Henry Wood]. *Zhixin mianbingfa* 治心免病法 [*A Method for Healing the Mind and Avoiding Sickness*]. Translated by Fu Lanya 傅蘭雅 [John Fryer]. Shanghai: Gezhi shushi, 1896.

Xia Xiaohong 夏晓红. *Jueshi yu chuanshi: Liang Qichao de wenxue daolu 觉世与传世: 梁启超的文学道路* [*Awakening the People and Passing Down to Posterity: The Literary Path of Liang Qichao*]. Shanghai: Shanghai renmin chubanshe, 1991.

Xiong Yuezhi. *The Eastward Dissemination of Western Learning in the Late Qing Dynasty*. Translated by Yanxin Chan, Jiao Li, and Junchao Wang. Vol. 1. Singapore: Enrich Professional Publishing, 2013.

Xiong Yuezhi 熊月之. *Xixue dongjian yu Wanqing shehui 西学东渐与晚清社会* [*The Eastward Dissemination of Western Learning in the Late Qing Dynasty*]. 2nd ed. Beijing: Zhongguo renmin daxue chubanshe, 2011.

Xiyang Wanchang 夕阳晚唱. 'Jiu Nanlaibeiwang de wenti xiangqide 就"南来北往"的问题想起的 [Nanlaibeiwang's Question Has Reminded Me of Something]'. *Xici hutong 西祠胡同*, 27 August 2012. http://www.xici.net/d175186744.htm.

Xiyang Wanchang 夕阳晚唱, and Wan Zhaoyuan 万兆元. Email. 'Qingjiao 请教 [An Inquiry]'. Email, 23 October 2012.

Xu, Fengna. 'Shaping "A Community of Shared Future for Mankind": New Elements of General Assembly Resolution 72/250 on Further Practical Measures for the PAROS'. *Space Policy* 44–5 (2018): 57–62.

Xu Guangqi 徐光啟. *Xu Guangqi ji 徐光啟集* [*Collected Writings of Xu Guangqi*]. Beijing: Zhonghua shuju, 1963.

Xu Guangren 徐光仁. 'Ping Luo Mengwei xiansheng du renmin minzhu zhuanzheng duiyu Kang Youwei de pingjia 评罗孟韦先生"读人民民主专政"对于康有为的评价 [A Review of Mr Luo's Essay Entitled "Reading Mao's Comment on Kang"]'. *Huanan shiyuan xuebao 华南师院学报*, no. 2 (1959): 78–84.

Yan Fu 严复. *Yan Fu ji 严复集* [*Collected Works of Yan Fu*]. Edited by Wang Shi 王栻. Vol. 1. Zhonghua shuju, 1986.

Yan Fu 严复. *Yan Fu ji 严复集* [*Collected Works of Yan Fu*]. Edited by Wang Shi 王栻. Vol. 4. Zhonghua shuju, 1986.

Yang, C. K. *Religion in Chinese Society: A Study of Contemporary Social Functions of Religion and Some of Their Historical Factors*. Berkeley: University of California Press, 1961.

Yang Quan 楊銓. 'Jialileiwo zhuan 加里雷倭傳 [A Biographical Sketch of Galileo]'. *Kexue 科學* [*Science*] 1, no. 1 (1915): 99–105.

Yang Quan 楊銓. 'Zhanzheng yu kexue 戰爭與科學 [War and Science]'. *Kexue 科學* [*Science*] 1, no. 4 (1915): 353–58.

ZDLDG 中国第二历史档案馆 [China Second Historical Archives], ed. *Zhonghua minguoshi dang'an ziliao huibian* [*ZMDZH*] *3: wenhua 中华民国史档案资料汇编 3: 文化* [*Compilation of Archival Materials Pertaining to the History of the Republic of China 3: Culture*]. Nanjing: Jiangsu guji chubanshe, 1991.

Zeng Yi 曾亦. *Gonghe yu junzhu: Kang Youwei wanqi zhengzhi sixiang yanjiu 共和与君主: 康有为晚期政治思想研究* [*Republic versus Monarchy: A Study of Kang's Political Thought in His Later Years*]. Shanghai: Shanghai renmin chubanshe, 2010.

Zhang Jian 张剑. 'Cong "kexue jiuguo" dao "kexue jianguo" 从"科学救国"到"科学建国"[From "Saving the Nation" to "Building Up the Nation" through Science]'. *Ziran bianzhengfa tongxun 自然辩证法通讯* [*Journal of Dialectics of Nature*], no. 3 (2016).

Zhang Junmai 张君劢. 'Renshengguan 人生观 [The Philosophy of Life]'. In *Kexue yu renshengguan 科学与人生观* [*Science and the Philosophy of Life*], by Zhang Junmai 张君劢 et al., 30–37. Shenyang: Liaoning jiaoyu chubanshe, 1998.

Zhang Junmai 张君劢. 'Zailun renshengguan bingda Ding Zaijun 再论人生观并答丁在君 [Another Discussion on the Philosophy of Life and My Replies to Ding]'. In *Kexue yu renshengguan 科学与人生观* [*Science and the Philosophy of Life*], by Zhang Junmai 张君劢 et al., 30–37, New. Shenyang: Liaoning jiaoyu chubanshe, 1998.

Zhang Songzhi 张颂之. 'Kongjiaohui shimo huikao 孔教会始末汇考 [A Review of the History of the Confucian Religion Association]'. *Wenshizhe 文史哲* [*Journal of Chinese Humanities*], no. 1 (2008): 55–72.

Zhang Xiang 张翔. 'Cong li gongli zhixue dao yi datong lijiao 从立公理之学到以大同立教 [From the Learning of Universal Principles to a Religion Based on the Great Unity]'. *Zhexue dongtai 哲学动态* [*Philosophical Trends*], no. 3 (2015): 13–22.

Zhang Xiang 张翔. 'Ruxueshi xushu de fenduan yu Kongzi zhiyi de bifushi quanshi 儒学史叙述的分断与孔子之义的比附式诠释 [The Demarcation of the History of Confucianism and the Analogous Interpretation of the Meaning of Confucius]'. *Zhongguo zhexueshi 中国哲学史* [*History of Chinese Philosophy*], no. 6 (2019): 120–28.

Zhang Xiuhua 张秀华. 'Kexue yu zongjiao guanxi tanjiu de xinjinlu: dui jinnian Zhongmei kexue yu zongjiao xueshu huiyi de fenxi 科学与宗教关系探究的新进路——对近年中美科学与宗教学术会议的分析 [New Paths for Inquiry of the Relation between Science and Religion: A Review of the China-America Academic Conferences on Science and Religion in Recent Years]'. *Tsinghua daxue xuebao 清华大学学报 (哲学社会科学)* [*Journal of Tsinghua University (Philosophy and Social Sciences)*], no. 3 (2013): 88–95.

Zhang Yong 张勇. 'Kang Youwei de "zuowei" jiqi xiandu 康有为的"作伪"及其限度 [On Kang Youwei's "Forgery" and Its limits]'. *Lishi yanjiu 历史研究* [*History Studies*], no. 6 (2005).

Zhang Yongle 章永乐. *Wanguo jingzheng: Kang Youwei yu Weiyena tixi de shuaibian 万国竞争: 康有为与维也纳体系的衰变* [*Nations as Rivals: Kang and the Decline of the Vienna System*]. Beijing: Shangwu yinshuguan, 2017.

Zhang Zhigang 张志刚. *Zongjiao Zhongguohua yili chensi 宗教中国化义理沉思* [*Sinocization of Religions: A Theoretical Study*]. Beijing: Zongjiao wenhua chubanshe, 2017.

Zheng Kuangmin 郑匡民. *Liang Qichao qimeng sixiang de dongxue beijing 梁启超启蒙思想的东学背景* [*The Japanese Intellectual Background of the Enlightenment Ideas of Liang Qichao*]. Shanghai: Shanghai shudian, 2003.

Zheng Zhiming 郑志明. 'Kang Youwei de zongjiaoguan (shang) 康有为的宗教观 (上) [On Kang Youwei's Religious Views I]'. *Ehu 鹅湖* 23, no. 3 (1997): 17–28.

Zheng Zhiming 郑志明. 'Kang Youwei de zongjiaoguan (xia) 康有为的宗教观 (下) [On Kang Youwei's Religious Views II]'. *Ehu 鹅湖* 23, no. 4 (1997): 22–31.

Zhong Wendian 钟文典. *Guangxi tongshi 广西通史* [*A Comprehensive History of Guangxi*]. Vol. 2. Nanning: Guangxi renmin chubanshe, 1899.

Zhongguo kexueshe 中國科學社, ed. *Kexue tonglun 科學通論* [*General Discussions on Kexue*]. Shanghai: Zhongguo kexueshe, 1934.

Zhou Min 周敏. 'Zouxiang renlei mingyun gongtongti: yige bijiao wenhua de shijiao 走向人类命运共同体：一个比较文化的视角 [Towards a Community of Shared Future for Mankind: A Perspective from Comparative Cultural Studies]'. *Shanghai jiaotong daxue xuebao 上海交通大学学报* [*Journal of Shanghai Jiaotong University (Philosophy and Social Sciences)*] 27, no. 126 (2019): 68–75.

Zhou Weichi 周伟驰. *Taiping tianguo yu Qishilu 天平天国与启示录* [*Taiping Heavenly Kingdom and Revelation*]. Beijing: Zhongguo shehui kexue chubanshe, 2013.

Zhouyi zhengyi 周易正義 [*The Verified Meaning of the Book of Changes*]. Beijing: Peking University Press, 1999.

Zhu Weizheng 朱维铮. *Qiusuo zhenwenming 求索真文明* [*In Search of True Civilization*]. Shanghai: Guji chubanshe, 1996.

Zhu Xi 朱熹. *Sishu zhangju jizhu 四書章句集注* [*Interlinear Analysis of and Collected Commentaries on the Four Books*]. Beijing: Zhonghua shuju, 1983.

Zhu Yitian 朱忆天. *Maixiang jindai wenming guojia de tansuo: Kang Youwei houqi sixiang yanjiu 迈向近代文明国家的探索：康有为后期思想研究* [*Towards a Modern Civilized Country: A Study of Kang's Later Thought*]. Shanghai: Shanghai renmin chubanshe, 2016.

Zhuang Cunyu 莊存與. *Weijingzhai yishu: Xici zhuan lun 味經宅遺書：繫辭傳论·* [*The Unpublished Writings of the Weijingzhai: A Treatise on the Xici of the Book of Changes*]. Yanghu zhuangshi, 1882.

Zhuang Shidun 莊世頓 [R. Johnston]. 'Zhuang Shidun jun yuanshu 莊世頓君原書 [Letters from Mr. R. Johnston]'. *Kongjiaohui zazhi 孔教會雜誌* [*The Confucian Association Monthly*], no. 3 (1913).

Zhuo Xinping 卓新平. 'Zhongguo zhishijie dui zongjiao yu kexue guanxi zhi lun 中国知识界对宗教与科学关系之论 [Discourse on the Relationship between Science and Religion among Chinese Intellectuals]'. In *Qiao: kexue yu zongjiao 桥：科学与宗教* [*Bridging Science and Religion*], edited by Kang, Phee Seng, Ted Peters, and Gaymon Bennett, 2nd ed., 230–44. Beijing: Zhongguo shehui kexue chubanshe, 2007.

Index

Printed in the United States
by Baker & Taylor Publisher Services